POLICY-MAKING IN THE EUROPEAN UNION

POLICY-MAKING IN THE EUROPEAN UNION

THIRD EDITION

Edited by
Helen Wallace and William Wallace

OXFORD UNIVERSITY PRESS · OXFORD

Oxford University Press, Great Clarendon Street, Oxford OX2 6DP

Oxford New York

Athens Auckland Bangkok Bogota Buenos Aires Calcutta
Cape Town Chennai Dar es Salaam Delhi Florence Hong Kong Istanbul
Karachi Kuala Lumpur Madrid Melbourne Mexico City Mumbai
Nairobi Paris São Paolo Singapore Taipei Tokyo Toronto Warsaw
and associated companies in
Berlin Ibadan

Oxford is a registered trade mark of Oxford University Press

Published in the United States by
Oxford University Press Inc., New York

First published 1996
Reprinted in paperback 1996, 1997, 1998

British Library Cataloguing in Publication Data
Data available

Library of Congress Cataloging in Publication Data
Data available

ISBN 0-19-878128-8
ISBN 1-19-878129-6 (Pbk)

Printed in Great Britain
on acid-free paper by
Bookcraft (Bath) Ltd
Midsomer Norton, Avon

PREFACE

This is both a new book and the third edition of an old one. It follows the pattern established in *Policy-Making in the European Communities* (1977), extended and developed in the second edition of 1983. Its seventeen chapters are, however, entirely rewritten, with references to the earlier case-studies as appropriate. The merged Communities had become in general usage *the* European Community (EC) by 1983; with chapters on the second and third pillars established under the Maastricht Treaty on European Union (TEU) included in this volume, the title now reflects our coverage of the broader European Union (EU).

The processes of EC/EU policy-making have become much more complex in the twelve years since the second edition. Two revisions of the constitutional structure set up in the Treaties of Paris and Rome have been negotiated and ratified. The EC of 9 of 1977 had been joined by Greece in 1981, with Spain and Portugal in mid-negotiation for accession when the last edition went to press. The EU expanded from twelve to fifteen in January 1995, with the shadow of further enlargement to the east and south already falling across its policy process, as several of the chapters in this volume make clear. The role of the European Court of Justice (ECJ) has expanded further and become more controversial. The European Parliament, after four direct elections and two treaty revisions, now plays a more active part in many areas of EU policy.

The policy agenda itself has broadened, to such an extent that several governments have attempted to claw back from Brussels policies and powers which have shifted from state or sub-state authorities to the EU level. To provide the student of European integration with a sense of the diversity of policies and of patterns of policy-making which were channelled through the EU institutions in the early 1990s, the number of case-studies has now grown to fourteen, from nine in the first edition and eleven in the second. Familiar issues of distribution and redistribution, the common market, agriculture, competition, external trade, monetary policy, and foreign policy are retained; energy policy, which was covered in the first edition, is brought back in. These cases are complemented by studies of issue areas in which the salience of the EU arena has grown significantly during the last decade, or in which the EU's approach has markedly changed: environmental regulation; the social dimension, one of the most sensitive dossiers negotiated in the TEU; justice and home affairs; and the Community's response to the transformed position of its eastern neighbours from 1989 onwards. Some of the best examples of how European policy is made and managed concern individual products. The first and second editions took the sugar regime as their demonstration case; this edition includes a study of the banana regime.

The academic literature on west European integration, limited in the mid-1970s and as regards theory largely American, is now extensive. It is also substantially European. The consolidated bibliography at the end of this volume provides the reader with a guide to the recent—as well as the classical—literature, in English, German, and French. The choice of three German contributors for this volume is a partial reflection of the very considerable contribution which the German academic community has made to studies of European integration over the past decade. We chose for this edition not to retain a survey chapter on theories of integration, partly because Carole Webb—our co-editor in the last two editions—has now moved on to other responsibilities, and partly because we have chosen instead to open the volume with our own analytical approach. We have missed Carole sorely in preparing this edition.

The eighteen authors of this volume, seven of whom contributed to the second edition, are drawn from five EU countries and from two non-member states. As before, the task of drawing together a collective study has been eased by working with people who are friends as well as colleagues; five are also former or current students of the editors. Informal ties and shared extraneous interests help to hold together transnational academic networks, as well as transnational official and interest-group networks: conversations around this book have touched upon model railways, toy soldiers, historical novels, and (of course) children, as well as on the intricacies of national and Community policy-making in different fields.

A preliminary meeting of contributors was held in the margins of the second world conference of European Community Studies' Associations in Brussels in May 1994. Several draft chapters were presented to a panel at the biennial conference of the American European Community Studies' Association in Charleston in April 1995. We are grateful to Tim Barton for his patience and encouragement at Oxford University Press, to Alasdair Young for his thoroughness and tolerance beyond reasonable limits in consolidating the bibliography, to Julie Smith for sustained and sustaining help at several stages, to Eva Østergaard-Nielsen for help in preparing the manuscript for the press, to Edward Wallace for his technical skills and tolerance in converting computerized texts into a single manuscript, and to

Harriet Wallace for her determined efforts to make our scientific metaphors more rigorous.

H. W., W. W.

London,
June 1995

CONTENTS

PART ONE: PROCESS AND INSTITUTIONS

PART TWO: POLICIES

PART THREE: CONCLUSIONS

DETAILED CONTENTS

FIGURES

BOXES

TABLES

ABBREVIATIONS

ACP   African, Caribbean, and Pacific countries
AFNOR   Association française de normalisation
APEC   Asia Pacific Economic Cooperation
AWACS   Airborne Warning and Control System

BAT   best available technology
BATNEEC   BAT not entailing excessive economic costs
Benelux   Belgium, Netherlands, and Luxemburg

CAP   common agricultural policy
CCP   common commercial policy
CEECs   countries of central and eastern Europe
CEN   European Committee for Standards (abb. from French)
CENELEC   European Committee for Electrical Standards (abb. from French)
CEP   common energy policy
CEPR   Centre for Economic Policy Research
CFSP   common foreign and security policy
CFSR   Czech and Slovak Federated Republic
cif   prices including cost, insurance, and freight
CIREA   Centre for Information, Discussion, and Exchange on Asylum (of JHA,
           abb. from French)
CIREFI   Centre for Information, Discussion, and Exchange on the Crossing of
           Internal Borders (of JHA, abb. from French)
CIS   Commonwealth of Independent States (ex-USSR)
CMEA   Council for Mutual Economic Assistance ('Comecon')
COPA   Confederation of Professional Agricultural Organizations
Coreper   Committee of Permanent Representatives
Coreu   Correspondant Européen (European Communications Network of
           EPC)
CP   Contracting Party
CSCE   Conference on Security and Cooperation in Europe
CSF   Community Support Framework

DG   Directorate General (of European Commission)
DGB   Deutscher Gewerkschaftsbund
DIN   Deutsche Institut für Normung
DM   Deutschmark

DOM	départements d'outremer
EA	Europe Agreement
EAGGF	European Agricultural Guidance and Guarantee Fund
EBRD	European Bank for Reconstruction and Development
EC	European Community
EC6	EC of Six (Belgium, France, Germany, Italy, Luxemburg, and the Netherlands)
EC9	EC of Nine (Six plus Denmark, Ireland, and UK)
EC10	EC of Ten (Nine plus Greece)
EC12	EC of Twelve (Ten plus Portugal and Spain)
EC15	EC of Fifteen (Twelve plus Austria, Finland, and Sweden)
ECB	European Central Bank
ECJ	European Court of Justice
ECO	European Cartel Office
Ecofin	Council of Economic and Finance Ministers
ECSC	European Coal and Steel Community
ecu	European currency unit
EDC	European Defence Community
EdF	Électricité de France
EDF	European Development Fund
EDU	European Drugs Unit
EEA	European Economic Area
EEC	European Economic Community
EFTA	European Free Trade Association
EIB	European Investment Bank
EIS	European Information System
EMI	European Monetary Institute
EMS	European Monetary System
EMU	economic and monetary union
EP	European Parliament
EPC	European Political Cooperation
ERDF	European Regional Development Fund
ERM	exchange-rate mechanism
ERT	European Round Table of Industrialists
ESA	European Space Agency
ESC	Economic and Social Committee
ESF	European Social Fund
ETUC	European Trade Union Confederation
EU	European Union
Euratom	European Atomic Energy Community
Eurocorps	multilateral European force, expanded from Franco-German brigade in 1991
Eurogroup	European group within Nato from 1970
Europol	European Police Office

FCO	Foreign and Commonwealth Office
fob	prices free-on-board
FoC	Forum of Consultation (of WEU)
FTA	Free Trade Agreement
FTAA	Free Trade Agreement for the Americas
G7	Group of 7 (western economic powers)
G24	Group of 24 (member states of OECD)
GAC	General Affairs Council
GATS	General Agreement on Trade in Services
GATT	General Agreement on Tariffs and Trade
GDP	gross domestic product
GDR	German Democratic Republic
GNP	gross national product
GNS	Group on Negotiations in Services (of GATT)
GPA	Government Purchasing Agreement
GSP	General System of Preferences
IDP	Integrated Development Programme
IEA	International Energy Agency
IEM	internal energy market
IEPG	Independent European Programme Group
IGC	Intergovernmental Conference
IIA	Inter-Institutional Agreement
IMF	International Monetary Fund
IMP	Integrated Mediterranean Programmes
IPR	intellectual property rights
ISO	International Standards Organization
IT	information technology
JHA	justice and home affairs (third pillar of EU)
K4	committee of senior officials for JHA
LDCs	less developed countries
MCA	monetary compensatory amount
MEP	member of the European Parliament
Mercosur	Common Market of the Southern Cone
MFA	Multi-Fibre Arrangement
MFN	most favoured nation (in GATT)
mtoe	million tonnes of oil equivalent
MTR	Mid-Term Review of Uruguay Round
NACC	North Atlantic Consultative Council
NAFTA	North Atlantic Free Trade Agreement

Nato	North Atlantic Treaty Organization
NGO	non-governmental organization
NICs	newly industrialized countries
NTB	non-tariff barrier
OECD	Organization for Economic Cooperation and Development
OEEC	Organization for European Economic Cooperation
OPEC	Organization of Petroleum Exporting Countries
OSCE	Organization for Security and Cooperation in Europe
PFP	Partnership for Peace (of Nato)
Phare	Poland and Hungary: Aid for the Restructuring of Economies (later extended to other CEECs)
piatnika	grouping of five successive Council presidencies
QMV	qualified majority voting
QR	quantitative restriction
QUAD	GATT/WTO group of four: Canada, EU, Japan, and USA
R&D	research and development
RDP	Regional Development Plans
SCA	Special Committee on Agriculture
SEA	Single European Act
SEM	single European market
SIRENE	Supplementary Information System of Schengen
SIS	Schengen Information System
SME	small and medium-sized enterprises
SPD	Single Programming Document
T&C	textiles and clothing
TACIS	technical assistance (of EU) for CIS
Tempus	Trans-European Mobility Programme for University Studies
TEN	Trans-European Network
TEU	Treaty on European Union
TNC	Trade Negotiations Committee (of GATT)
TOM	territoires d'outremer
Trevi	Terrorisme, Radicalisme, Extrémisme, Violence, Information (agreement on police cooperation)
troika	grouping of three successive Council presidencies
UCLAF	Unité de Coordination de la Lutte Anti-Fraude
UK	United Kingdom
UN	United Nations
UNECE	United Nations Economic Commission for Europe
UNICE	Union of Industries of the European Community

UR	Uruguay Round
US, USA	United States
VAT	value added tax
WEU	Western European Union
WG	working group
WTO	World Trade Organization

LIST OF CONTRIBUTORS

DAVID ALLEN	*University of Loughborough*
MONICA DEN BOER	*European Institute of Public Administration, Maastricht*
ANTHONY FORSTER	*St Hilda's College, Oxford*
JANNE HAALAND MATLARY	*University of Oslo*
MICHAEL HODGES	*London School of Economics and Political Science*
BRIGID LAFFAN	*University College, Dublin*
STEPHAN LEIBFRIED	*University of Bremen*
PAUL PIERSON	*Harvard University*
ELMAR RIEGER	*University of Mannheim*
ULRICH SEDELMEIER	*Sussex European Institute*
ALBERTA SBRAGIA	*University of Pittsburgh*
MICHAEL SHACKLETON	*European Parliament*
CHRISTOPHER STEVENS	*Institute of Development Studies, Sussex*
LOUKAS TSOUKALIS	*University of Athens and College of Europe*
HELEN WALLACE	*Sussex European Institute*
WILLIAM WALLACE	*London School of Economics and the Central European University*
STEPHEN WOOLCOCK	*London School of Economics and Political Science*
ALASDAIR R. YOUNG	*Sussex European Institute*

EDITORS' NOTE

We have chosen for the sake of simplicity to refer generally to the European Union in this volume, except where the references are specific to pre-Maastricht events or where the 'Community pillar' is pertinent, or where one of the three original Communities is under discussion.

PART ONE: PROCESS AND INSTITUTIONS

CHAPTER ONE

POLITICS AND POLICY IN THE EU: THE CHALLENGE OF GOVERNANCE

Helen Wallace

West European political integration remains an experimental process, driven by both ideas and interests as policy-makers grapple with issues beyond the capacities of nation states. Assessing both the process and its outputs demands insights from a variety of social science disciplines. The impetus for transnational policy cooperation was provided by the geopolitical context (until 1989), by responses to globalization, and by the evolution of the west European welfare state; the vicissitudes of economic cycles and the pull of national loyalties exerted contrary pressures. European governance has thus depended on the swings of a pendulum between opposing forces. Sustainable common policies have emerged where policies have been devised that command a consensus of ideas and interests. But European governance repeatedly has to compete for primacy with other levels, a source of inherent instability which permeates the institutional processes of the EU. This competition for primacy opens up opportunities for a wide variety of actors to influence the policy process, while at the same time depriving it of the traditional anchors of legitimacy.

Introduction

This volume is intended as a guide to the policy process of the European Union (EU) and its roots in the European Community (EC). It aims to explain how European policies are made, the characteristics of the intermediating institutions, and their impact on the development of west European integration. This opening chapter offers a road-map of the Community policy process, and like any map it reflects the predilections of the map-maker in its choice of distinguishing features. The chapter also identifies ways in which policy-making has varied over time and differs according to issue areas.

During the thirteen years since this volume's predecessor appeared much has changed. 'Eurosclerosis' in the early eighties was followed by a period of hyperactive 'Europhoria', with the Single European Act (SEA) of 1987, the internal market programme for 1992, and ambitious plans for both economic and monetary union (EMU) and a common foreign and security policy (CFSP). Yet by the early 1990s, paradoxically as the EC was subsumed within the framework of a more ambitious European *Union* agreed at Maastricht in 1991, the integration model was under fire as endogenous and exogenous factors combined to throw into question what had appeared to be the givens established since the early 1950s.

Much has been written about whether integration has produced cumulative and irreversible transfers of policy competences from countries to the European level. It has become fashionable to suggest that integration has hit a critical threshold beyond which it would need to be a fully-fledged polity and not only an accepted arena for some shared policy regimes. Our last edition concluded that the then EC was 'more than a regime and less than a federation' or polity. This edition picks up that question again and seeks answers that fit with the altered circumstances of the 1990s.

As before, we have tackled the question by examining both big policy issues and 'everyday' issues; indeed we believe that the process can be understood only by understanding both kinds of cases (Moravcsik 1991; J. Peterson 1995a). Our selection of case-studies is in part a reprise of those that were covered in previous editions, bearing in mind that many of the cornerstones of EC policy activity have stayed in place (the budget, competition policy, agriculture) or been reinforced (the single market, structural policies, monetary policy, and foreign policy). We have referred back to earlier editions for comparison or to point up differences of substance or of interpretation. The accounts in this edition are wholesale revisions and focus predominantly on recent and current developments in each policy area.

In addition this new edition covers issue areas that have become more vigorous or more debated over the past decade, in particular environmental policy, social policy, internal security, and relations with eastern Europe. Two chapters—on bananas and the Uruguay Round—focus on particular episodes and the interaction of internal and external factors in defining policy. It remains our contention that to generalize about the policy process demands sustainable analysis across a variety of issue areas. The reader is invited to draw her own conclusions as to whether such generalization is plausible or whether the variations between policy arenas are more striking than the similarities. Our own judgements appear in the concluding chapter to the volume. We have not repeated the commentary on the theoretical debate as it stood in the early eighties. For that readers should glance back at Carole Webb's introduction to the second edition (1983). Instead we offer a rather different way into the discussion of integration that takes account of both intervening analytical approaches and the large changes in the context of integration.

Finding a starting-point

Figuring out how European policies are made has never been a straightforward task for either the practitioner or the commentator. Nor has the passage of time made the process easier to understand. The experiment in political and economic integration in western Europe that led to the creation of the EC now has an accumulation of over forty years' experience in testing and retesting the ground for the development of shared solutions to shared policy problems. That experience has been shaped by economic and political cycles, sometimes encouraging collective European responses, often yielding fierce disagreements. Our focus in this volume is particularly on politics and policy, but we acknowledge at the outset the necessity of getting to grips with the economics and law of European integration. However, we equally insist on the impact of politics on economics and on the law.

Always the 'real' economy has cast a long shadow, influencing what was feasible and what policy prescriptions would suit the economic circumstances of different areas within the EU. From the outset the Community experiment was tied to the development of shared policies to govern economic activities. Economic policies have thus to be judged by their economic impact, by the fit between policy problem and the offered policy solution. None the less the pursuit of common policies has been conditioned by the broader political context, which leads decision-makers to moderate their judgements of policy options in terms of the political results that they may achieve as well as their economic impact. Beyond the policy-makers stand the recipients of policy, some of whom judge only the economic impact, others of whom are more interested in the political consequences.

The policy process has been set in a distinctive legal framework, which has embedded rules and norms that not only develop policy but bend the political relationships among the participants in the process. European law from early on made certain kinds of policy more feasible than others, and not always quite in the way that the policy-makers had originally intended. Indeed in some respects European law has acquired quasi-constitutional elements with a surprisingly firm foundation of shared jurisdiction (Weiler 1991; Burley and Mattli 1993). This feature of European integration has long been understood by European scholars, who have generally acknowledged that the law has been a powerful instrument of policy-making and rule-application. What seems less certain is how far we can assume this as a constant, as we see lines of resistance developing to some of the tentacles of European law.

The role of institutions has been crucial, institutions both in the narrower sense of particular institutional formats and mechanics and in the larger sense of political relationships and behaviour. Over the years we have seen periods of intense investment in EC institutions, designed to equip the Community system to fulfil its functions more effectively, interspersed with periods of institutional disinvestment or drift, under pressures to limit the autonomy of 'Brussels'. In the next chapter we set out the main features of the institutional system as a backcloth to the illustrations in the case-studies.

The EC, later the EU, has changed significantly over time. The geopolitical context has shaped the policy process and, as we can now observe with the benefit of hindsight since 1989, it has been a potent source of political cohesion. In the past the engagement of successive US governments was very important, some would say determining, in encouraging west European cooperation, as well as sometimes in constraining its endeavours (W. Wallace 1994a). The lightened impact of the USA may now have reduced the pressures for European integration, though competition with the USA still sometimes meets or provokes robust European solidarity (e.g. the USA–Euratom Treaty disagreement of 1992–5). For forty years or so the menace of the Soviet Union, both ideological and military, provided elements of cohesion among the participating west European polities. Its removal as an adversary has left west Europeans somewhat confused as to where their common interests and shared ideas lie.

The early shared ideas and collective interests of the period following the Second World War were rooted in geopolitical stabilization, efforts of economic reconstruction, the development of the welfare state, and the embedding of modernized liberal democracies. European integration seemed a useful adjunct for the predominant economic and political élites and could broadly sustain wider public support in the countries that originally signed up for membership (E. Haas 1958; Lindberg 1963). But the membership has changed, with successive enlargements bringing in new member states, in some of which the symbolism of reconstruction was absent. In any case by the 1990s it has become much less clear how a collective 'modernization project' might be defined, even though Europeans are still under pressure to define their collective identity *vis-à-vis* the rest of the world (Garcia 1993; H. Wallace 1994). The early integration theorists flagged strongly this feature of the EC and the way it permeated some of the choices of policy priorities. One question to bear in mind in reading the case-studies that follow is which European policies currently serve 'modernization' goals and which have become instruments of 'conservation' (Scharpf 1988).

One should be wary of over-reacting to economic and political cycles—much of what the EU now does resembles what the old EC did, and the stresses and strains currently visible in many ways recall the push and pull between region, country, Europe, and globe as competing levels of governance that have always characterized European integration. Some sections of earlier editions of this volume thus remain apposite. Yet some fundamentals have changed, and three particular factors stand out.

- *Enlargement:* the EU now consists of fifteen members, absorbing the old schism between integrationists and loose cooperators from western Europe that had marked the early period after the Second World War. It faces pressing demands for an eastern expansion to overcome the other and graver European schism between communism and liberal democracy (Michalski and Wallace 1992).
- *Scope:* the stated scope of European shared policies has grown ambitiously, only to encounter serious disputes relating to both doctrine and

ultimate goals. This contrasts with the founding period, in which the creation of the EC could rest on a cross-class and to some extent cross-country consensus about a particular socio-economic paradigm and the political methods that might give it transnational expression (Lindberg 1963).

- *Consensus*: for many years the EC was the beneficiary of a permissive consensus by which pliant publics allowed economic and political élites a relatively free rein in defining the scope, methods, and functions of European integration. This no longer holds: the 'democratic deficit', perhaps better 'legitimacy deficit', is a stringent reproach to the 'Community method' of building agreement and makes it much harder to create or sustain collective policies or to identify European public goods (Franklin *et al.* 1994).

We add two further considerations that we assess to be of much greater importance than we recognized in the previous two volumes: the role of *ideas* and the impact of *policy reappraisal* in the participating countries. By ideas we mean the range of intangibles that impact on politics, economics, and society and inform policy choices: values, symbols, ideologies, doctrines, and knowledge (Hall 1986; E. Haas 1990; Goldstein and Keohane 1993; Weiler 1992). By policy reappraisal we mean major changes of direction within one or several member states (for whatever reason) which change the basis from which policy is defined or negotiated collectively—the impact of the Green movement in northern Europe, especially Germany, the socialist turn-round on economic policy in France in the early 1980s, the arrival of Thatcherism in the UK, the collapse of Christian Democracy in Italy, and so on.

It is our contention that both the movement of ideas and policy redirection within the member states have been underrated features of the European policy process. Two examples recur in this new edition: the impact of shifting and contrasting views about the welfare state, and the emergence of market liberalization as a predominant theme in enough member states and the Commission to propel new policies (Jobert 1994). These debates have started to be addressed in the burgeoning literature in comparative politics on the welfare state and regulation, as well as in the international relations literature on epistemic communities (networks of professionals with recognized expertise and an authoritative claim to policy-relevant knowledge) (P. Haas 1992; Rieger and Leibfried 1995).

The search for an analytical chart

There is a bewildering range of analytical approaches to European integration as a political phenomenon and as a policy arena. In intellectual cycles successive generations of social scientists and historians have grappled with the challenge of explaining, even trying to predict, the integration process, its dynamics and statics. To get a good handle on the discussion really does require an effort to distil elements drawn from a number of disciplines and perspectives.

In the 1950s and 1960s the intellectual high ground was firmly occupied in both economics and political science. In economics those who provided a clear statement of the different steps from free(d) trade to economic union were arguing with the grain of what common sense and observed behaviour seemed to support (Balassa 1962; Pelkmans 1984). The question as regards the EC was over how far down this particular road EC policy-makers would seek to travel, with little contestation of the underlying model— Keynesianism with additives of Beveridge, Erhard, and Monnet. Yet surprisingly little attention was given by mainstream economists to empirical analyses of European economic integration. The territorial specificity of the EC held no great intrinsic interest in analytical terms, except for the applied economist interested in a particular range of policies. In the mid-1980s it was left to the Commission to fund studies (as it turned out mostly by consultants) on the economic impact of market integration—the Cecchini studies (1988)—because so little had been produced by the economics profession in a form that would aid the policy-maker.

This changed dramatically with the advent of the 1992 programme, the rekindled debate on economic and monetary union (EMU), and subsequently the emergence of central and east European countries (CEECs) struggling to establish market economies. In response the economic literature on European economic integration—both theoretical and applied—has burgeoned. This literature has, however, done surprisingly little to redefine the basic analytics of economic integration. Where theories derived from economics and mathematics *have* been posited as making a larger contribution to the understanding of the policy process is in the wave of literature on rational choice (for examples see Bueno de Mesquita and Stokman 1994 or Widgrén 1994).

In political science the neofunctionalists (E. Haas 1958; Lindberg 1963) had the initial advantage, both because of the intellectual coherence of their arguments and because the alternative paradigm of realism, in which the state remained stubbornly resilient (Hoffmann 1966), was too resonant of the *status quo ante* for the innovative. By the 1970s uncertainty was beginning to creep in, at least within the political science community. It became fashionable to knock down the arguments of neofunctionalism and to produce counter-evidence. A persistent doubt lingered over whether the EC was an example of an international organization or an embryonic state, thus confusing the criteria for assessing the results of the EC's existence. Political scientists were on the whole honest enough to admit that they did not know the relationship between political and economic factors of integration and disintegration, though much was being written about the international political economy (for overviews of the theories see Pentland 1973; Webb 1983). Much of the political science literature therefore became heavily descriptive and often narrowly focused on institutions.

Lawyers, in contrast, were always at the forefront of efforts to enhance our understanding of European policy-making and for a long time held by far the best cards. It rapidly became clear that Community law was a new species with distinct characteristics, remarkable tenacity, and an uncanny capacity to define and to redefine the ground on which policy-makers negotiated, policy-implementers administered, and entrepreneurs did business (Joerges 1994). Indeed it became commonplace to argue that the law itself, rather than economic logic or political choice, was the single most important motor of integration, as interpreted by both national courts and the European Court of Justice (ECJ) (Cappelletti *et al.* 1986). Thus, at least for countries marked by respect for law and for the pronouncements of the courts, pioneering European law was both an independent variable and an important resource to be exploited by the skilled. This remains so; the development of strong Community law has apparently been little marked by cycles of more and less intense integration. It remains to be seen how sustainable this would be, faced with either more determined cheating or defaulting, or by a political climate of retrenchment from integration.

Historians started out at an obvious disadvantage in assessing the new EC. A minority looked for earlier parallels in federal history—the USA, the *Zollverein*, Austro-Hungary, Switzerland, and so on. Most of what passed for history was biographical or contemporary commentary without the benefit of historical method or systematic evidence. This unavoidable absence of historical perspective perhaps made it too tempting for the contemporary social scientists to over-emphasize the specificity and novelty of the EC integration model. As records from the 1950s and 1960s have become available the gap is being filled, often disconcertingly (Küsters 1982; Milward 1984; Hogan 1987; Milward 1992). The received wisdom about the early phase of EC construction as a kind of heroic period is now being met by a barrage of critical analysis, much of which seeks to portray the EC as always much more 'normal' and less creative; it may after all simply have been the preferred instrument of some policy-makers and policy-influentials to project and to protect their interests, when necessary, at the European level.

One further introductory point needs to be mentioned. The early analytical literature on EC integration was dominated by Americans: the founding of the EC had much to do with US sponsorship (both governmental and corporate), and the US federal model permeated much of the debate about the yardsticks for 'successful' integration. For most of the 1970s and 1980s the literature was primarily European, often not in English, and much more empirical than theoretical. Since the late 1980s and the resurgence of American interest in European integration there has been a more vigorous competition of approaches to the subject, though it remains hampered by language barriers. In particular, the rich literature in German (more perhaps, on some estimates, than in all other Community languages combined) and the burgeoning literature in Spanish are too little read outside their originating countries.

It was against the confused debate about integration theory in the early 1970s that the first edition of this volume was conceived in 1973 as an attempt to describe and to characterize what the EC policy process produced and how it produced what it did. Our starting presumption was that we would not necessarily find a single underlying pattern or single map, but that to identify the range of patterns would at least be useful and might help to explain some of the evident discontinuities. By the late 1980s the study of European integration had changed dramatically, essentially as a function of the apparent change in the object of our attention. The advantage for the student is that there is now a much richer literature on European integration, both empirical and theoretical, and from a wider range of disciplines. There is now too an extensive *demi-monde* literature, driven by the consultancy boom, with handbooks to guide the practitioner through the often arcane and always complicated byways of European policy-making.

The student is thus faced with a huge array of explanations of how European policies are made and needs coordinates to find a way through the intellectual maze. A persistent debate remains whether European integration should be seen as a subset of international relations or as an external projection of country-based polities, the turf of comparative politics specialists (Bulmer 1984; Hix 1994). But to resolve this would still leave choices within each discipline between competing paradigms. It is not our ambition to settle this debate, though we believe it to be misconceived. Our preference, as will be argued in more detail later, is for a 'seamless web' approach, that is to say that modern governance, at least in western Europe, involves efforts to construct policy responses at a multiplicity of levels, from the global to the local. The European arena constitutes points of intense interface and competition between levels of government and between public and private actors (Scharpf 1994*a* and 1994*b*; Kohler-Koch and Woyke 1995). What interests us is what clusters of factors generate an agreement that on specific issues the EU level should predominate as the preferred policy arena.

The swinging pendulum of policy cooperation

We have taken the *pendulum* as our metaphor for characterizing European governance, introduced here and spelled out in more detail in the conclusions. The pendulum oscillates between two magnetic fields, the one country-based and the other transnational, each with its own inducements and repelling features. Depending on the varying strengths of these two fields there will sometimes be a propensity for transnational policies to be adopted and sometimes national ones. If the magnetic fields at both levels are weak, then no coherent policy at all may emerge—it is tempting, but misleading, to think the default remains national policy. Neither the transnational level nor the country-level of governance is uncontested and neither provides a comprehensive policy capability. Indeed the transnational level may be articulated through European integration or other international fora. The country level is primarily oriented around the central core of the national polity, but most west European countries also contain smaller magnetic fields at local and regional levels. The term pendulum is simply a metaphor for the movement in the process of European integration, sometimes regular, sometimes erratic, sometimes sustained, and sometimes stationary.

Our model is first summarized and then the elements are explained in more detail. It rests on three contextual premisses:

- that the west European state is politically inadequate;
- that globalization has a significant impact; and
- that the (west) European region has specific features.

It assumes that these lead to choices between national policy and various forms of transnational collaboration, choices subject to endemic political and policy competition, thus inherently unstable or fluctuating in the sense of not necessarily settling down into a sustained and regular pattern. We argue that this competition is played out as a function of the interplay of *ideas*, *interests*, and *institutions* and focused on the choices made by a variety of actors about how to respond to issues. In the context of European integration these issues

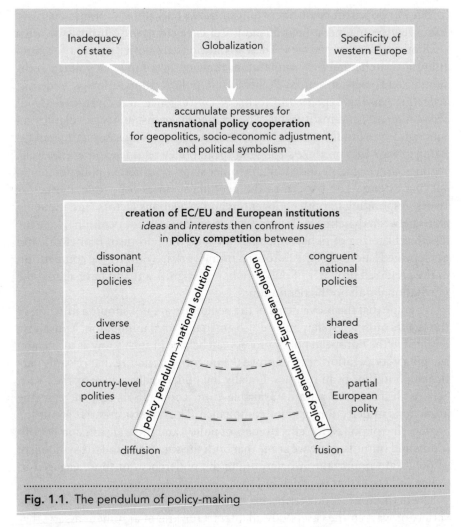

Fig. 1.1. The pendulum of policy-making

are particularly clustered around *geopolitical stabilization, socio-economic adjustment,* and *political symbolism.*

Where ideas and interests (we insist on both) are broadly congruent, that is to say shared across and within the member states, or complementary, that is mutually compatible and mutually reinforcing, there is a propensity to adopt collective policies and for the transnational arena to be accepted as the preferred form of governance. To the extent that the congruences and complementarities cluster within the magnetic field of the EU, transnational governance is pursued through the European integration process. But other transnational fora may exert stronger attractions, as Nato has in the security arena for most of the period since the Second World War.

The development of distinctive *institutions*, backed by supranational law, has given the magnetic field of the EC greater strength in most issue areas than other international fora. These institutions are both a source of competition and a channel for competition, drawing in a range of political, economic, and societal actors. Political and policy competition is endemic, reflecting varying degrees and forms of convergence and divergence about ideas, interests, and institutions. These latter make instability or volatility an inherent feature of the model. Hence the pendulum swings, sometimes oscillating in ways that produce strong European policy regimes, sometimes weak regimes, and engaging individual member states or different policy actors to varying extents. The pattern of the pendulum varies over time and across issues, responding to both endogenous and exogenous factors, and inducing shifts between dynamic and static periods or arenas of cooperation. None the less the clustering of policy outcomes through the European framework and the repeated 'success' of the European framework in winning the competition as the preferred provider of public policy are significant and have altered the gravitational pull on the pendulum.

We argue that the movements of the pendulum are not simply a function of the factors outlined so far, but are also amenable to manipulation. Thus those engaged in the European policy process can behave in such a way as to encourage policy cooperation or to improve the incentive structure that rewards participation in European regimes. Equally the policy-makers may by their own actions decrease the rewards available from cooperation and increase the attractions of defection or poor compliance. Thus questions arise about the engineering of policy, the effectiveness of policy, and the transmission of both inputs and outputs. Here we argue that both the substance and the symbolism of policy are important in determining the preferences of those engaged in policy and the extent to which European policy is legitimated.

The definitions of our terms are set out below and they are explored in a variety of ways in the case-study chapters. Our line of argument asserts that European integration is in a broad sense a political process composed of a multiplicity of interacting factors. These do not necessarily generate a unidirectional or cumulative pattern of shifts from the country to the transnational level of governance. Indeed there is scope within our approach for policy regimes to emerge in different mixtures of country-based or transnational elements. It is also consistent with our model that sub-national policy regimes or focuses of political activity may at some times, or in some countries, or for some issues, win the competition as the preferred level of governance. Some would go further and argue that the extent of cross-level governance is such as to blur the boundaries between levels (Scharpf 1994*a*)—in our terms overlapping magnetic fields.

In this sense our model implies that the diffusion of governance may just as well be an outcome as fusion (for the fusion thesis see Wessels 1992). In other words, we do not need to assume that the process must result in either the European arena or the state 'winning' the political and policy competition once and for all . Nor do we need to assume that all member states will be tied into the process in identical ways. None the less by observing the patterns of the competition and the propensities to practise cooperation or conflict we can reach some judgements as to the likelihood of one or another field of governance to predominate and as to the scope for the competition to be steered in one or another direction. Thus the policy process can itself make a difference to the outcomes.

Our insistence on three clusters of policy issues as relevant to European integration—geopolitical stabilization, socio-economic adjustment, and political symbolism—is deliberate. These clusters, specified below, are all important and, we argue, the European arena of policy cooperation has been driven by and dependent on its acceptability across the three clusters. They deliberately include issues of both 'high' and 'low' politics and deliberately exclude some areas of policy as much less relevant.

The inadequacy of the state

We start from the simple premiss that no one set of political authorities can command sufficient political and economic resources to project policy authoritatively across all the main arenas of public policy. The period following the Second World War saw a vast extension of the powers of the state and the provision of public goods. Thus the functions of government, irrespective of level, and more and more of the life of the individual citizen and of hitherto private activities became subject to state activity. As functions increased, so states, or their governing élites, found themselves under pressures of increasing demands and extended their own ambitions to respond.

One means of responding was to draw on resources from transnational cooperation as instruments of domestic policy. Indeed the scope for doing this may well have served to disguise the inadequacy of the state and paradoxically to preserve the notion that the state could remain effective as the primary source of most public goods as well as the primary point of political affiliation. We note also that individual states are 'inadequate' in different ways and thus that transnational policy cooperation may serve quite different functions in different countries, depending on which gap in local policy provision is being addressed; in the UK the structural funds of the EC filled, first, an EC budgetary gap and, later, a central government funding gap, while in Ireland they contributed to a strategy of infrastructure and investment renewal and in the cohesion countries an offset for the pressures of the single market (A. Smith 1995).

More recently the conventional wisdom has been that the welfare state subsequently became overstretched, hence the debate has been joined on the retraction of public goods and the reduction of the powers of government. The symbolism of state sovereignty and the resilience of national political identities none the less persist. It could even be argued that the reduction of public powers at the country level has led to a reflex to demonstrate that the state remains alive by reclaiming control over some policy arenas that had been 'delegated' to transnational cooperation, by renationalizing or repatriating them.

Globalization

Globalization makes the traditional state level of governance particularly inadequate for a number of important issue areas, including—and this is troublesome—some policy issues traditionally defined as core prerogatives of statehood or the stuff of high politics. The traditional arguments about the impact of globalization relate to the interdependence of economic transactions and the processes driven by new technologies, all operating with increased intensity over the past twenty years. Typically these trends have been offset by the appearance that the state retained an important role in external security and defence policy. But this has been true in only the limited sense of the formal reservation to the state of the decision on whether or not to go to war. The reality is, however, one in which security interdependencies have also grown, and the definition of what constitutes security has altered to embrace the increasing impact of transnational factors. Nor does the state any more exercise monopoly control over who resides within its boundaries or the rights of individual residents or the lawfulness of their activities.

Globalization has also enabled a variety of societal actors to have access to an international arena as an extension of or an alternative to the state arena (Kapteyn 1995). It has generated new kinds of policy issues and new forms of political interaction. In particular it often means that some economic actors have reduced or negligible loyalty to the state or any one state. Thus a British-based company will now reach its judgement on monetary policy differently depending on whether its business is conducted more in dollars or more in European currencies, not in relation to some purported UK interest. Political élites, in contrast, for the most point are condemned to success or failure in terms of their impact within an individual state. We tend therefore to the view that 'the state' has limited meaning, as the focus of 'state preferences' or 'national interests'. This does not prevent those who remain bounded by the state from asserting a 'national interest' defence of their reduced political responsibilities—on the contrary.

Thus globalization in the economic and security arenas and in the new 'global issues', energy supplies, ecology, and so on, induces needs for transnational cooperation. But these pressures coexist with policy arenas and political activities confined by traditional political territory.

The specificity of the west European region

Neither the inadequacy of the state nor the phenomenon of globalization are specific to Europe. However, we argue, both have an unusual intensity in the west European region. It is a region with long, albeit interrupted, traditions of transnational interaction and cooperation, both formal and informal, going back over centuries.[1] The breakdowns of transnational cooperation in Europe have produced wars still close enough in the collective memory to induce a profound commitment to policies of stabilization and a willingness to develop regimes of sustained collective security. In western Europe these regimes have included both the militarily aligned and the neutral countries, a factor not always well understood. Arguments about the specificities of these regimes, which continue, should be seen more as about means than about ends.

Since the Second World War the preoccupation with collective security and geopolitical stabilization has been marked by the concern to stabilize intra-west European relationships, especially *vis-à-vis* Germany; it was also a function of the superpower confrontation (Herbst *et al.* 1990). We argue that, even though European integration was formally focused on issues of political economy, it was part of a wider pattern of relationships that included these geopolitical issues.

Moreover, European integration can also be seen as a distinct west European effort to contain the consequences of globalization. Rather than be forced to choose between the national polity for developing policies and the relative anarchy of the globe, west Europeans invented a form of regional governance with polity-like features to extend the state and to harden the boundary between themselves and the rest of the world.

Western Europe is an overcrowded region in terms of both population and number of distinct states, no longer able to solve the pressures of overcrowding by exporting portions of the population to other parts of the globe. West European states have been particularly ambitious in expanding the scope and cost of public policy, while simultaneously seeking to maintain competitive economies. The welfare state, in its several variants, remains an essentially west European phenomenon, partly because it draws on deep social and political roots and an embedded political culture, and partly because it was the form of 'modernization' or socio-economic adjustment adopted in the period following the Second World War (Kaelble 1990; Flora 1993). Because this choice of modernization was internationally constrained, its adoption was intimately bound to the development of external policies and thus to transnational cooperation.

A third distinctive feature of western Europe has been the prevalent commitment to forms of liberal democracy, not as a monopoly of west Europeans, but as a strongly preferred system to be defended against competing models in the immediate neighbourhood. Thus both those west Europeans who inhabited long-democratic states and those who had memories of previous non-democratic experience have retained an active concern: to embed democratic practice in Germany and Italy, remembering the experience of the 1930s; to rescue countries like Spain and Portugal from dictatorship; and to conserve liberal democracy from the ideological competition with communism at their borders.

The competition with state socialism was particularly intense and structured in Europe, impinging on political, economic, and security issues for all the west European countries on the liberal democratic side of the line, as well as in the countries with more authoritarian regimes. It is impossible to understand the history, the policies, and the institutional forms of the EC without taking these factors into account. The EC was intended, and is widely perceived as having succeeded, to contribute to democratic stabilization. In a way it was necessarily democratic, because it was a club of west European democracies. The EC was bound to be open to enlargement to include European countries that were setting out on a path of redemocratization (Lepsius 1991).

Thus, we suggest, the political symbolism of this shared concern has been a powerful factor in inducing certain patterns of transnational cooperative behaviour. Political symbols have been weighed against more material concerns in the determination of policy preferences. What is less clear is how far these particular symbols and concerns still weigh in the balance or whether they remain equally potent for political opinion across western Europe, or how far they are still attached to the EC/EU as a vehicle for their assertion. The redemocratization of western Europe is taken by some to be sufficiently 'given' not still to require an intense investment in its transnational articulation. Also 1989, by marking the end of serious ideological competition, may have weakened the impact of political symbolism and of the case for democratization eastwards.

This shared attachment to liberal democratic values and to exporting them across the continent has, however, always been accompanied by differences between west European countries and by the resilience of distinctive political cultures from country to country (Schulze 1994). In the early period of west European integration these differences had a relatively small impact at the transnational level. More recently the differences have been magnified and more influential in explaining lines of resistance to integration or differences of interpretation about the scope and limits to integration. The counter-symbolism of national identity has probably gained more appeal as a result of waning fears of nationalism in western Europe, in contrast to 1945, and also as a partial response to globalization. It has also coincided with efforts by some European politicians to increase the level of European political integration.

The agenda of cooperation

European governance has embraced a broad agenda of policy cooperation. One consistently important item on that agenda has been *geopolitical stabilization*, both the provision of external security and the embedding of 'safe' democracy. For these purposes a twin-track set of arrangements was developed in the 1940s and 1950s, partly European, through what became the Western European Union (WEU), and partly transatlantic, through Nato. The EC did not become an accepted arena for the direct provision of security, though it made an indirect contribution in two senses: first, by making stronger the multilateral links and commitments between its members, especially between Germany and its former adversaries; and, secondly, by encouraging economic stabilization. The changed context since 1989 has reopened the question of whether or not the EU should become the direct provider of geopolitical stabilization for its existing members and its eastern associates as the primary arena for a common security and defence policy.

The second core area of policy cooperation at the European level has been that of *socio-economic adjustment*. Here too the EC had to compete for a share of relevant policy responsibilities, both with the state and with other international fora. The founding treaties gave the three European Communities distinct policy tasks in relation to post-war reconstruction, the liberalization of markets, and corollary sectoral and horizontal policies. The precise policy agenda has changed over time, reflecting changes in the impact of globalization, in the character of industrial society, and in the definition of new public policy concerns.

Broadly, however, a division of labour was struck between the more social elements, and in particular the core of the welfare state, as primarily country-based, and the more economic elements, of which substantial parts were concerted at the European level. The one striking and important exception was agriculture, as is argued in Chapter 4, where the welfare of producers was tied to the European level of governance. But the division of labour was not tightly drawn, leaving persistent grey areas between the national and the European levels of policy and thus creating policy competition over the control of these grey areas. Also, differences in the character of the welfare state in individual countries produced different preferences and expectations of European policy cooperation in this area (Streeck 1992).

The main explicit scope of Community policy has lain in this area of socio-economic adjustment. It has often been accompanied by the language of 'Community preference', that is to say a privileged position for intra-Community interests, with shared identities being built around shared policy regimes for agriculture, some industrial sectors, trade instruments, monetary stability, product standards, and so on. But the orientation and focus of socio-economic policy have altered over time, reflecting changes in the economy, in industrial structures and competitive performance, in societal characteristics and demands, and in ideas and doctrines. We have seen over the past decade efforts to expand the scope of European policy to include, at one end of the spectrum, monetary and macro-economic policies and, at the other end, social policy broadly defined, from labour markets to issues such as health. Both efforts at policy expansion have been deeply controversial and remain contested.

The core emphasis on socio-economic adjustment has been interpreted as requiring European regulation, though differently defined across the decades (Héritier et al. 1995). But this core area drew in its train an unavoidable entanglement with issues of distribution and redistribution (Peffekoven 1994). There has thus been competition between countries, regions, sectors, and socio-economic actors for the benefits of redistribution and to avoid undue burdens. Some of the competition about distribution is played out around explicit EC policies with more or less planned distributional consequences. Other contests are over the distributional impacts of policies which have different primary purposes, such as trade and market liberalization. The arguments over distribution have recurrently been amongst the sharpest that have affected the development of collective policies. Their sharpness has, if anything, increased as the pressures on distribution within the member states have also become more acute since the 1980s and as redistributive policies have been thrown into question by neo-liberal ideas and market-emphasizing social and economic policies.

The third core function of European integration has concerned the *political symbolism* attached to the EU as a club of west European democracies. This has permeated institution-building and the development of law within the EU; it has defined which countries were eligible for membership; it has generated some specific common rules and policies; and it has impacted on the definition of external policies. Some policy developments can be seen as attempts to appropriate more intense political symbols to support European governance—the social dimension and the elements of EU citizenship are both examples of this. But the limits to the resonance of European political symbols can be seen in the continuing power of political symbols attached to the polities of the member states. These in turn constrain policy cooperation in areas where symbolism is a crucial ingredient to be weighed alongside judgements about substantive rewards and burdens from cooperation (Schnapper 1992).

Endemic political and policy competition

Thus, as we have seen, the pressures have been intense for transnational policy cooperation to compensate for the inadequacies of the state, to respond to globalization and to do so in a distinct form that reflects historical experience and shared political values in western Europe. The agenda of that cooperation has over the decades included geopolitical stabilization, socio-economic adjustment, and the projection of political symbols, as well as everyday practical or functional policy cooperation.

But, as we have also seen, the forces that have induced cooperation have coexisted with factors that have inhibited it. Thus there has been a recurrent and, we have argued, endemic political and policy competition. This competition has been played out in several ways:

- The EU has had to compete with other forms of transnational cooperation to establish a predominant position: with Nato and WEU in the defence arena; with the European Free Trade Association (EFTA) for some countries; with wider groupings, such as the Organization for Economic Cooperation and Development (OECD), on some issues to engage a broader range of countries; or with specific sectoral organizations, such as the European Space Agency (ESA).

- The EC has had constantly to compete with country-based alternatives for the development of policy and for the achievement of political and economic objectives.

- Within the framework of the EC the allocation of powers has been the subject of competition, precisely because the EC and more recently the EU still do not constitute a polity based on a clear specification of powers, scope, and authority.

Ideas

It has become commonplace to argue that integration has been a function of the interaction and definition of interests, a point we develop further below. However, it follows from what has already been asserted that *ideas* have played a very important part in defining the grounds for cooperation and for sustaining cooperation even when the direct consequences may be uncomfortable. The neofunctionalists already had recognized that there was a shared body of ideas that bound together coalitions of élites prepared to invest in integration. Their definitions were particularly linked to the notion of integration as a process of transforming and modernizing the political economies and social impacts of public policy across the participating countries, through what Leon Lindberg (1963) has called a cross-class and cross-country compact. One key element in this was the large extent of political space occupied by a form of collusion between Social and Christian Democrats in the founder-member states of the EC. This shared ground provided an underpinning consensus of a quasi-ideological (rather than post-ideological) character. It enabled much of the discussion on the specifics of policy to be determined on apparently more technical and substantive grounds.

The underestimation in later years of the impact of ideas was perhaps in part because often they were implicit rather than explicit. When ideas are taken to be held in common the participants may not feel the need to keep restating them, especially if they are not the subject of vigorous contestation from opposing viewpoints. As Marcus Jachtenfuchs (1993) has argued, in an excellent essay on the need to bring ideas back into the analysis, a range of intangibles to do with beliefs, values, and world views, as well as to do with prevailing normative preferences, are necessary parts of the explanation of European integration. Joseph Weiler (1992) has articulated the same point cogently in his insistence on the extent to which certain shared values have underpinned the development and application of European law in the member states of the EC. Indeed it is hard to explain the embedding of European law without taking some shared value set as at least implicit and sometimes explicit in the formulation of distinct legal doctrines with often powerful economic, political, and social consequences.

The new debate on the role of epistemic communities is a different and also fruitful way into the discussion of the role of ideas (P. Haas 1992). In some senses it is a reprise of the Monnet approach to European integration. For Monnet and his colleagues it was important to provide a framework through which 'the brightest and the best' could be enabled to pioneer new ideas for collective and supranational policies. It was a strategy designed to permit the shared European policy arena to predominate by producing better-informed policy guidance than the normal process of politics, the model on which the French Planning Commissariat had also been developed by Monnet (Winand 1993; Duchêne 1994). The underpinning assumption was that knowledge-based policy communities could be built and were likely to be persuasive in developing policy.

In the early period of the EC this form of expertise was cultivated in the High Authority, and later the Commission, and in its choice of external inter-locutors around the initial policy agenda. Indeed it can be argued that an epistemic community was also built around the concept of European inte-gration as a policy model, which drew in alongside the more technocratic exponents of the method certain political and economic élites as well as more independent experts. In a relatively small EC, in which relatively tight-knit policy communities could be built, there were special opportunities for these groups to be bound together by shared ideas as well as by a shared concern with a particular policy sector, and for these to take root as an important influence on policy outcomes.

Versions of this phenomenon can be found in the subsequent history of the EC, especially in 'new' policy arenas, the environmental perhaps being the most amenable, as scientific knowledge and specialized policy communities began to develop and to find a response from European-level policy-makers (Héritier 1994). Indeed it is hard to explain the initial development of European environmental policy without acknowledging the force of ideas ahead of the definition of substantive interests as a determinant of policy, with the 'leaders' pulling along the laggards, as Alberta Sbragia argues in Chapter 9.

Another example of the role of ideas in defining the policy process is the emergence of new thinking about the role of liberalized markets and the shift to a regulatory approach to collective policy (Pelkmans 1990; Majone 1996). In the mid-1980s a variety of factors combined to refocus thinking and even-tually decisions on the 'common market' core of west European economic integration. The paradigm on which policy was based altered, as several case-studies in this volume illustrate. The new paradigm had to compete, and still does, with an earlier paradigm which assumed a more beneficial role for intervention and some deliberate distortions of market forces, often defended as legitimate Community preferences. Significantly, the new para-digm started to undercut some of the old consensus on the welfare state, although without completely replacing it. This has led to the paradox of two important sets of ideas coexisting, but sometimes conflicting, in the reshap-ing of public policy at both European and member-state levels.

Interests

All analysts agree that interests and self-interests have repeatedly played an important part in the European policy process (Moravcsik 1991 and 1993). But there is considerable disagreement on which kinds of interests carry most weight and on how and why interests are defined in the way that they are. We note that interest-based analyses of cooperation are increasingly being explored in rational choice and game-theoretic explanations of collective action and conflict. Such analyses rest their arguments on the primacy of self-interest and calculated strategies for achieving defined goals. Depending on which kinds of actors are assumed to be pre-eminent and which kinds of interests are considered to predominate, such theories then offer systematic and repeatable assessments of the likely pattern of outcomes from the games that are played around the development of European policies.

We acknowledge throughout this volume the importance of interests and self-interest in influencing the behaviour of actors in the European arena. However, we also argue that some interests have been developed around the establishment and sustaining of that policy arena as such, that is to say that some policy positions reflect views about the policy and political utility of the EU as a forum. Also we believe that we can identify cases in which ideas are modifiers of interest-definition and calculation. We argue that the diffuse nature of the European policy arena enables an unusual range of actors to intervene and thus that the 'n-person' character of bargaining across multiple issues limits the applicability of game-theoretic explanations.

We make thus six assumptions about the interests that bear on European governance.

- *A plurality of organized interests* is engaged in efforts to influence the policy process and, where their propagators can, operates at whichever level of governance they believe to be the most receptive to their preferences and demands. Whether or not corporatism used to be a feature of the EC model (a debatable point), pluralism is now established (Streeck and Schmitter 1991). However, there are distortions in the process such that access for interests may be relatively open, but opportunities for influence are biased to favour some interests more than others. In any case the high level of involvement and activism of many different organized interests should not be confused with definite influence on outcomes (Young 1995).

- *Interest satisfaction* is a crucial feature of the process. As the neofunctionalists also argued, the development and sustainability of European policies requires both a basis of common interests and the satisfaction of individual interests. Rewards on substance help to generate agreement on the system as a whole.

- *National interest* is a misleading and mostly unhelpful term. It is frequently invoked by member governments in Community bargaining to elevate the salience of the positions that they defend. But member governments, albeit representatives of member states, are composed of winning parties with partisan preferences. To assert something as a national interest suggests either that a nationally negotiated consensus exists or that there is some clear and quasi-objective stake for a country as a whole.

- *Nested games* (most actors are simultaneously involved in a whole network of games) play an important part in defining which interests rise to the surface (Tsebelis 1990). The competition of interests within states is juxtaposed with the competition of interests deriving from the international arena.

- *Ideas may define interests.* Over the years German political opinion has attached high normative value to the consolidation and democratization of European institutions, and German governments have been prepared to let such considerations influence their views on specific substantive issues, even to the point of accepting relatively high material costs (Katzenstein 1987). The 'shadow of the future' does weigh in the balance and includes for other member governments also a commitment to systemic goals (Keohane 1986; Axelrod and Keohane 1988).

- *Rational choice theories are of limited utility* in helping us to understand the definition and defence of interests in the Community. Actors often adopt satisficing rather than optimizing strategies, and the interplay of ideas and interests makes it hard to identify preferences as systematically and consistently as rational-choice analyses demand, especially in a context of complex multilateralism (Ruggie 1993).

Institutions

We deliberately did not start our account of the European policy process with its institutions, although many accounts of European policy-making do precisely that. One of the most distinctive features of European integration is that it has been built around an unusual pattern of institutions. These have formed the skeleton of the European policy process, becoming an enduring framework as policies have developed, as personalities have come and gone, and as the interplay of interests has evolved. The institutional skeleton has been state-like in certain important respects, creating the impression and the expectation that the politics and policy process would resemble those conventionally associated with the state (Schuppert 1994). Indeed the preoccupation with the 'democratic deficit' rests partly on the assumption that these European institutions should be subject to tests of accountability and representativeness similar to those applied to the individual member states.

Our argument here is threefold. First, the aspiration to establish state-like institutions has permeated their design and their operating practices. The Commission has tried in many respects to behave like a governing executive; the Court has sought to become a constitutional court, and perhaps succeeded; and the Parliament has endeavoured to acquire traditional parliamentary powers.

The Council is the institution that is hardest to equate, which brings us to our second point, namely that precisely because the EC was not established as a state its institutions were in crucial respects not state-like. Indeed the invention of the European arena of governance might reasonably be expected to have developed different institutional features and modes of operation. It is this that has led to so much insistence on the EU as *sui generis*. In this sense the institutions have had and retain an experimental character, managing public functions traditionally associated with the state, but having to rely on different political resources, forms of authority, or substitutes for political territory. Thirdly, the institutions have a provisional character and have been repeatedly subject to review, both in the formal sense of treaty reforms and informally, as participants in the policy process have sought to foster particular institutional roles and relationships.

Further details on the institutions are set out in the next chapter. There has been an unusual process of institution-building at the European level, both the formal institutional structures and the more informal patterns of institutionalized behaviour (Quermonne 1994). These institutions have had distinguishable consequences for the modes of building collective policies, both as constraints and facilitators. But, secondly, institutional development and, in particular, institutional reinforcement have been intimately tied to policy agreements that appeared to require institutional changes. The linking of institutional change to policy change in the SEA is a vivid example of this.

We might go further and suggest that significant institutional changes are improbable unless linked to changes in the level of agreement about policy. Thus a more intense consensus on policy would ratchet institutional change, as is in theory intended by the agreement to construct on EMU, while dissensus on policy could unravel some institutional commitments, as perhaps illustrated by the recent emphasis on subsidiarity, voting rules in the Council, and questioning of ECJ judgements.

We note also three specific points about European institutions that are not generally stated. First, the institutional impact of the European arena in the polities of the member states varies considerably. The Community policy process is articulated through quite different institutional settings in different countries—to take an obvious point, through a broadly consensual framework in a country like the Netherlands, and through an adversarial framework in a country like the UK. Secondly, some features of the institutions of the member states are absent at the European level, in particular the programmatic competition that characterizes choices between parties as electable or not and also the interest aggregation and adjudication functions of parties. Thirdly, the partial or incomplete coverage of European institutions makes the rules of access to the institutions very unclear and more unclear than in most established polities. It is partly for this reason that 'policy networks', currently amongst the most attractive and rewarding areas of study, have become so important (Héritier 1993).

Issues

It is the issues of policy that test the capacities and reveal the features of the European policy process. It is only when faced by choices over issues that policy-makers and policy-influencers are forced to consider whether the European arena is likely to be more or less productive and more or less acceptable than the available possible alternatives. As issues develop policy-makers and policy-influencers inject ideas about the issues, thus defining the range of potential policy responses, and begin to formulate their preferences as to how to set policy. As and when issues are channelled through the intricacies of the European institutions they acquire specific characteristics.

The case-studies in this volume deal with specific and varied issues or clusters of issues. They are collected precisely in order to demonstrate the range and the diversity of issues on the Community agenda and to illustrate the variety of responses from the European policy process. They show the importance of understanding the substantive content of each set of issues and reveal great differences in the ways that issues get on to the policy agenda for possible European resolution. Some issues have been driven by the Community itself, some by the emergence of a new body of ideas, some by specific constellations of organized interests or member states, some by events, some by external pressures, some by a combination of several impulses. It is by examining a variety of issues that we can start to specify the ways in which ideas, interests, and institutions interact and are intermediated and negotiated to produce outcomes both positive and negative—that is, sometimes generating policy agreement and sometimes controversy.

Inherent instability

There is an inherent instability in the European policy arena. By this we mean that it is rarely certain that the outcome of the policy dialogue will produce a clear and consistent line of policy amenable to a sustained collective regime. In other words European policy regimes are conditional rather than definitive, a consequence of the continuing fluidity of the political setting as less than a polity, pulled between the political territories of the member states and the pressures of global and European influences.

But we do not intend to state by our use of the term 'instability' that everything is uncertain. On the contrary, as cases in the volume show, the actual record of policy development may accumulate policy authority at the European level and weight the European as against other contenders as the most attractive policy provider and on a long-term basis. We note also that this may emerge in unintended or unexpected ways. The social dimension, covered in Chapter 7, reveals particularly clearly the emergence of a distinctive European social regime, but substantially different from what has traditionally been seen as social policy. The policy process is unstable and unpredictable in that it varies over time, across countries, and across issue areas. Wolfgang Wessels (1992) has made a compelling argument that these characteristics are as much a result of fluctuations in the policy processes of the member states as a reflection of European integration. He views the country-derived fluctuations as paving the way for a 'fusion' of country and European policy processes. We are more agnostic in suggesting that diffusion may as well be the result.

Policy engineering

Most polities provide opportunities for the policy engineer, skilled at manipulating the policy process in pursuit of particular policy objectives or interest satisfaction. We understand policy engineers as groups of policy-makers who are concerned more with forcing policy change or policy replacement than with maintaining the *status quo* and who try to mobilize and manipulate policy-building resources in order to sustain policy change and policy replacement. These engineers may be the politicians and bureaucrats associated with the traditional policy processes of countries, but they also include those who can succeed in inserting new forms of policy-influencing, whether as a complement or a competitor to the traditional ones.

The European arena, not quite a polity, and with endemic competition for influence and advantage to a particularly large extent, allows special scope for the policy engineer to tip the balance of the competition. The experimental or transformational character of European integration has provided new institutional opportunities for leverage and invited policy engineers with a 'mission' to develop policy strategies in ways that would be harder in more established and traditional settings—'leaders' versus 'laggards'. The élitist character of the forum has been an added factor, decoupling the policy engineer more abruptly from democratic scrutiny than is at least the appearance within the member states.

Obviously the European Commission has been a particular pivot of influence for the policy engineers, but opportunities also arise for nationally based policy engineers to exploit the interface between national and European governance and also for non-governmental actors to act unusually directly as policy-makers and not just policy-influencers on occasion (Eichener and Voelzkow 1994). Whether this is to be seen as a creative or a manipulative process is a matter of judgement, but we should at least recognize the importance of policy engineering as less subject to political intermediation and often more sector-specific than is generally the case within the member states. In this context the roles of policy networks and of advocacy coalitions (Sabatier 1988) are especially important political resources to be mobilized in support of policy development.

Effectiveness of policy

We argue that tests of policy effectiveness play a particularly important part in the European policy process. This is for two different but related reasons. First, European policies, and especially their sustainability, depend on establishing credibility as a more 'efficient' alternative to other options or as a means of filling gaps in national policy provision. But by not being embedded in a secure form of political authority this credibility is always open to challenge, a challenge that can be withstood to the extent that European policies deliver substantive results that are valued by the recipients.

Secondly, the base of legitimacy for European policies is fragile, because of the partial character of institutional underpinning and weak mechanisms of conventional parliamentary accountability. Questioning of legitimacy may be reduced by policy resonance and by the building up of alternative support bases with the proponents and clients of policy. But support bases that are issue- or measure- or sector-specific are more often contingent than systemic, thus more dependent on results-based gathering of support.

There is evidence from across a range of policies of both negative and positive feedback resting on statements of the purported ineffectiveness or effectiveness of policy content. Moreover, policy gains are frequently claimed by member governments, while policy failures are readily attributed by national politicians to the European level. In the European case both the credibility of the process itself and the outcomes of implementation are hazardous. The sustainability of European-level policy cannot be taken for granted unless policy problem and policy outcome mesh together. Policy success can enhance the policy setting; policy failure can undermine it. The policy engineers, who may not be very visible, may not get the credit for the success, but only the criticism for failures, since national political leaders are only too willing to claim the successes and to distance themselves from the failures.

We therefore need to identify ways of assessing policy effectiveness at the European level: all the debate on subsidiarity since its formulation in the Spinelli draft treaty has talked of policies that can be 'better' or 'more effectively' or more 'appropriately' delivered at the European level (Sinn 1994). We list here, for exploration across the case-studies, some questions about effectiveness:

- policy authority clearly established at European level and not elsewhere, such that no alternative is available, and effectiveness then has to be judged by results;
- policy based on rules that can be and are backed by tough European law, thus providing firm compliance;
- policy backed by resources and their distribution at the European level, thus building in incentives for the beneficiaries;
- policy that may change behaviour by relevant actors, hence the argument that the key test of the internal market is not the implementation of the legislative programme, but whether business behaviour has altered in ways that promote industrial restructuring, more competitive production, etc.;
- policy based on 'really good ideas' developed at the European level, thus effectiveness tied to the integrity of the policy remedy;
- policy based on an equilibrium point (Pareto-optimal or Nash equilibrium) where everyone relevant is as well off as possible and where alternative policy arenas would not satisfy these conditions, and thus it is the best alternative to no agreement (Fisher and Ury 1982);
- policy may not be excellent, but alternatives are worse, i.e. second best;
- policy effectiveness within the issue area may be questionable, but policy serves symbolic or setting goals and thus in this sense effective.

Transmission systems

The formulation and implementation of policy depend on the transmission systems through which policy demands are articulated, interpreted, and find practical responses. In the European arena the transmission systems tend to emphasize the technical over the political, or at least political considerations are often implicit rather than explicit. The 'technicity' of policy formulation tends to make details more evident than underlying principles, and the importance of legal interpretation as a tool tends to produce language that is quite impenetrable for the lay person. Policy proposals and rules, because of both their subject-matter and their terminology, are not readily amenable to parliamentary scrutiny or commentary in the general media. It is a debatable point whether this is so intended by policy-makers or an unintended consequence. Either way the recent preoccupation with finding means to improve 'transparency' is a revealing signal of concern at the political results of operating a seemingly closed process.

The transmission systems that handle inputs into the policy process essentially consist of agencies from the member states, expert groups, and alert organized interests. Political parties play a minor role and so do parliaments, except to the extent that the EP is now emerging as a potential agenda-setter (Tsebelis 1994); national parliaments have played a negligible role in initiating policy demands, except on rather rare issues of acute national sensitivity.

The transmission systems that handle the outputs of policy are similarly restricted and mainly built around interactions between the Commission, implementing bodies in the member states (sometimes at local level), and the directly affected clients of policy (A. Smith 1995). Transmission systems vary between sectors, reflecting the segmentation typical of the European policy arena. None the less they include a wider range of opinion and actor participation than is typical of international policy cooperation. It might also be argued with some plausibility (though we lack good empirical studies on this) that networks of key activists from the west European Christian Democrat and Social Democrat parties have played a significant role in conditioning the European policy environment and in facilitating some important collective agreements.

The restrictiveness of participation seems marked if compared to what happens (or is supposed to happen) within a traditional polity. The predominance of bureaucrats also seems striking. On the other hand, policy networks are more widely drawn around European policy issues and differently configured from national policy networks, thus permitting different combinations of policy-influencers.

In addition, however, we note that differences of access points to the European policy arena favour some organized interests more than others. Historically the lopsided formal institutions of the EC gave special access to those organized interests that could 'capture' the ear of the Commission or member governments. These have varied according to issue area: bankers and financiers in the monetary arena, business interests, including firms, in areas impacting directly on industry, the farming and food processors as regards agriculture, and so on. Such groups have been important elements in the policy communities that have emerged and, as far as we can tell, have been quite influential as regards policy outcomes, perhaps more important in the European arena than at the national level because of the absence of political parties as primary forces in the process.

In most policy arenas there has been a competition between interests, although again of a particular kind. The environmentalists and the scientists have found an effective niche at the European level with those services in the Commission dealing with the environmental issues, as they have in some member states. Yet elsewhere business interests have been more prominent in defining the environmental policy options, either by their influence on particular governments, for example the British (Héritier 1994), or by their skill in defining an issue as not primarily environmental, as happened with the initial definition of the case for Trans-European Networks (the TENs).

However, the European policy process has become more politicized by both the greater involvement of the EP in policy, through its strengthened legislative powers, and by the eruption of more controversy about integration as such. Thus we have seen the relative grip of organized interests weaken and the opportunities for countervailing interests open up (note for example the animal-rights movement, the anti-roads groups, and the anti-biotech groups, and also the 'consumer' on national food preferences). A key question about the European policy process is whether the bias of influencing opportunities is different at the European level from that within member countries.

Negotiation as the predominant policy mode

The European policy process has been peculiarly dependent on negotiation as a predominant mode of reaching agreements on policy and of implementing policies once agreed. Much of the literature is misleading in suggesting that the model is either a negotiation model or something else. The analytical question is what characterizes the negotiating process, not whether it exists. Neofunctionalists identified one form of cumulative and solidarity-inducing negotiation as the characteristic 'Community method' (Lindberg and Scheingold 1970). Realists accept the existence of negotiation, but assert that states as the predominant actors can and do reserve the right and capability to set firm limits to what is negotiable and *in extremis* to exit from a negotiation. The liberal institutionalists build much of their argument on the opportunity of political leaders to manipulate negotiations in their 'double-edged' diplomacy in order to gain advantage (Putnam 1988; Evans *et al.* 1993).

So the interesting questions about European negotiations are:

- whether the essence of the Community model is really only about negotiation, as distinct from other forms of political interaction;
- what characterizes the process of negotiation and what special features, if any, are to be found in the Community case;
- what bearing the process of negotiation has on the character and quality of the policy outcomes;
- which participants exercise power within the negotiations and why; and
- what the incentive structure for cooperation is and what would induce defection, or—more commonly in our context—non-compliance.

None the less, although we have asserted that a negotiating process lies at the heart of the policy process more generally, we can make some observations about changes in its character and centrality at different periods. In the early phase of the EC participation in the process was limited, a technocratic mode was predominant in defining policy, and political leaders had many opportunites to 'collude'. Subsequently negotiations became in some ways encapsulated and often hostage to 'veto-groups', the phenomenon that Fritz Scharpf castigated as the 'joint decision trap' (1988).

Later the negotiating arena expanded in scope and membership, more policy actors as well as more countries, and as the fora for negotiation on EC issues proliferated. In this sense pluralism ran riot, making negotiations more open in their outcomes and making it harder to establish the determinants of the results. More recently negotiations have become more bounded by the circumspection of many participants. Positions have become more defensive and the windows of opportunity for agreement have become narrower, as the arguments for restraining the scope and the authority of the collective arena have been more loudly voiced.

Another development has occurred. Part of the reason for the predominance of negotiation at the core of the process had to do with the cartel of élites that dominated the negotiating fora and the interests that lay behind them. That cartel has been 'threatened' by the impact of other forms of policy influence. These include the imposition of policy through the courts, both ECJ and national, and the emergence of a form of parliamentarism at the EC level. Irrespective of whether the EP provides legitimation of European executive decisions, it certainly interferes with the negotiating process. It can, and sometimes does, overturn the results of negotiation in and around the Commission and the Council.

The debate joined at Maastricht about the Community pillar versus the two new 'intergovernmental' pillars is partly an argument between the notion of European integration as a developed policy and political process and the limitation of the European policy process to what is negotiable between governments as the predominant actors. This can be differently expressed as a competition between the Commission and the Council for institutional predominance. We return to this set of questions in Chapter 2.

Substance and symbolism of policy

It is our underlying contention that actors engage in seeking policy cooperation through European means for both substantive and symbolic purposes. Factors that relate to both weigh in the definition of goals, the calculation of interests, and the development of negotiating strategies; they explain where the coalitions are formed and where the deals are struck. Some actors are preoccupied more by symbolism than substance, though substantive arguments tend to predominate much of the time for probably the majority of participants in the process. But on occasion quite important substantive interests may be relegated to a subordinate role in determining preferences. The symbolic goals of participants often cluster around preferences for more national control or more scope for the European arena, i.e. on the issue of integration itself.

Moving the pendulum

As we have argued, the policy pendulum may be set in motion by a number of different impulses. The length and direction of the swings of the pendulum varies, sometimes in steady motion, sometimes resting more on the side of European policy outcomes, sometimes veering towards the retention of the national policy arenas. The shape of the movements differs between issues and policy sectors and the pull of the national polity is stronger for some countries than for others. Even motion depends on a similarity of policy demands from the participating countries and transnational policy networks. But it does not depend on identity of policy demands or necessarily imply convergence of needs or demands. We argue rather for the importance of congruence or compatibility of policy needs and demands, in which European policies may satisfy quite different ideas or interests for different countries or groups.

The institutions of the EU represent an effort to provide a casing to constrain the pendulum, and thus to channel its movements. It is through the interactions within the institutions that ideas and interests are negotiated, allowing winning coalitions to emerge or blocking coalitions to obstruct. The relative solidity of the institutional casing imposes distinctive structure on the negotiating process and helps to consolidate the results at the European level. But the sustainability of policy agreement depends on the congruence of ideas and interests and on the effectiveness or utility or value-added of the outcomes, as well as on the institutional channels through which they are expressed.

Note

1. Dorothy Dunnett's Niccolò saga makes good reading on this for the 15th c., the first being *Niccolò Rising* (London: Penguin, 1986).

Further Reading

For general accounts of early attempts at theorizing about European integration see Pentland (1973) and Webb (1983), or for the original analyses of neofunctionalism E. Haas (1958) and Lindberg (1963) and the *riposte* of Hoffmann (1966). For the American theorizing that followed the Single European Act see Sandholtz and Zysman (1989), Moravcsik (1991), Sbragia (1992b), and Schmitter (1992). For European analyses see Marks (1993), Scharpf (1994a), Schnapper (1992), W. Wallace (1990b), and Wessels (1992). To understand the role of the law in the integration process see Burley and Mattli (1993) and Weiler (1991), and of economics see Tsoukalis (1993).

Burley, A.-M., and Mattli, W. (1993), 'Europe Before the Court: A Political Theory of Legal Integration', *International Organization*, 47/1: 41–76.

Haas, E. B. (1958), *The Uniting of Europe: Political, Social, and Economic Forces, 1950–1957* (Stanford, Calif.: Stanford University Press).

Hoffmann, S. (1966), 'Obstinate or Obsolete: The Fate of the Nation State and the Case of Western Europe', *Daedalus*, Summer: 862–915.

Lindberg, L. N. (1963), *The Political Dynamics of European Economic Integration* (Stanford, Calif.: Stanford University Press).

Marks, G. (1993), 'Structural Policy and Multilevel Governance in the EC', in A. W. Cafruny and G. G. Rosenthal (eds.), *The State of the European Community*, ii: *The Maastricht Debates and Beyond* (Boulder, Col.: Lynne Riener, 1993).

Moravcsik, A. (1991), 'Negotiating the Single European Act: National Interests and Conventional Statecraft in the European Community', *International Organization*, 45/1: 19–56.

Pentland, C. (1973), *International Theory and the European Community* (London: Faber).

Sandholtz, W., and Zysman, J. (1989), '1992: Recasting the European Bargain', *World Politics*, 42/1: 95–128.

Sbragia, A. (1992*b*) (ed.), *Euro-Politics: Institutions and Policymaking in the 'New' European Community* (Washington, DC: Brookings Institution).

Scharpf, F. W. (1994*a*), 'Community and Autonomy: Multi-Level Policy-Making in the European Union', *Journal of European Public Policy*, 1/2: 219–42.

Schmitter, P. C. (1992), 'Interests, Powers, and Functions: Emergent Properties and Unintended Consequences in the European Polity', unpublished paper.

Schnapper, D. (1992), 'L'Europe, marché ou volonté politique?', *Commentaire*, 60.

Tsoukalis, L. (1993), *The New European Economy: The Politics and Economics of Integration*, 2nd edn. (Oxford: Oxford University Press).

Wallace, W. (1990*b*) (ed.), The Dynamics of European Integration (London: Pinter).

Webb, C. (1983), 'Theoretical Perspectives and Problems', in H. Wallace, W. Wallace, and C. Webb (eds.), *Policy-Making in the European Community*, 2nd edn. (Chichester: John Wiley and Sons).

Weiler, J. H. H. (1991), 'The Transformation of Europe', *Yale Law Journal*, 100/8: 2403–83.

Wessels, W. (1992), 'Staat und (westeuropäische) Integration: Die Fusionthese', *Politische Vierteljahresschrift*, Sonderheft 23/92: 36–61.

CHAPTER TWO

THE INSTITUTIONS OF THE EU: EXPERIENCE AND EXPERIMENTS

Helen Wallace

Institutional powers and relationships were contested in the original EC design and throughout subsequent reforms; they have varied across policy sectors and over time in complex and inconsistent ways. These reflect the interplay of ideas, interests, and policy substance. A succession of different policy modes with different institutional configurations has marked the history of European integration. The Monnet mode of partnership was challenged by the Gaullist mode of negotiation. Segmentation of policy-making became endemic, reflected in the entrenched structures of the Commission and the Council. The Single Act was characterized by the cooption of wider political and economic élites, but the ambitious aims of

Maastricht produced yet more varieties of cooperation rather than a tidy construction. The outcomes for the main institutions—Commission, Council, Parliament, and Court—are assessed against the backcloth of impending challenges to the Community model of governance.

Introduction

Most features of the European policy process are incomprehensible without an understanding of the special setting of the European Union (EU), its institutional rules and norms, and institutional behaviour.[1] The particular pattern of institutions makes some policy outcomes possible and others impossible, and it imprints special policy modes on European governance. Indeed we have argued that the institutions provide a critical catalyst in the process of policy formation. It has often been asserted that the EU institutions are *sui generis*; we prefer to describe them as in some crucial senses experimental. Over the past decade there has been an almost continuous history of deliberate institutional change, with more envisaged at the 1996 Intergovernmental Conference (IGC), following three decades of informal institutional evolution. Enlargement will prompt further institutional changes beyond the purely mechanical and thus the institutional *status quo* is almost certainly not an option or a probable outcome.

In this chapter we emphasize this experimental feature and the malleable character of European institutions. It is a key point of our argument that policy-makers are themselves involved in crafting the institutions and in bending them to suit their own purposes—and not only their substantive policy purposes. Precisely because European policy-making questions the adequacy of national institutions, just as it probes the scope for and consequences of new forms of transnational governance, those involved often have institutional objectives *per se*. They may accept particular policy outcomes because of their institutional consequences and may even reject policy outcomes that would favour their substantive policy interests because they do not wish to accept the institutional implications.

A contested point of departure

Variety and complexity of institutional behaviour were the two features that we emphasized explicitly in earlier editions of this volume. We argued that institutional behaviour was quite significantly different in one sector from another, engaging policy interests in different kinds of ways and producing different kinds of outcome. Unless these variations were well understood the policy analyst was likely to misunderstand what drove particular policies, and it was imprudent to assume that behaviour developed in one policy arena would necessarily be replicated in another. Indeed we suggested that different policy arenas had their own institutional dynamics, an observation that is largely supported in studies of policy-making within countries. This variety remains a feature of European policy-making and is illustrated throughout the case-studies that follow.

Complexity was, we argued, also an endemic feature of the European institutions. The devil was generally to be found in the detail. Thus the good student had to burrow deep into a subject to dig out as much detailed empirical evidence as possible and to build up thereby a portrait of the politics and policy that related to the particular policy field. As one sharp critiquer observed to this author once, we were pointing a way for the traveller who needed to be familiar with each individual village, but we were perhaps not equipping the traveller to understand the wider terrain. The complexity remains, so much so that 'transparency' has become a key contemporary theme in the EU institutional debate, with practitioners under increasing pressure to explain and to justify the way in which they reach decisions. Another reflection of this concern is the call for 'simplification' of the institutional processes and rules of the EU.

In previous editions we did, however, chance our arms at some generalization, in particular that the then European Community (EC) was a part-formed polity. The institutional system was, we asserted, state-like, even if some of the attributes of a fully-fledged polity were denied, precisely because the EC had to coexist with the still relatively robust and embedded polities of the member states. Thus the last edition concluded that the EC was much more than a regime, the then currently fashionable term in international relations. Where we remained agnostic was on the question of whether the EC was developing into a federal or quasi-federal entity, though we acknowledged this as a possibility.

Beneath this line of argument, however, lay an assumption that we would now contest. The notion of the EC as 'less than a federation, but more than a regime' implied that there was a linear progression from one to the other and

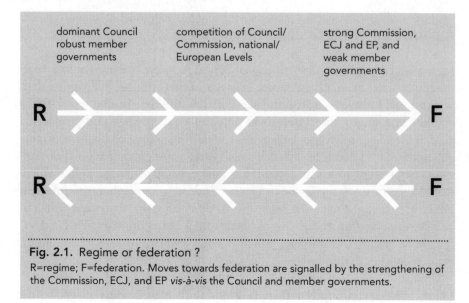

Fig. 2.1. Regime or federation ?
R=regime; F=federation. Moves towards federation are signalled by the strengthening of the Commission, ECJ, and EP *vis-à-vis* the Council and member governments.

that this was expressed institutionally (see Fig. 2.1). Though we were prudent enough not to assume that a shift along the line was inevitable, and though we indicated that different policy arenas were located at different points on the line, the potential for 'progress' along the line was assumed. The progress would be marked by a combination of increased collective policy and 'strengthened' common institutions that increasingly resembled those of a state.

Expressed in conventional institutional terms, a predominant Council, composed of resilient national governments, would characterize a point on the line closer to R, while a predominant Commission and European Court of Justice (ECJ), with a strong European Parliament (EP), would show evidence of a point closer to F. The remarkable strength of European law and of the ECJ helped to point the arrows towards F. In contrast, the pull of national sovereignties and of national political territory would turn the arrows to point in the other direction, a tempting conclusion in mid-1995, with the post-Maastricht emphasis on increased 'subsidiarity' (or 'repatriation', in the discourse of British conservative politicians).

The *pendulum* metaphor set out in Chapter 1 precisely throws into question the assumption of linear development. Instead we have argued that the tussle between levels of government is reflected in a persistent institutional competition. Our revised assumption is that European policy-making rests on a reconfiguration of governance and a diffuse and shifting allocation of institutional roles. The participating countries are engaged in a transnational policy process through collective decision-making. The scope and reach of the collective institutions can be larger or smaller (see Fig. 2.2); the member states (here simplified as **A**, **B**, and **C**) have a durable character, but have accepted a degree of shared governance (**S**). The extent of shared governance may vary over time, less in Fig. 2.2(*a*) and more in Fig. 2.2(*b*), or between countries, as in Fig. 2.2(*c*).

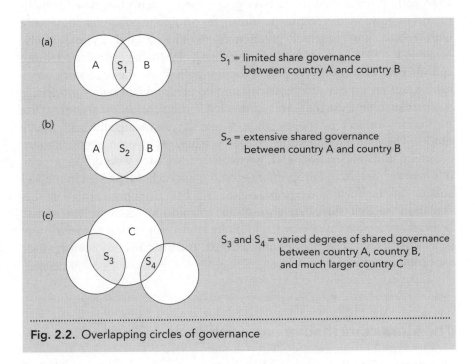

(a)

S_1 = limited share governance
between country A and country B

(b)

S_2 = extensive shared governance
between country A and country B

(c)

S_3 and S_4 = varied degrees of shared governance
between country A, country B,
and much larger country C

Fig. 2.2. Overlapping circles of governance

These overlapping circles of governance do not produce exclusive jurisdictions or a clear division of powers between national and EU levels, but, on the contrary, overlapping areas of institutional authority. This permits examples of strong institutional powers at the European level, but with weak policy impacts, and also of strong policy emanating from weak institutional mechanisms. It also means that the European institutions look very different depending on the angle from which they are viewed.

In the sections that follow we have plotted changes in the institutional patterns over time. We aim to show the coexistence of different institutional patterns, as well as a succession, in order to emphasize the competition between these different patterns. The point we want to stress here is that particular episodes and their accompanying policy developments have been installed at different periods as the key illustrations of the prevailing policy model and of the way that the institutions work. In the 1960s it was agriculture and the customs union that set burgeoning supranationalism as a frame of reference for the analysts (Lindberg 1963). In the 1970s it was the gap between stated goals and demonstrated achievement that appeared to show stagnation in integration. In the mid-1980s the hyperactivism around the single-market project, cohesion policies, and environmental programmes provoked a flurry of new analyses that uncovered multi-level games, policy networks, new forms of regulation, and increased private influences on public policy, all competing with robust governmental preferences (Sandholtz and Zysman 1989; Moravcsik 1991; Majone *passim*). The early 1990s have generated more agnosticism, with monetary policy, foreign policy, and the regulatory arena exhibiting confusingly different traits and no predominant policy model.

These successive portraits are represented in this volume. We argue that the institutions impact on policy and *vice versa*, producing different policy modes. These we have identified as: partnership, negotiation, segmentation, cooption, and cooperation, each particularly associated with different periods in the evolution of European integration, but each still present in the discussion of the further development of the EU. In commenting on this evolution we have three reference-points in mind for assessing the consequences of the policy modes: what ideas are embedded in each mode; the way in which interests are expressed and satisfied; and the capability of each mode for problem-solving.

The Monnet method of partnership

The essence of the distinctive institutional framework created for the EC was a form of systematic *partnership* between those who could speak for the individual member states, the economic actors who would be directly affected by the results of the new system, and a new group of European public servants who would develop the collective agenda. Its inspiration was drawn from the contribution of Jean Monnet and his associates to the development of the EC (Duchêne 1994). The core of the partnership consisted of the 'tandem' relationship between the Council and the Commission. Each had its function in this partnership and the object was to devise institutional practices that would produce net value-added for all concerned in terms of specific policies, institutional stability, and political synergy. The hope was that this would become a functioning equilibrium, dependent on the Commission, as proposer of legislation and implementer of agreed policies, and the Council, as the negotiating forum in which representatives from member governments would fight their corners, feed in their preferences, and legitimize the decisions reached. This 'tandem' was moderated to an extent by the interventions of the European Parliament, interventions both symbolic and sometimes substantive, but rarely defining or definitive. The backbone of compliance and 'good behaviour' was provided by the ECJ, facilitating enforcement, providing those affected by European legislation with opportunities to assert their rights, and sometimes insisting that the legislators had intended more than they themselves recognized.

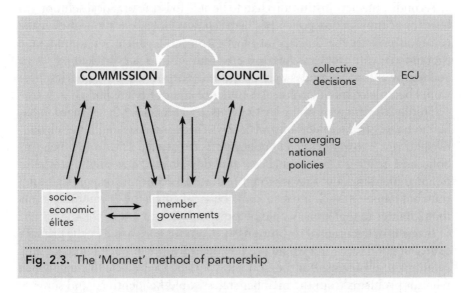

Fig. 2.3. The 'Monnet' method of partnership

The Monnet method (see Fig. 2.3) in effect underpinned the analysis of the neofunctionalists (E. Haas 1958 and Lindberg 1963) and continues to influence policy-making and the debate about political integration. It is part of what informs the current nostalgia for the presumed intimacy of the 'old' Community of the Six. Common European institutions were the vehicle through which shared European public goods were to be defined and delivered. They would coexist with the institutions of the participating countries and in time perhaps displace them as the primary focus of political activity, at least in those issue arenas where they had significant authority. The institutions would focus the interactions of key economic and political élites from the member states and in time create new or successor élites for whom the Community arena would be the preferred and predominant arena. The model is straightforward to describe, but its operation rested on a kind of collusion behind the scenes, assuming a permissive consensus from supporting publics, the Council to deliver national electorates and the Commission to deliver satisfied clients.

The institutional process, anchored essentially by this interaction of the Commission and the Council, would achieve its purposes if it could succeed in three important ways. First, it must be able to stimulate good ideas and attract talented policy engineers. That is to say it needed to draw in the 'brightest and the best' as the sources of innovation, modernization, and cosmopolitanism. At the time benign technocracy was the image. We might now talk of epistemic communities as the necessary underpinning for enlightened policy designs (P. Haas 1992).

Secondly, the new institutions had to be able to bring practical solutions to bear on acknowledged policy problems. It was not a question of devising common policies for the sake of having common policies or technocratic aggrandizement: problems had to be solved, and policies backed by tough and legally enforceable rules were the means. Without effective policies there would be little chance of political confidence and policy credibility.

Thirdly, the interests of the participants, and especially the engaged élites, had to be satisfied in a demonstrable way. Those interests might be substantive and relate to very concrete policy preoccupations. They might be symbolic and relate to values, beliefs, and doctrines. They might attach to national identities, but they might also attach to economic or political function. But if enough such interests could be satisfied via the Community arena then loyalties to that arena would become stronger.

In terms of the original institutions the Commission would be the primary source of good ideas and would provide the transmission belt for generating solutions to problems. But the Council would be the vehicle through which relevant problems from the member states would be identified and some of the relevant interests would be articulated. Other relevant interests would be fed through to both the Commission and the member governments and perhaps through formal consultative bodies for social and economic partners. Hence the Commisssion/Council tandem was surrounded by a plurality and diversity of other stakeholders in the Community policy arena, both socio-economic interests and national policy-makers. 'Normal' politics in the sense of parliamentary politics were to be kept at arm's length, not so much because the aim was autocracy for the clever policy engineers as because parliamentary politics were judged to have failed several important European countries within living memory. Socio-economic interests and their representatives were early identified as more important to engage than the conventional 'political forces', thus implying a form of corporatism (Streeck and Schmitter 1991).

What the treaty design did was to sketch, but only to sketch, the basis for endowing the EC with the capabilities to satisfy these three aims of good ideas, problem-solving, and interest-satisfaction. The sketch did, however, add in some stabilizers, which with hindsight can be identified relatively clearly. The Commission and the ECJ were given the potential for a substantial degree of institutional autonomy and leverage. They would have to earn this by their own effective performance (Cappelletti *et al.* 1986). The member states were given a degree of reassurance by the way in which the Council was structured and the various formulae for reaching decisions and making common rules, depending on the nature of the policy powers being developed.

The interests of the collectivity and a bias towards the collective over the particularist were partly to depend on the Commission and the ECJ upholding collective goals and interpretations of shared purposes and agreements. But collective interests and shared purposes were to be further bolstered by giving the smaller members among the founders, the three Benelux countries, somewhat enhanced status. This was precisely because these countries had the strongest possible reasons for investing in a strong shared framework that would circumscribe the temptations of the larger countries to use their greater economic and political muscle egoistically or hegemonically. Initially broader political endorsement for this new process would be indirect and come via the legitimation of the participating states. Over a period the European Assembly, later EP, might take on a more important role, but only if a basis for this could be defined and agreed, and, implicitly, if parliamentarians could earn their passage.

For a decade and a half EC governance did indeed rest on this Commission/Council tandem. The perhaps inevitable competition between Council and Commission, however, became a dramatic confrontation in 1965, which marked the subsequent evolution of both institutions. Thereafter the Monnet method had to compete with other approaches to the use of collective institutions.

The Gaullist method of negotiation

The Luxemburg crisis of 1965 brought to the fore the limits of the Monnet method. Even though France stood to gain much in substance from the Commission's 1965 proposals (secured agricultural financing and budgetary 'solidarity'), the French president judged that the institutional price was too high: greater Commission autonomy, the beginnings of a real role for the EP, and clear agreement to use the treaty provisions on qualified majority voting (QMV) in the Council. Out of the protracted debate emerged a more restrained scope for the common institutions. The Commission was given less room for manœuvre; the development of the Parliament was stunted; and it proved very difficult to exercise the technical opportunity for QMV, with the weight of the 'Luxemburg Compromise' appearing to imply that the right to national vetoes was clearly embedded across the board. Only the ECJ escaped untouched from this confrontation, leaving it free to develop as the most independent and 'integrationist' of the institutions (see Fig. 2.4).

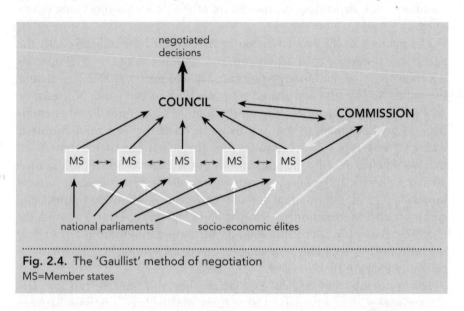

Fig. 2.4. The 'Gaullist' method of negotiation
MS=Member states

In the subsequent period there was a recurrent tension between the Monnet method, operating in areas where the turf had been won for the partnership mode, and the negotiation mode that took root as an alternative. The well-known face of the Gaullist method was the series of confrontations on 'high politics' issues, with the French government the most explicit defender of member-state powers. Indeed it was the prevalence of this image that began to undermine neofunctionalism and to make intergovernmentalism a more common descriptor of the Community's institutions. Yet more important, we argue, was the shift to negotiation as the predominant policy-making method. Here the essence was not antagonism as such, but rather positioning, a guarded approach to policy extension and a sharp insistence on entrenching the interests of particular clients of policy in the policy rules and in the institutions.

With the benefit of hindsight we can see rather clearly the imprint of the new mode on the institutions. By the early 1970s the Council had spawned a great substructure of committees and working groups, from the Committee of Permanent Representatives (Coreper) outwards and downwards. This came to lock armies of national officials into recurrent negotiation with each other and to change the equilibrium point in the Council/Commission relationship in favour of a greater weight for the Council. All of these committees in some senses mimicked the Council in putting to the fore the defence of 'member state' (more accurately member government) interests, even though at working-group level some of the habits of functional technocracy survived. The negotiation structures also led to a more rigorous calculation by the participants of their relative positions *vis-à-vis* each other and *vis-à-vis* their own statuses *quo ante*.

Nor was this negotiation mode confined to the Council. It permeated the Council/Commission relationship, as the Commission participants inside Council discussions came to act more tactically *vis-à-vis* the Council as a whole and individual member governments. Within the Commission negotiation between commissioners and between services became characteristic, as the Commission itself came to operate as less of a college. It is to the early 1970s that belongs the emergence of the *cabinets* of commissioners as the vehicle for 'pre-negotiations' between the commissioners, acting in a rather veiled way to filter decisions for the college, as Coreper did for national ministers.

The Commission also began to proliferate committees for consultation before policy was drafted and for the management of policy after it was agreed. It was only after these committees had proliferated that the odd piece of Eurojargon was coined—'comitology'—to identify the phenomenon and then to turn it into a term of art for the *cognoscenti*. It was only later that the issue of accountability of these committees began to be voiced. Interestingly, for many practitioners on the inside track the main concern was about the acceptability and legitimacy of the Council committees, while those a little further removed presumed that the Commission-sponsored committees were the more contestable.

We should also note that the negotiation mode made it quite hard for socio-economic interests to get a handle on the policy process. The Monnet method had invited and indeed won the engagement of certain economic élites. The Gaullist method forced socio-economic interests back on to the use of more traditional national channels, since what became vital was to insert their points into the national negotiators' briefs early and persuasively. Since those briefs were written mostly in national capitals, through the burgeoning and increasingly complex systems of national policy coordination, this emphasis on national channels of influence was wholly rational in those areas of policy where the negotiation mode predominated.

Segmentation of policy-making

The pattern of policy-making that developed during the 1970s was essentially more and more segmented between policy arenas. Different policy fields took on different characteristics; these reflected the varying attributions of policy competences to the EC and the differential abilities of particular policy communities to exploit opportunities that would increase the scope for collective action. It is precisely this development that we covered extensively in earlier editions of this volume.

Within the Commission individual directorates general (DGs) took on increasingly distinct personalities and developed different working methods (see Table 2.1). They shaped varied relationships with their external interlocutors and behaved differently in their dealings in the Council and *vis-à-vis* member governments. In part these variations reflected differences in the tasks that they had been assigned: DG IV (competition) had a regulatory style; DG VI (agriculture) had a clientelist style; DG III (industry) had an interventionist style, and so on. These were reflected in individual 'organizational cultures' and working methods. Some of these differences flowed naturally from policy content, but the net effect was to fragment the Commission as an institution. Internal coordination became more and more of a problem, in spite of the efforts of the General Secretariat to correct it. Individual commissioners began to develop their own fiefdoms, the character of which was also shaped by the personalities of individual office-holders. No president of the Commission seemed able to grasp the institution as a whole.

Those close to the process were well aware of the costs and benefits of this segmentation. On the plus side the scope was there for individual areas of policy to blossom: the Commission played a pioneering role in developing transnational environmental regulation, as it did in developing what became the research framework programme. On the minus side were the diminished cohesion of the Commission as a collective institution and the lack of management controls over its working practices. Various efforts, internal and external, were made to counteract the segmentation, but with very limited results (see, for example, Spierenburg *et al.* 1979; Sutherland *et al.* 1992; and Metcalfe 1992). It became clear that the Council and the member governments were not much moved to help the Commission to improve its internal management and cohesion, perhaps not surprisingly. Rather more puzzling was the reluctance of commissioners and senior officials to take on the task themselves.

A similar story can be told about the Council. Early on the notionally

Table 2.1. Structure of the European Commission: 1995

DG I*	External economic relations
DG IA	External political relations
DG II	Economic and financial affairs
DG III	Industry
DG IV	Competition
DG V	Employment, industrial relations, and social affairs
DG VI	Agriculture (and Veterinary and Phytosanitary Office)
DG VII	Transport
DG VIII	Development
DG IX	Personnel and administration
DG X	Information, communication, culture, and audiovisual
DG XI	Environment, nuclear safety, and civil protection
DG XII	Science, research, and development (and Joint Research Centres)
DG XIII	Telecommunications, information market, and exploitation of research
DG XIV	Fisheries
DG XV	Internal market and financial services
DG XVI	Regional policies
DG XVIII	Credit and investments
DG XIX	Budgets
DG XX	Financial control
DG XXI	Customs and indirect taxation
DG XXII	Education, training, and youth
DG XXIII	Enterprise policy, distributive trades, tourism, and cooperatives

Secretariat General
   Forward Studies Unit
   Inspectorate-General
   Spokesman's Service
   Joint Interpreting and Conference Service
   Security Office
   Statistical Office
   Translation Service
Legal Service
Informatics Directorate
Consumer Policy Service
European Community Humanitarian Office (ECHO)
Euratom Supply Agency
Office for Official Publications of the European Communities

Attached agencies

European Foundation for the Improvement of Living and Working Conditions
European Centre for the Development of Vocational Training (CEDEFOP)
EUROBASES (EC Database)

* DG: Directorate General.

single Council had in practice fragmented into a variety of specialist Councils—General Affairs (GAC, composed of foreign ministers), Ecofin (composed of economic and finance ministers), and so on (see Fig. 2.5). These took on different personalities, different rhythms of work, and different approaches to

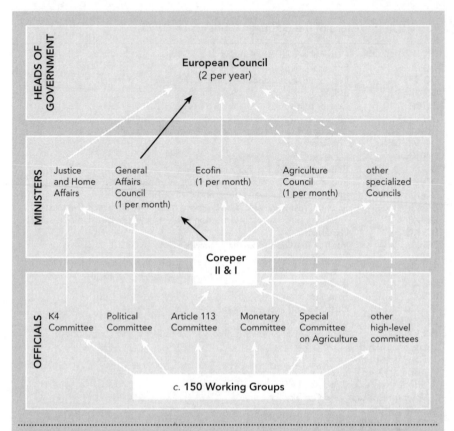

Fig. 2.5. The structures of the Council

Justice and Home Affairs (JHA), K4 Committee and Political Committee formally incorporated only after Maastricht.
Coreper II (Ambassadors) prepares Councils and topics that are more political;
Coreper I (Deputies) prepares the more technical and financial issues
Article 113 (EEC) Committee deals with trade issues

bargaining, some bearing the Monnet imprint, others the Gaullist imprint. Coordination became a problem within the Council, even though Coreper in principle filtered most decisions for ministers. The more senior Council formation of foreign ministers was supposed to be able to coordinate the more specialist Councils, but had limited success. The Council presidency, with its six-monthly rotation, offered some scope for assiduous and well-focused member governments to impose some order and coherence on the management of business, but not reliably and predictably so. The Secretariat General of the Council also played some part in coordination, but only up to a point.

The results, as with the Commission, were both positive and negative.

Individual Councils and their supporting committees were able to develop policy and practice quite extensively and to avoid too much disturbance from other Councils. This enabled the Agriculture Council, for example, to acquire very considerable policy autonomy. The downside was that issues could become blocked within a particular Council and the connections between related policy developments were often not well made. In addition the powerful impact on the Council of both the British budgetary quarrel and the external oil shocks contributed to an image of the Council in which blockages looked more characteristic than movement.

In some policy areas—monetary and foreign policies are the obvious examples—the institutions as such were barely engaged, policy cooperation depending much more on informal circles of consultation. It is tempting to call this 'intergovernmentalism', in that the classic establishment of clear Community competence and distinct treaty and institutional rules were absent. But intergovernmentalism is too rigid a term for the engagement of policy élites from the member states in the process and the structuring of dialogue that was achieved. There was a process of *engrenage* or locking in, and the costs of defection were significant, even if the scope for penalizing defectors was less than in the classical Community framework.

It was with these varied experiences that the decision was taken to hold meetings of heads of state and government as the European Council from 1974 onwards. This intensified the involvement of the most senior politicians from member governments, a reflection of the salience of the EC for them. A different policy mode almost emerged; a Schmidt–Giscard method of Franco-German propulsion made some running during the 1970s as a primary source of initiatives, best illustrated in the steps leading to the creation of the European Monetary System in 1979 (Ludlow 1982; Morgan and Bray 1984). But both the European Council and the Franco-German couple had rather uneven results in developing strategic decision-making. For those observers who saw the tip of the iceberg the image of stagnation was powerful. The fact that so much responsibility for seeking to impose coherence was thrown back on to national policy coordination within member governments contributed to the impression of an increasingly intergovernmental character to the EC. Fritz Scharpf (1988) castigated the prevailing policy mode of the period as 'the joint decision-trap'.

Within the classical Community framework some elements of institutional 'conservatism' began to be revealed. Both the Commission and the Council began to act defensively and to become prone to institutional rigidities. Within the Commission, as standard operating procedures came to be established, and the orthodoxies of 'appropriate' behaviour came to be acquired, the search for new policy proposals became increasingly predefined by precedent. This had the advantage of settling policy-makers into repeat-able routines and habits, precedent being a powerful argument for encourag-ing the acceptance of Community authority and extensions of policy rules. The terminology of the *acquis communautaire* exemplifies this as the induce-ment to fall into line. The *acquis* was shorthand for the accumulated obliga-tions and commitments agreed under the treaties and legislation of the EC over the years. By definition it was conservative, encouraging Commission officials to limit their policy proposals to established approaches to policy, even when they did not bear fruit. The history of industrial policy and of har-monization affecting non-tariff barriers in the 1970s both illustrate this.

Similarly, negotiation in the Council became caught in a pattern of cau-tious consensus-building. In part this was a legacy of the Commission/de Gaulle confrontation of the mid-1960s, in part a function of the treaty deci-sion-rules on voting which left the exercise of qualified majority voting tech-nically possible but politically difficult, and in part it was a product of a policy context that produced many substantive obstacles to agreement. In any event on many issues saying 'no' was easy and individual member governments were easily induced to say 'no', even if isolated; and saying 'yes' mostly implied only limited incremental change.

None the less in the late 1970s and early 1980s efforts were made to shift the institutional behaviour. The Commission started to look for different ways of changing policy by institutional innovation. The Davignon strategy for developing the R&D programme was one example, in which he sought very deliberately to build a new policy coalition with leading IT companies. Those dealing with market harmonization began to craft what became the 'new approach' (see Ch. 5). In managing the structural funds the Commission began to probe the scope for more direct links with regional authorities in the member states (see Ch. 8). In the Council during the early 1980s majority voting was used more frequently and began to change habits, indeed to reassure doubting governments that this need not be so dramati-cally disturbing as they feared. Actual recourse to the Luxembourg Compromise was barely mooted. The productivity of the Council as a leg-islative machine increased.

The Single Act and the cooption method

The Single European Act (SEA) took most analysts unawares and led to a flurry of new and often surprised assessments of Community policy-making. Our contention throughout this volume is that the SEA built on already existing foundations and consolidated rather than invented a new policy method that we have here called cooption. We set this out in summary here; examples are to be found in many of the following case-studies. As Fig. 2.6 shows, the Single Act method was a more complicated evolution of the policy process, reflecting the disjointed incrementalism of the EC more generally.

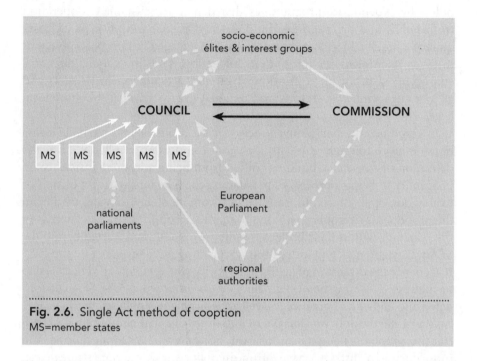

Fig. 2.6. Single Act method of cooption
MS=member states

Essentially three developments coincided in the early 1980s; we deliberately do not pass judgement here on their relative order or influence.

- The political economy debate altered, as neo-liberalism came into competition with the old Christian/Social Democrat consensus. This led some governing parties to start advocating a different policy mix at both national and European levels.

- Big firms, with increasingly transnational bases, began to look to the European policy process to produce different operating rules and thus to channel different and rather focused policy demands towards both national and Community policy-makers (Cowles 1994).

- The Commission began to develop different and more strategic relationships with various of its policy clients. The net result was to coopt the outside clients of policy as the architects and engineers of policy.

They were helped by the enabling circumstance of a weakening of traditional policy relationships at the state level. This also released transnational networks or advocacy coalitions which mobilized arguments and support for collective European policies. Enlargement during the 1980s also added countries with new policy demands that significantly increased the size and range of the Community budget. This in turn extended the reach of Community policy by expenditure programmes that demanded sustained cooperation from public and private organizations across the Community. As a consequence a much wider range of interests were engaged in European policy, as much in its implementation as in its formation. Thus the parallel phenomena occurred of an intensification of interest articulation in Brussels and greater engagement of sectoral and regional interlocutors at the more local level.

In response, the bargaining process in the Council also altered. Voting became more frequent explicitly, but more important implicitly. Thus the definition of consensus-building shifted from being a wooden insistence on unanimity to being a process of persuading the reluctant to shift position or to abandon opposition. This resulted less from insistence on votes than from the proponents of a policy working harder to accommodate the concerns of the reluctant within their preferred policy. This shift had started before the SEA was negotiated, and indeed was an enabling factor even before the 1985 IGC, hence the extension of qualified majority voting (QMV) did not need to be addressed as an issue of big principle. In other words, the SEA consolidated and embedded practice, as well as enlarging its application.

After 1985 the policy process went into overdrive. The combination of substantive proposals being taken forward, not only the 1992 programme, and more purposive discourse gave European policy-making a more strategic feel and a more effective image. Many commentators have attributed this 'success' specifically to the leadership role of Jacques Delors as President of the Commission from January 1985 (C. Grant 1994; Ross 1994). To deny his impact would be absurd, and he was the first Commission President since Walter Hallstein (Loth *et al.* 1995) to make such a big impression on both the Commission and its interlocutors, as well as on wider public opinion. Indeed both Delors and some other commissioners came to be very closely identified with particular policies and their successful advocacy, a factor in enhancing the credibility of the Commission as an institution. None the less much of the new dynamism within the European institutions was a product of the other factors summarized above and illustrated in the case-studies that follow.

Another important point needs to be stressed. The cooption method worked very effectively at the meso-level of policy and in providing a frame of reference for the micro-level of policy. In both cases there were transna-

tional groups to be mobilized, clients to be satisfied, and a loosening of constraints as member governments retracted the reach of national public policy. However, during the 1980s policy achievements were much more limited on issues of high politics—the fields of monetary, foreign, and internal security policies.

Maastricht: competing policy methods

The Treaty on European Union (TEU) had grand policy ambitions and muddled institutional results. Optimism, built on experience of the Single Act method, led to expectations that comparable successes could be achieved in policy arenas of high politics which had few of the features that had facilitated the burst of policy activity during the 1980s. Monetary, foreign, and internal security policies are the property of rather closed groups of policy-makers, with rather limited scope for marshalling wider coalitions and characterized by practice and precedent rooted in the defence of national political territory. The proponents of reform sought to bring these policy areas within the Community framework and to subject them to the same institutional rules and norms.

The negotiations during the IGC dealt with an array of institutional propositions. A key issue became whether or not to maintain a single primary institutional framework which would revolve around the interaction of the Commission and Parliament and also give clear positions to both the Parliament and the ECJ (Cloos *et al.* 1994). The outcome was to retain the inherited institutional pattern for the range of policy issues that it already covered, but to extend those policy competences somewhat by enlarging their impact in areas cognate to those where the EC was already active. In addition, the European Parliament won the right to influence legislation more extensively under the new 'co-decision procedure'.

However, for the new 'high politics' issues of foreign and security policy, and justice and home affairs, the price of their inclusion within the scope of what would thereafter be the Union was that they would be subject to different and weaker institutional regimes. These would be organized under a second and a third 'pillar' for more 'intergovernmental' cooperation, with the main decision-making to be conducted through the Council and European Council. The Commission would participate, but with a rather light role compared with the first pillar; the Parliament would have very limited opportunities to comment on second and third pillar work; and the Court was more or less excluded from them. A long discussion took place over where to put economic and monetary union (EMU), whether as a separate pillar or within the first. The latter view prevailed, although EMU is subject to special institutional rules as regards the independence of the eventual European Central Bank, and in specifically catering for only some member states to be fully engaged.

Thus the established institutional pattern was put into explicit competition with a new and more loosely cooperative mode (see Fig. 2.7). In one sense it could be argued that this simply formalized the *status quo ante*, since, as Chapters 15 and 16 show, the two new pillars built on established circles of intergovernmental cooperation. However, to formalize in this way and to reject the orthodox institutional framework was not a casual decision. It reflected serious reticences on the part of some member governments about the Europeanization of these policy areas. This conditionality for some, notably the British government, rested on the hope that this intergovern-

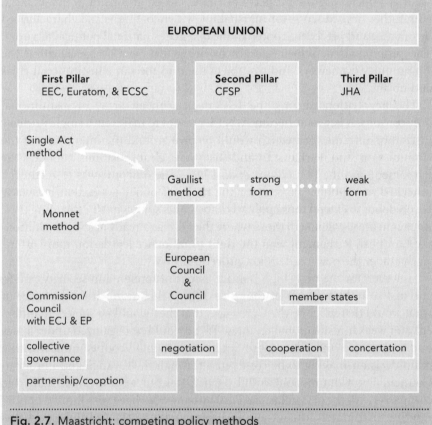

Fig. 2.7. Maastricht: competing policy methods

mental mode would defeat calls for any extension of the more supranational or 'centralizing' features of the first pillar.

The institutional outcomes

Each of the main institutions in the policy process has thus altered its behaviour as part of this changing pattern.

The Commission

The Commission has had everything to play for in terms of acquiring policy credibility and political territory (Edwards and Spence 1994). It became a complex organ and spawned a variety of direct relationships with the clients of its policies, with experts, and with interlocutors in the member states. The ugly term 'comitology' for its formal relationships with interlocutors implies a heavy procedural bureaucratization. In crucial respects the image is misleading, since the Commission has also cultivated, often rather effectively, close and productive relationships with experts and has often been quicker to respond to new policy problems than the also bureaucratized and often conservative or tradition-bound processes of governance within the member states. Epistemic communities have developed around the Commission and generated new policy ideas. New opportunities for exercising influence on policy definition have also been provided by the new relationships increasingly exploited by pressure groups (Mazey and Richardson 1993a) and by regional-level public authorities and organizations concerned with local economic development (Keating 1992). One unexplored empirical question is how far these relationships contributed to the positive expectations of big firms from the single-market programme or to the lack of expressed anxiety from the regions about winners and losers from the single market.

On the other hand, the Commission's concern to get close to its clients also drew it into very close relationships with some of them. Here the Commission behaved more like a traditional government in acquiring the baggage of expectation of vested interests. Examples abound in familiar areas—agriculture, sensitive industries, and the like. Also, though this phenomenon is also underexplored, some individual big firms were able to establish extremely close relationships with the Commission and to exert considerable leverage on policy: the R&D programmes are one example, and the increasing resort to contingent trade protection, through anti-dumping measures and voluntary restraint agreements, is another much criticized case (Cowles 1993).

To produce a systematic audit of the Commission's performance goes beyond the scope of this chapter, but some broad observations might be substantiable. The Commission has shown a significant capacity in some, but not all, policy sectors to develop new ideas and to feed them through into policy proposals or agenda-setting for the EC/EU. Indeed sometimes the Commission is criticized for being too innovative, even eccentric in its suggestions. The obverse is also probably true, namely that where the Commission has been lacking ideas or out of touch with the relevant epistemic communities, its policy proposals have been quite vulnerable. Though proposals have still to be bargained with the Council, those that lack intellectual cogency are particularly vulnerable to erosion.

Ideas are all very well, but of little use unless capable of being transformed into practicable policies. Here the Commission suffers from a 'Catch 22' structural obstacle. Necessarily the Commission is operating a long way away from the coal-face and with little direct experience of 'hands-on' policy delivery. This sometimes frees it to be innovative, even idealistic, and often to be rather detached and relatively 'objective'. But its credibility is always open to challenge precisely because of its 'irresponsibility'. The member governments in contrast have deep experience and ingrained policies, as well as direct political responsibility to electorates. But they also have the conservatism of habit and precedent and the weight of local patronage to defend. When problems are 'solved' as a result of European policies, member governments are quick to claim the credit. When problems are posed, or seem to be posed, by European policies, governments have often been quick to scapegoat the Commission. Thus it is hard for the Commission to win credit, but easy for it to attract blame.

As for the satisfaction of specific interests, the Commission has become recognized in its areas of active competence as a necessary interlocutor. The growth of lobbying in Brussels attests to the importance that a wide variety of special interests attach to trying to get the Commission to adopt their preferences and to produce proposals that reflect their concerns. Indeed in some cases special-interest representatives find a readier ear in Brussels than in capitals (Mazey and Richardson 1993a). But here too the link between channels of influence and institutional support for the Commission is elusive. In relatively few areas does the Commission exercise monopoly policy powers and thus special interests divide their attention between the Commission and member governments. In any case special interests are fickle in their affiliations, since their concerns often have more to do with individual issues than systemic relationships. Another complication is that the Commission lacks the political resources to mediate between competing interests, because its policy competences are constrained and it has no direct political mandate.

The Council

The Council is both a European institution and the prisoner of the member states, or perhaps rather of the member governments. Such collective identity as it has developed is fragile and always vulnerable to competition between the member governments, as well as competition with the Commission. The remarkable growth of the Council with its vertical segments and horizontal layers reveals two paradoxical trends. On the one hand, the extension of policy authority to a process of European governance demands more and more of the energy of delegates from the member governments in the Council, better Councils, with ministerial participants and thousands of national officials. The intensity and intimacy of their involvement is unparalleled at the transnational level. Yet, on the other hand, this engagement is also an effort to control the content and direction of European policies and to subject the Commission to continuous and detailed scrutiny. It also reveals what Alberta Sbragia called the resilience of 'political territory' as a determinant of behaviour (Sbragia 1992a).

Thus the Council in its various formats is always there to second guess the Commission and often to deploy counter-arguments. The bargaining behaviour within the Council, from working groups of national experts to the elevated gatherings of ministers with political responsibility, is geared not to the production of bright ideas but to the translation of ideas into palatable decisions. Thus the Council rarely discusses general principles as distinct from detailed texts. Hence the new transparency code agreed after the Edinburgh European Council of December 1992 talks of opening up Council sessions to public gaze for discussions of principles, but the Council has not been able to deliver on its own promise in this respect. Although the Council and especially the European Council frequently produce statements of intent with big ideas for collective action, they rarely produce detailed decisions that focus on the big ideas. Individual governments are also often more concerned to present the nuanced points on which they have won a concession from others in the Council than to present the core idea of common policy development.

In their defence many participants in the Council would argue that it is precisely the task of the Council to turn the more or less bright ideas of the Commission into practical policies. It is precisely because they each have their feet on the ground within the member states that they concentrate on points of feasibility and operational specifics. Of course there is substance in this defence, but there is also a distortion. Members of the Council are primarily concerned with the impact of policy within their own countries or with the impact of policy on other member states in so far as their own necessarily parochial concerns are affected. It is thus easier to view the Council as a loose composition of distinct national delegates than as a body oriented towards identifying the most appropriate collective policy. It was for this reason that Fritz Scharpf talked of a 'joint-decision trap' created by the Council following a 'bargaining' mode of behaviour rather than a 'problem-solving' mode (Scharpf 1988).

In the past, analyses of the Council tended to blame the unanimity decision-rule for this joint-decision trap. The need to tie all participants into the same decision both enhanced the power of veto groups and concentrated negotiation on adding complexity to policy, in order to satisfy a wide array of special pleading arguments from each member state. Since the early 1980s, and more visibly since the SEA, QMV has become a more frequent decision-rule, making the Council much less dependent on formal consensus. This has enlarged the scope for the Commission to devise proposals that can get past the opposition of a few member governments. It has also enabled groups of governments with a clearly shared policy preference to rally to a particular side of the debate, whether as a propulsive majority or a blocking minority. Thus on some issues it has become possible to change the direction of collective policy and to escape from the 'decision trap'. Even on complex package proposals the Commission has found it possible in this changed environment to make its own preferred proposals 'yesable' to a surprising extent. The Delors-1 and Delors-2 packages on budgetary issues are striking examples. None the less QMV can also have the effect of repeatedly isolating some governments in a marginal minority and of diminishing commitments to those decisions to which a government has not assented.

In the Council governments thus contribute more by way of problems than by way of ideas, and the process has become geared more to restraining than to innovating. Yet this is an over-simplification. New ideas are sometimes fed in by governments or groups of governments (Garratt 1992). The Germans generated much of the push for environmental measures in the EC; the French instigated much of the debate on improving technological capacity; the British pressed earliest for market liberalization; the Spanish argued for the concept of EU citizenship. Thus new policy options can flow from specific governments and then be adopted; successful adoption has generally required both entrepreneurship by the Commission and coalition-building in the Council.

One of the key functions of the Council is to deliver satisfaction for particular interests. The language of Council bargaining is permeated with the vocabulary of what is or is not deemed to be in the interests of the individual member states or some projection of these as wider Community interests. Here we need to be clear as to what is meant. The Council consists of representatives from the member states (actually, those who participate are delegated by member governments). Though they use the vocabulary of national interest, they frequently articulate more narrowly based interests, the partisan interests of particular parties in office, the sectional interests that command the ear of particular ministers or ministries, or those domestic interests that governments choose to promote—for example, favoured firms or favoured regions. There are interests within the member states that may not be much drawn into consideration by governments. So for both the Commission and the Council the spectrum of interests engaged is partial. In the case of the Council this is liable to produce discontinuities as electoral cycles bite on the process.

The European Court of Justice

The impact of the ECJ on the European policy process is undeniable and well embedded (Burley and Mattli 1993). Our concern here is, however, limited to the ECJ's impact on ideas, problem-solving, and interest satisfaction, and to the consequences for the competition between the Council and the Commission. The way in which the ECJ has been a source of ideas is by elucidating general principles of European law; the examples are impressive—direct effect, proportionality, equality, legal certainty, fundamental rights, and so on. Successive ECJ rulings have made a tangible difference to the bases on which EC rules have been defined and applied; they have on the whole enlarged the ground on which European public policies rest and strengthened the capacities of European institutions, especially the Commission, to act. Moreover, as Joseph Weiler has argued, the ECJ has endorsed a set of values to underpin European governance (Weiler 1992).

Rulings from the ECJ have also solved practical policy problems by clarifying an issue that was the cause of controversy within the Council/Commission dialogue. The most famous example by far was the *Cassis de Dijon* judgement, which enunciated the principle of 'mutual recognition' (1979). Though its direct scope is not as wide as posterity has claimed, this did open up a new avenue for collective rule-making and thus underpin the 'new approach' to harmonization and market liberalization (see Ch. 5). Other judgements have been more controversial and have been viewed quite differently as problem-solving or problem-creating, depending on the perspective. The Barber judgement (1990) and subsequent line of cases (several decided in September 1994) on equal rights for men and women in occupational pension schemes (actually going back to an earlier 1976 ruling) settled what had otherwise been ambiguous or restrictive in the practice of employers and pension schemes (see Ch. 7). But these judgements also created problems by requiring from some employers substantial and costly adjustments, so much so that the Maastricht IGC was lobbied hard and successfully by several big employers to take a limiting decision, in itself a curious example of direct special-interest lobbying of the European Council.

Through the ECJ and the opportunity for direct litigation 'individuals' have been able to seek legal redress and clarification of their entitlements and obligations. Whose interests have been satisfied by this opportunity is less easy to state. One caveat to be noted is that you have to prove that you are an interested party to have *locus standi* and plead your case with the ECJ. Firms directly affected obviously find this easier to demonstrate than, say, the consumer vaguely defined. So the bias is in favour of access to litigation for the direct economic actor. The cases brought have been directed at both the Commission and member governments, as well as against other economic actors. As for issues of free movement of workers and equal treatment of workers, individual has meant individual workers, beyond whom stand cohorts of comparably situated individuals. Most cases have complained at the application of Community law by governments or firms within the member states.

Until recently it has been broadly argued that the ECJ's contribution has been to extend the reach of Community law, to sharpen its definition, and to reduce the scope for member states to vary legislation on issues within the orbit of Community competence. In this sense the ECJ might be argued to have been a political actor and not just a source of jurisprudence, or a least to be an active influence on the definition of policy options. The fact that the legal services of the Council and the Commission are such important parts of both organizations stands witness to the encroachment of ECJ rulings on the freedom of manœuvre of policy-makers and politicians. Equally, the use to which ECJ rulings have been put by both policy-makers and politicians, from both national and European institutions, shows also that Community jurisprudence provides opportunities as well as constraints (see Ch. 6 and Ch. 7). Moreover, the ECJ impacts by developing doctrine and ideas and not just interpretation and enforcement of individual cases. However, recent judgements from the ECJ are beginning to reveal greater prudence. The judgement on how the EU is represented in the new World Trade Organization, discussed in Chapter 12, shows an erosion of the integrity of the first pillar of the framework.

There has been resistance, both legal and political, to the ECJ. National courts have not always accepted easily either the supremacy of European law or the judgements of the ECJ on individual issues. National politicians do not always welcome the constraints that the ECJ sets on them. Nor does the Commission always have an easy ride in proving its general interpretations or specific cases. Recently resistance to the ECJ has become more focused and more tenacious. The ratification of Maastricht generated considerable contestation of the reach of Community law and criticism of the 'activism' of the ECJ. The ruling on the TEU from the *Bundesverfassungsgericht* in October 1993 (Winckelmann 1994) rang warning bells, suggesting that acceptance of EU decisions and of ECJ rulings could not be taken for granted. The subsequent ruling from Karlsruhe in January 1995 on the Community regime for the import of bananas (see Ch. 13) is another shot across the bows of Community jurisprudence on issues at the heart of the agricultural and trade policy regimes. Indications that some member governments might want to limit the influence of the ECJ by treaty amendment at the next IGC are another signal of pressures to alter the balance between the institutions.

The European Parliament

In the early phases of European integration the EP lacked influence on policy development and was more decorative than effective. Gradually, since its acquisition of first budgetary powers and then some legislative powers, the EP has inserted itself much more directly into the institutional processes. Indeed the EP was perhaps the largest net beneficiary of the institutional changes in the TEU, having already won ground in the SEA. Some commentators would now argue that the overall result has precisely been to alter the institutional balance within the EU in favour of the EP, even though its powers remain nominally less than those of national parliaments *vis-à-vis* national executives and even though it is commonly argued that a democracy deficit persists.

How is this reflected in terms of our questions about ideas, problem-solving, and interest satisfaction? The opportunity of the EP to generate ideas in the form of a political programme has been hampered by its lack of role in determining the composition of the executive branch and the relative weakness of party discipline. None the less, two sizeable and increasingly well-organized party groups—the Socialists and the European Peoples' Party—do now operate from broadly shared political platforms and often support policy preferences that coincide with those platforms. They expect to cooperate more closely with commissioners and Council members from their political families. MEPs also regularly weigh into the institutional debate in support of strengthened powers for not only the EP but also other 'collective' European institutions. As George Tsebelis has recently argued, the EP is exploiting its existing leverage to influence 'agenda-setting' within the EU (Tsebelis 1994). The requirement for assent from the EP on some important policy issues, such as enlargement and external agreements, has extended the influence of the EP and provided some scope for linkage with its own ideas about the development of the EU. Some specific new ideas can be generated by the EP voting budget lines and demanding programmes to use them, as it did in arguing for what became the Phare democracy programme.

The EP plays less of a role in practical problem-solving, in any case a less obvious parliamentary task. None the less, EP amendments to and modifications of policy proposals have rather surprisingly often been incorporated. In a parliamentary body where expertise is a cultivated attribute by many members, the 'expert' MEP has scope to introduce practical proposals as well as broad comment. In some very precise instances there has been a direct partnership between MEPs and the Commission or Council in the management of a particular issue—one striking example was the role of a few MEPs in facilitating the absorption of the former East Germany into the then EC (Spence 1991).

As for specific interests and the role of the MEPs in promoting them, the dog that has barked remarkably rarely is 'national interest'. Votes or position-taking on national and cross-party alignments are infrequent and indirect. We know that lobbyists and special interests pay increasing attention to MEPs and also that the recent applicants for accession from EFTA countries worked hard to convince MEPs of the merits of their case, though we know less about the correlation between lobbying and policy results. Regional interests have beaten a path to the EP and do look for opportunities to persuade 'their' MEPs to promote their cases, much as do their American counterparts *vis-à-vis* the US Congess. Tales circulated of regional side-payments offered to induce EP assent from doubtful 'southern' MEPs for the EFTA enlargement in May 1994. It remains to be seen where and how the new Committee of the Regions will locate itself in the process.

Current trends

The original institutional design in the founding treaties reflected several different approaches and objectives, inviting the protagonists of these different views to prove their preferences. It is not only that political institutions and processes are generally subject to change in response to circumstances and experience, but also that volatility was an in-built feature of the European integration model. There was and is no clear separation of powers. There is no firm allocation of policy powers between European and member-state institutions. But the ambiguities that result create tensions as well, leave gaps, produce some duplication, and provide opportunities. The discussion of subsidiarity is one result of the ambiguity; another is the confusion about how to handle the so-called democratic deficit; a third is the slide into regulation as a primary instrument of policy (Majone 1993); yet another is the issue of where responsibility lies for carrying out common decisions—the implementation deficit (Sutherland *et al.* 1992) and the phenomenon of policy concealment or policy by stealth—touched on by several authors in this volume.

The shift from EC to EU marks an attempted shift from a collective policy framework to a collective polity; it thus demands both satisfied policy clients and an explicit vesting of political authority in EU institutions. The difficulty of achieving this is revealed by the contestation of 'Maastricht', both its content and its legitimacy. The original institutional design was light on legitimation mechanisms, much reliance being placed on member governments indirectly to confer legitimacy by being seen to endorse and to sponsor collective policies, and on the European and national legal systems to ensure due process and legality.

Enlargement has been a major complication, making it harder to tell what would or could otherwise have been the pattern of institutional and policy development in the EU. Some incline to argue that the inherited institutions would work well with a smaller number of willing and similar member states. For this view institutional dysfunction is a reflection of some combination of too many and too heterogeneous members to be accommodated (Wallace and Wallace 1995).

On the other side of the debate are ranged those who argue that the problems lie in the institutional methods themselves and their policy consequences. The need to improve 'effectiveness' has emerged as a key theme for the 1996 IGC and is generating a heterodox range of proposals, many calling for radical change. For example, some commentators have argued for a radical downgrading of the Commission (González Sánchez 1994 and Vibert 1994), some for an explicit alteration of power relationships between the member states (Schäuble and Lamers 1994; Weidenfeld 1994) to create a directing inner group, others for an explicitly federal constitution.

The policy credibility of European governance has always been subject to challenge from critics in the member states and from those who feel their interests are insufficiently taken into account. Institutional adaptations have tended to make the process more complicated, not more straightforward. The Commission has been long acknowledged as needing reform (Spierenburg *et al.* 1979), yet no agreement has yet been reached on how to reform it. Over time it has acquired its own conservatisms, suffered from overload and under-resourcing, and probably been 'over-lobbied'. This has fragmented its sense of collective purpose. These weaknesses have been successfully offset in those areas and at those periods that the Commission has been able to produce policy leadership and persuasive argument. The Council's members have ensured the persistent centrality of 'their' institution and imposed a heavily bargained style of policy decision-making. But they have not made the Council into an institution capable recurrently of producing policy leadership, though the European Council from time to time does attempt this. The increased influence of the Parliament on policy has added layers to the process, but has not provided a firm base for legitimizing European policies. Much has been left to rest on legal authority, as distinct from political cogency.

In this tough environment for European governance much is left to hang on institutional performance *per se*, and institutional relationships are left to bear much of the strain. Perhaps inevitably the Commission is expected to carry a disproportionate share of the imputed political responsibility. We should therefore not be surprised that proposals are now circulating that seek not to carry out reforms of the Commission to make it more effective but rather to reduce its functions and influence (Vibert 1994). Nor is it surprising that some member governments should assert that more intergovernmentalism is the necessary antidote.

When in addition a fourth enlargement to include some EFTA countries and the prospect of an eastern enlargement bear on the redefined EU, the case for finding a different balance between and within the institutions seems irresistible. The arguments running most strongly in this debate are thus about how to adapt the core Council/Commission relationship, about the respective roles of national and European parliaments, and about the 'balance of power' between different member states. The debate, as another IGC is prepared, seems therefore to stress the 'macro-political' choices about institutions, in particular between supranationalism and intergovernmentalism, rather than what institutional forms would favour the development of specific and effective collective policies. As this volume goes to press the initial diagnoses of institutional performance are being prepared for the Reflection Group that is considering options for the IGC and should produce valuable insights into the workings of the institutions (Commission 1995c; Council 1995). Perhaps so politicized a discussion was unavoidable, given the dramatic change in geopolitical context and stage of evolution of the European model of governance. It is, however, a high-risk conclusion to draw and it leaves much less scope than in previous periods of the European integration experiment for the balance of proof to be provided by substantive policy outcomes. Nostalgia for the old Community and the old inner core almost certainly will not deliver agreement on a successor model of governance. But to forget the original corner-stones of finding good ideas, proposing solutions to common problems, and satisfying particular interests would be to reject the lessons of experience. The challenge is to identify modern functional equivalents and to devise institutional adaptations that will encourage them to flourish.

Note

1. This chapter draws on H. Wallace, 'Die Dynamik der Europäischen Institutionengefüges', in Jachtenfuchs and Kohler-Koch (1995).

Further Reading

For general background consult Dinan (1994) and Keohane and Hoffmann (1990) and for earlier developments H. Wallace (1983b). On the Commission see Edwards and Spence (1994), Grant (1994), and Ross (1995). On the Council see de Bassompierre (1989), Hayes-Renshaw and Wallace (1996) and Ludlow (1995). On the ECJ see Neville Brown and Kennedy (1994). On the European Parliament see J. Smith (1995) and Westlake (1994). On socio-economic interests and lobbying see Kohler-Koch (1994) and Mazey and Richardson (1993b). For analyses of the policy process see Jachtenfuchs (1995), J. Peterson (1995a), and Scharpf (1994a).

Bassompierre, G. de (1988), *Changing the Guard in Brussels: An Insider's View of the EC Presidency* (New York: Praeger).

Dinan, D. (1994), *Ever Closer Union* (London: Macmillan).

Edwards, G., and Spence, D. (1994) (eds.), *The European Commission* (London: Longman).

Grant, C. (1994), *Delors: Inside the House that Jacques Built* (London: Nicholas Brealey).

Hayes-Renshaw, F., and Wallace, H. (1996), *The Council of Ministers of the European Union* (London: Macmillan).

Jachtenfuchs, M. (1995), 'Theoretical Perspectives on European Governance', *European Law Journal*, 1/2: 115–33.

Keohane, R. O., and Hoffmann, S. (1990), 'Conclusions: Community Politics and Institutional Change', in W. Wallace (ed.), *The Dynamics of European Integration* (London: Pinter, 1990), 276–300.

Kohler-Koch, B. (1994), *The Evolution of Organised Interests in the EU*, Paper presented at the IPSA Congress, Berlin, August.

Ludlow, P. (1995) (ed.), *L'Équilibre européen: Études rassemblées et publiées en hommage à Niels Ersbøll* (Brussels: CEPS).

Mazey, S., and Richardson, J. (1993b) (eds.), *Lobbying in the European Community* (Oxford: Oxford University Press).

Neville Brown, N., and Kennedy, T. (1994), *The Court of Justice of the European Communities*, 7th edn. (London: Sweet and Maxwell).

Peterson, J. (1995a), 'Decision-Making in the European Union: Towards a Framework for Analysis', *Journal of European Public Policy*, 2/1: 69–93.

Ross, G. (1995), 'Assessing the Delors Era in Social Policy', in S. Leibfried and P. Pierson (eds.), *European Social Policy: Between Fragmentation and Integration* (Washington, DC: Brookings Institution, 1995).

Scharpf, F. W. (1994a), 'Community and Autonomy: Multi-Level Policy-Making in the European Union', *Journal of European Public Policy*, 1/2: 219–42.

Smith, J. (1995), *Voice of the People: The European Parliament in the 1990s* (London: RIIA).

Wallace, H. (1983b), 'Negotiation, Conflict and Compromise: the Elusive Pursuit of Common Policies', in H. Wallace, W. Wallace, and C. Webb (eds.), *Policy-Making in the European Community*, 2nd edn. (Chichester: John Wiley and Sons, 1983).

Westlake, M. (1994), *A Modern Guide to the European Parliament* (London: Pinter).

PART TWO: POLICIES

Brigid Laffan and Michael Shackleton

The EU's budget, though limited in scale, has a significant political impact: budgetary argument hinders other policies, as in the 1980s; budgetary peace helps other policies to progress. The introduction in 1970 of Community 'own resources' made the budgetary process a testing-ground of institutional powers and position, especially for the European Parliament. The Delors-1 debate then marked a critical threshold, as an example of a classic package-deal that was firmly tied to the 1992 programme, the SEA, and the need to bind in the less prosperous member states. Similarly the Delors-2 debate was a key ingredient of the agreement that underpinned the TEU. None the less the budget tests the Community process to its limits, as the problem of controlling fraud reveals. As the EU contemplates EMU and grapples with enlargement, big questions loom about the sustainability of the budgetary bargains.

Introduction

Historically, budgets have been of immense importance in the evolution of the modern state and they remain fundamental to contemporary government. The purpose of this chapter is to enter the labyrinth of EU budgetary procedures so as to unravel the characteristics of budgetary politics and policy-making in this evolving political order. Where EU money comes from, how it is spent, and the processes by which it is distributed are the subject of intense political bargaining. Budgets matter politically because money represents the commitment of resources to the provision of public goods. Making budgets necessarily involves political choices about the allocation and distribution of scarce resources among the member states, regions, and social groups within those states.

The politics of making and managing budgets have had considerable salience in the evolution of the EU. A number of factors contribute to the importance of the budget. First, the search for an autonomous source of public finance for the original EC was critical in building a Community that went beyond a traditional international organization. Second, budgetary issues have inevitably become entangled with debates about the role and competence of individual EU institutions and the balance between the European and the national levels of governance. Third, budgetary flows to the member states are highly visible; 'winners' and 'losers' can be calculated with relative ease. National politicians regard EU money as 'bringing home the bacon' or conversely as losses to the national coffers. Hence budgetary politics are more likely than rule-making to become embroiled in national politics and national electoral competition.

Questions about the purpose of the budget and the principles that govern the use of public finance in the Union are linked to wider questions about the nature of the EU and its evolution as a polity. Changing ideas about the role of public finance in integration shape the policy agenda in areas such as EMU, regional policy, and social policy. Analysis of budgetary politics casts light on the relationship between political and economic integration. Financial resources are an important means of applying political cement to market integration. Put another way, the budget is a useful yardstick by which to measure positive integration. The size and distribution of the EU budget has implications for the operation of a vast range of policies. In post-1989 Europe the EU budget is an important indicator of the capacity of the Union to meet the challenge of external and not just internal solidarity. The process of managing, and not just making budgets, also raises questions about the management capacity of EU institutions, particularly the Commission. All the institutions, and in particular the Court of Auditors, are paying increasing attention to fraud in the budget and searching for better ways to protect the financial interests of the EU.

A thumbnail sketch of the budget

In the early years of the Community, the budget was a financial instrument akin to those found in traditional international organizations. The budget treaties of 1970 and 1975 led to a fundamental change in the framework of budgetary politics and policy-making.[1] First, the treaties created a system of 'own resources', which gave the EC an autonomous source of revenue. 'Own resources' consisted of three elements: customs duties, agricultural levies, and a proportion of the base used for assessing VAT in the member states, up to a ceiling of 1 per cent. The system of 'own resources' removed the budget from direct national control and strengthened the capacity of the EC to develop independent criteria for the allocation of financial resources. One of the basic principles of the 'own resources' system was that it should apply to all member states, regardless of their size, wealth, the pattern of EC expenditure, or their ability to pay. This was to cause increasing difficulty in the years ahead.[2] Second, the European Parliament (EP) was granted significant budgetary powers, including the right to increase, to reduce, or to redistribute expenditure in areas classified as 'non-compulsory' expenditure, to adopt or reject the budget, and to give annual discharge, through a vote of approval, to the Commission for its implementation of the budget. The 'power of the purse' gave the EP leverage in its institutional battles with the Council of Ministers and allowed it to promote autonomous policy preferences. The Council of Ministers was no longer the sole budgetary authority, although it retained the last word in the legislative field. Third, the 1975 Treaty provided for the creation of the independent Court of Auditors to enhance account-ability in the budgetary process and the management of EC money. (See Box 3.1.)

The emergence of the budget as a genuine instrument of EC public policy was constrained by a basic factor which still shapes EC finances. The EC budget was and remains a small budget; it is small in relation to Community GNP and small in relation to the level of public expenditure in the member states. The 1994 budget represented 1.19 per cent of Community GNP and just 2.4 per cent of public expenditure in the member states (Commission 1993b). Moreover, although the budget has little macro-economic significance for the Union as a whole, it is very important for those of the smaller member states

Box 3.1. The European budgetary cycle and rules

Articles 199 to 209 lay down the financial provisions governing the EEC Treaty, with Article 203 establishing the precise timetable and procedure for making the budget each year:

The Commission initiates the budgetary cycle by presenting the Preliminary Draft Budget to the Council by 1 July (though usually by 1 May).

The Council adopts a draft budget by 5 October of the year preceding its implementation. (The financial year starts in January.) The Council meets with the EP in a conciliation meeting before actually adopting the draft budget.

The EP has 45 days to complete its first reading of the draft. It is entitled to propose modifications to compulsory expenditure, expenditure needed to fulfil the Community's legal commitments (essentially agriculture guarantee spending), and amendments to non-compulsory expenditure. Its control over non-compulsory expenditure is limited to increases within a 'margin of manœuvre', which is equal to half the 'maximum rate of increase', a percentage determined each year by the Commission on the basis of the level of economic growth, inflation, and government spending.

The Council has 15 days to complete its second reading of the draft budget. The Council has the final word on compulsory expenditure but returns the draft to the EP, indicating its position on the EP amendments to non-compulsory expenditure.

At its second reading, the EP has the final word on non-compulsory spending within the limits of an agreed maximum rate of increase. After its second reading of 15 days, the EP adopts or rejects the budget. If it is adopted, the EP President signs it into law.

If there is no agreement on the budget by the beginning of January, the Community operates on the basis of a system of month-to-month financing, known as 'provisional twelfths', until agreement is reached between the two arms of the budgetary authority.

The Commission then has the responsibility for implementing the budget. The Court of Auditors draws up an annual report covering the year in question, and on the basis of that report the EP decides whether or not to give a discharge to the Commission in respect of the implementation of the budget. The discharge is normally given in the second year after the year in question.

that receive extensive transfers from the structural funds: between 1989 and 1993, Community funds represented a sizeable 11 per cent of total investment in Greece, 7 per cent in Ireland, and 8 per cent in Portugal. The continuing overall small size of the budget masks impressive increases in financial resources over recent years and a very significant extension of the Community's policy range (see Fig. 3.1).

The slenderness of EC budgetary resources highlights an important feature of the emerging European polity, namely, the significance of regulation as the main instrument of public power in the Union. The expansion of regulatory policies reduces the need for extensive fiscal resources at EC level and reflects a limited view of the role of public finance in integration. This view has not always been dominant. In the 1970s the acquisition of sizeable financial resources for the budget was widely seen as essential to integration, especially to economic and monetary union (EMU). It was anticipated that a larger budget would be necessary to deal with external shocks and fiscal stabilization, which member states could no longer deal with through management

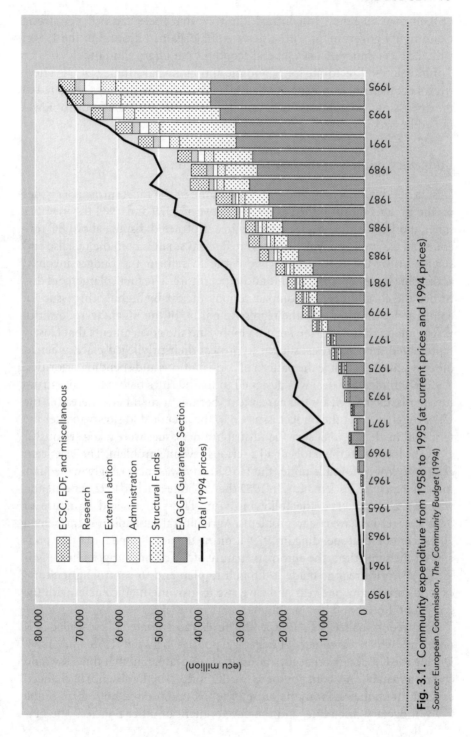

Fig. 3.1. Community expenditure from 1958 to 1995 (at current prices and 1994 prices)

Source: European Commission, *The Community Budget* (1994)

of their own currency. In contrast, the view that there can be 'very strong Community government with a slender purse' gained ground in the 1980s, with the ascendancy of a neo-liberal economic paradigm (Ludlow 1989). The Commission's Directorate for Economic and Financial Affairs (DG II) effectively kept the cohesion issue off the EMU agenda. Cohesion was relegated to the discussions on political union and its implications for the future of EMU not addressed.

Budgetary politics at the centre of the agenda

Despite its small size, the budget has been one of the most intractable issues on the EC agenda since the late 1970s. Between 1979 and 1984 the member states and EC institutions were locked in a protracted dispute about EC revenue and expenditure which contributed in no small way to the malaise and stagnation of the early 1980s. The battle to reform the budget involved addressing issues of both revenue and expenditure. The first enlargement disturbed the cosy budgetary compact among the Six by highlighting issues of equity, redistribution, and the regressive nature of the EC budget. Powerful vested interests, notably the farming lobby and the governments that closely represented their interests, sought to protect their privileged place, whereas others wanted to change the pattern of contributions and spending priorities. The 1970 budgetary deal was designed to fix the rules before the UK, structurally disadvantaged by the agreement, became a member. Thereafter the UK struggled to get the budget issue on to the agenda and slowly to alter the terms of the debate to ensure that distributional issues were taken seriously.

By the late 1970s the budget had a serious revenue problem. The 1 per cent ceiling on VAT, allowable under the 1970 Treaty, was increasingly inadequate to meet expenditure demands. In 1982 the VAT rate reached 0.92 per cent and was virtually breached the following year. The pattern and structure of expenditure also gave rise to problems. Agricultural expenditure represented 70 per cent of total spending in 1979, limiting the development of other policies. At the same time, the automatic nature of the common agricultural policy (CAP) price regime made it difficult to plan and to control agricultural expenditure in any one year and gave rise to growing inefficiencies with the growth of 'butter and beef mountains' and 'wine lakes'.

The predominance of CAP expenditure also accentuated the so-called 'UK problem', which dominated budgetary politics between 1979 and 1984. During the UK accession negotiations, it became apparent that the UK would end up as a major net contributor to the EC budget, in the absence of changed expenditure policies. From the outset, the UK had pressed for reform of the CAP, budgetary discipline, and the development of other policies. The new

Labour government (1974) renegotiated the terms of accession and got agreement at the Dublin European Council (1975) to a complex Financial Mechanism, designed to allow for rebates from the budget in the case of heavy gross contributions. However, this mechanism failed to work, bringing the question of distributional equity to the forefront of Community politics. The problem was seen to be structural rather than the result of chance consequences. Despite being one of the less prosperous member states, the UK became the second largest contributor after Germany. Hence in 1979, the new British prime minister, Margaret Thatcher, began to demand a structured rebate system which would guarantee the UK a better balance between contributions and receipts. The Commission and the other member states were loath to concede the British case at the outset. The Commission has always been reluctant to enter the discussion of the net financial flows to the individual member states, lest this encourage too narrow a calculation of the benefits of Community membership, and lead states to seek *juste retour*, that is to get out of the Community budget more or less what they put in.

Although Mrs Thatcher's confrontational approach was regarded as *non-communautaire*, the essence of the British case was conceded in the form of *ad hoc* spending programmes from 1980 onwards.[3] The continuing search for a long-term solution was underlined by the decision of the European Council in May 1980 to request the Commission to examine ways of solving the problem by restructuring the budget. In its subsequent Mandate Report the Commission advocated a reform of the CAP, an expansion of the structural funds, and reinforced budgetary discipline, notably for agricultural spending. The UK, not content with *ad hoc* arrangements, continued to press for structural reform, with the added leverage that the 1 per cent VAT limit had been reached. Any increase in the Community's revenue base required the unanimous approval of the member states.

These interlinked budgetary issues came to a head at the Fontainebleau European Council in June 1984 during the French presidency of the Council. President Mitterrand wanted the six-month presidency to revitalize the Community. As part of this process, he sought to remove the contentious budgetary issues from centre stage. At Fontainebleau, agreement was reached to increase the VAT ceiling from 1 per cent to 1.4 per cent in 1986 and to establish a mechanism for dealing with excessive British contributions on a longer-term basis. The agreement ameliorated but did not solve the financial crisis in the Community. The limits of the new VAT ceiling were reached in 1986, its first year of operation. The 1987 budget was balanced by the artificial device of delaying CAP payments. Resource shortage now replaced the UK problem as the main budgetary challenge.

Conflict between the Parliament and the Council

While the member governments were engaged in restructuring the budget, the European Parliament and the Council were involved in a continuing struggle over their respective powers on budgetary matters. The EP rejected the 1980 and 1985 draft budgets and the annual budgetary cycle was characterized by a continuous struggle between the two institutions, often referred to as the 'twin arms' of the budgetary authority. In 1982 and again in 1986 the Council of Ministers brought an action in the European Court of Justice (ECJ) for the annulment of the budget that the President of the Parliament had signed. The Council sought to limit the level of power-sharing with the EP as far as it legally could.

The Parliament, on the other hand, was determined to use the budgetary powers it had acquired in 1975 to enhance its position in the Community's institutional landscape and to further its policy preferences. This was done in three ways. First, the Parliament attempted to use its budgetary powers to gain some leverage in the legislative field. Although the EP gained budgetary powers, its role in legislation continued to be severely circumscribed. The EP took the view that the budget itself was a sufficient legal basis for using appropriations entered in the budget. Consequently it inserted additional budgetary lines designed to promote new Community actions, for example, in relation to aid to Latin America and Asia, which both the Council and the Commission resisted strongly. The Council of Ministers, on the other hand, argued that budgetary appropriations had to be underpinned by a separate legal basis over which it had effective exclusive control. Second, the Parliament used its amending power to increase expenditure to promote Community policies of interest to it, notably regional policy, transport, social policy, and education. Third, the Parliament used the annual budgetary cycle to expand the areas falling under what is termed non-compulsory expenditure and hence to have a larger volume of expenditure where it could use the margin for manœuvre available to it under Article 203 of the Treaty. (See Box 3.1.)

In view of these priorities, the EP tended to pay more attention to authorizing expenditure than to monitoring how it was spent. This was despite the fact that its new powers over expenditure were granted in the same year as it was given the sole right of discharge in relation to the Commission's implementation of the budget, as well as coinciding with the establishment of the Court of Auditors. Institutionally this imbalance was reflected in the fact that budgetary control was dealt with only by a subcommittee of the Budgets Committee until 1979. Thereafter the Budgets Committee tended to enjoy greater influence in the Parliament as a whole.

The sharing of budgetary authority and the divergent views between the EP and the Council on issues like the classification of expenditure and the maximum rate of increase inevitably made the annual budgetary cycle prone to conflict. The two institutions were forced to find ways of improving their collaboration. From 1970 onwards, various devices were developed to enable representatives of both institutions to meet to seek agreement. In June 1982, for example, the Council, the EP, and the Commission issued a political declaration laying down an agreed classification of expenditure as compulsory and non-compulsory. Despite the declaration, budgetary wrangles continued, with a basic absence of trust prevailing until agreement on the future financing of the Community was reached in 1988. This agreement served to limit and to channel the continuing differences between the two institutions.

Delors-1: a watershed in EC budgetary politics

The ratification of the Single European Act (SEA) in 1987 marked a relaunching of integration after the 'doldrums period' of the mid-1970s and early 1980s. Although the SEA appeared not to have overt implications for the budget, the new articles on 'economic and social cohesion' (Article 130, SEA) promoted by the Commission and the poorer member states proved a powerful peg for the Commission's strategy on redistribution. The Commission established a clear connection between the internal market process and the budget by launching a series of proposals on budgetary reform in two documents: *Making a Success of the Single Act* (1987a) and *Report on Financing of the Community Budget* (1987b). The proposals, known in common parlance as the 'Delors Package', were negotiated at the highest political level between June 1987 and February 1988, with the less prosperous states (Greece, Ireland, Portugal, and Spain) successfully linking the completion of the internal market to an increase in structural funds designed to reinforce 'economic and social cohesion'.

The Brussels agreement in February 1988 was a classical EC package deal. It combined measures for reinforced budgetary discipline, additional own resources, an expansion of the structural funds, and the maintenance of the UK budget rebate. The main elements were the following:

- an increase in the financial resources available to the Community, rising to a ceiling of 1.2 per cent of GNP by 1992;
- an extension of the system of 'own resources' to include a new fourth resource based on the relative wealth of the member states as measured by GNP;

- tighter and binding budgetary discipline to contain agricultural expenditure at not more than 74 per cent of the growth in Community GNP;
- a continuation of the complex Fontainebleau rebate system whereby the UK receives a reduction in its contribution to Community revenue equivalent to 66 per cent of the difference between its share of revenue provided and of total allocated expenditure; and
- a doubling of the financial resources available to the less prosperous areas of the Community as between 1988 and 1993.

The arguments on Delors-1 waxed and waned during the Danish presidency of the Council in the latter half of 1987. There were tense arguments on the overall size of the budget, the commitment of resources to the poorer regions, and the need to discipline CAP expenditure. The major split was between the poorer states fighting for a larger budget and the paymasters who sought to restrain the level of any such increases. At the same time the divergence between finance and agriculture ministers came into the open, with the latter doing their best to limit the extent of CAP reform. For a time, it looked as if no agreement would be found. At the Copenhagen meeting in December 1987, the European Council failed to find a solution and instead agreed to convene a special meeting in Brussels in February 1988 under the German presidency in order to overcome the deadlock. At that meeting Chancellor Kohl succeeded in brokering a deal which avoided the collapse of the Community's financial structure. In the end, Mrs Thatcher's assumption that Article 130 in the SEA was merely symbolic did not prevail, not least because Chancellor Kohl wanted to secure agreement and was willing to push for it, even though it meant a significant increase in German net contributions to the budget. The German government wanted to maintain the consensus on the internal-market programme. The UK government was caught by surprise because it had assumed that the Germans were in the austerity camp.

The Delors-1 package was accompanied by an Inter-Institutional Agreement (IIA) between the Commission, the Council of Ministers, and the EP, which entered into force in July 1988. It was based on agreement to a five-year financial perspective involving six categories of expenditure. The purpose of the agreement was to ensure that the Brussels decisions were not undone by a continuation of the unremitting conflict between the two arms of the budgetary authority. The three institutions agreed to respect the figures contained in each category of the financial perspective for all the years up to 1992. In return, the EP saw a substantial increase in non-compulsory expenditure underlined by a major commitment to the structural funds. Moreover, the agreement laid down that the figures for compulsory expenditure would not be revised in such a way as to lead to a reduction in the amount available for non-compulsory expenditure.

A number of features of the Delors-1 package illustrate just how far Community budgetary politics had shifted from the zero-sum bargaining of the early 1980s. The Community's policy process proved robust enough to produce a major agreement on what was traditionally a highly contentious issue on its agenda. The Brussels decisions represented a significant increase in EC expenditure: commitment appropriations were to increase by 16 per cent over the period of the agreement, rising from 44.1 to 52.8 billion ecus (in 1988 prices). The agreement to make decisions on budgetary discipline legally binding was part of the continuous search for CAP reform. The doubling of the structural funds for the poorer regions of the Community acknowledged that the benefits of market integration would be felt unevenly. Solidarity between the richer and poorer parts of the Community was affirmed as a 'value' in the political process. The goal of economic and social cohesion was embedded in the *acquis communautaire* in a manner that will be very difficult to dislodge. Spain and Portugal's accession to the Community had clearly enhanced the cohesion countries' bargaining power. The new fourth resource, based on the relative GNP figures for the member states, began to relate contributions to capacity to pay, something which had not been faced up to in 1970.

The Delors-1 package was a major negotiating success for Jacques Delors and the Commission; the Commission claimed with some justification that they had got 90 per cent of what they wanted. Moreover, the deal heralded a period of relative budgetary calm in the Community. The annual struggle to agree a budget was replaced by reasonable cooperation between the two arms of the budgetary authority. The consequence of the IIA was to make it possible for Council and Parliament to reach agreement as to the extent of the annual increase in non-compulsory expenditure. (See Box 3.1.) This removed a major source of budgetary conflict. Between 1988 and 1992 the President of the Parliament signed the budget into law each year on time and in accordance with the established procedure (Commission 1989*a*). Significantly, during the same period, the two arms of the budgetary authority managed to agree *five* revisions to the financial perspective to take account of new demands arising from events in east/central Europe, the Gulf war, and German unification. The normalization of relations was in stark contrast to the pervasive conflict before 1988. The inter-institutional struggle moved to the new legislative procedure of cooperation established by the SEA. Legislative power superseded budgetary power as the object of Parliament's desires in its search for an expanded role in the Community's policy process.

The next step: Delors-2

When the next treaty change was negotiated, it was once again accompanied by a new budgetary settlement. The political link between the SEA and Delors-1 was followed by a similar link between the Treaty on European Union (TEU) and the Delors-2 package. The Commission launched its Delors-2 proposals at the EP in February 1992, just five days after the TEU was formally signed, with its document *From the Single Act to Maastricht and Beyond: The Means to Match our Ambitions.* The Commission followed the formula successfully adopted for Delors-1 by proposing a medium-term financial perspective organized around a number of categories of expenditure. It envisaged that total spending would increase by some 20 billion ecus from 1992 to 1997, rising up to a ceiling of 1.37 per cent of Community GNP. Particular increases were earmarked for structural expenditure, further strengthening the commitment to the redistributive aspect of the budget, and for a new and separate category of external expenditure, reflecting the dramatic changes in central and eastern Europe and the former Soviet Union. Expenditure in this latter area was to grow from 3.6 billion ecus to 6.3 billion ecus by 1999.

The negotiations

The debate on Delors-2 was just as tortuous and controversial as the earlier debate on Delors-1. The member states grappled with their desire to reach agreement on the one hand, and with their determination that the terms of the agreement be as favourable as possible to their viewpoint, on the other. During 1992 the budgetary issue also became entangled in wider political issues. The heads of government met in Edinburgh in December 1992 against the backdrop of the TEU ratification crisis. The Danish 'no' of June 1992 was followed by turbulence in the ERM and the pressing demands from the EFTA states for accession negotiations, which added to the salience of the budgetary debate. Failure to agree to the Delors-2 package at Edinburgh would have heightened the sense of drift in the Community. The UK government, having been forced to withdraw from the EMS in September, did not want a failure at the European Council in Edinburgh, which would have reflected badly on its already troubled presidency.

The budgetary negotiations raised a number of tricky issues for the British presidency. The British government wanted both to protect the system of budgetary rebates and to contain expenditure at the lowest possible level. Its strategy was to conduct the negotiations on the future financing of the

Community within the framework of the Ecofin Council, because it felt that finance ministers would be more sympathetic than foreign ministers to its attempts to limit the growth of spending. This strategy did not work. In the lead-up to the Edinburgh meeting, the Delors-2 package was dealt with by a General Affairs Council (foreign ministers) on 9 November, by the finance ministers on 23 November, by a 'jumbo' meeting of both on 27 November, with a General Affairs meeting on 7 December just before the European Council. Ultimately it required the political authority of heads of government to agree the major elements of the package.

The elements of a deal

The Edinburgh Agreement reached conclusions on the financial resources that would be available to the EU in the 1990s and how this money should be spent. The Agreement had four main elements:

- the establishment of maximum levels of revenue and expenditure were established up to 1999, with the revenue ceiling maintained at 1.2 per cent of GNP for 1993 and 1994, but set to rise to 1.27 by 1999;
- expenditure was divided into six categories with specific financial allocations for each year in all six categories (unlike in 1988 when these details were settled after the Brussels Summit);
- the system of revenue-raising was slightly revised to take account of 'contributive capacity'; and
- the UK secured the maintenance of its abatement mechanism (A. Scott 1993).

The negotiations were dominated by the perennial conflicts about revenue and expenditure and their likely consequences for member-state contributions and receipts. The key issues revolved around the question of how much money, how it should be spent, and over what time-scale. The Commission was less successful during this round of negotiations in retaining the broad elements of its proposals and failed to get the agreement of the member states to the budgetary increases it sought. The Commission's proposal for revenue, for a ceiling of 1.37 per cent of GNP by 1997 (five years), was reduced to 1.27 per cent of GNP over a longer time-scale of seven years. The Commission's case was weakened by the fact that expenditure in 1992 represented 1.15 per cent of GNP, well within the 1.2 per cent established in the Delors-1 package. The UK presidency, in line with its well-established policy of austerity, argued that the existing GNP ceiling gave the Commission ample scope for the development of new policies. The presidency drafted a compromise document for the 'jumbo' Council meeting of 27 November which sought to restrict budgetary growth to 1.25 per cent of GNP by 1999, freezing the GNP ceiling at 1.2 per cent between 1993 and 1995. President Delors reacted angrily to the UK proposals by issuing a letter to the member states because he felt that the presidency compromise diluted the Commission's proposals in an unacceptable manner.

Two issues were most keenly fought over in respect of expenditure: namely, the extent of financial solidarity, and expenditure on measures designed to improve the Community's competitiveness. During the course of the Delors-1 package up to 1992, all the member states, with the exception of Greece, Spain, Portugal, and Ireland, saw their level of net contribution go up, with France joining Germany and the UK as a significant net contributor to the budget. A rise in the number of net contributors at a time of fiscal squeeze at the national level enhanced the determination of the austerity camp in the Council. However, the four cohesion countries continued to press for more money for redistribution, particularly in view of the prospective move to EMU. The Spanish prime minister, Felipe González, led the cohesion countries in very tough negotiations during the Edinburgh Council. He threatened to veto other agreements arrived at in Edinburgh, such as the agreement on Denmark's relations with the EC, unless a deal was achieved on the budget. In the closing stages of the Council, at a breakfast meeting, Chancellor Kohl and President Mitterrand agreed to endorse increases in the budget that went beyond the UK presidency compromise of November. This left the British prime minister with little option but to accept an agreement that allowed for higher spending than he would have wished. However, Mr Major managed to protect the British rebate, which had been questioned during the earlier negotiations. Six member states, including Germany and France, had opposed the principle and substance of the UK refund in November. Those opposing the rebate were unwilling to press their case at Edinburgh as they recognized the importance of a budgetary agreement and the potential impact of the issue on the Maastricht ratification debate, due to resume in the House of Commons in the New Year.

The cohesion countries had reason to be very satisfied with the terms secured in Edinburgh. The agreement envisaged a 41 per cent increase in expenditure on 'structural operations', with an effective doubling in financial flows to the poorer parts of the Community. Furthermore, it was agreed to finance a new instrument, the Cohesion Fund, for the four cohesion countries. Apart from structural spending, the Commission had limited success in getting agreement to additional internal expenditure. The Commission proposed that expenditure on internal policies grow at a faster pace than external expenditure. However, the member states were unwilling to see a substantial growth of expenditure in areas like research and development, transport networks, and telecommunications. Agriculture, by contrast, proved far less contentious than it had been in 1988. It was readily agreed that the general guideline established in 1988, whereby CAP expenditure would increase at a rate not exceeding 74 per cent of the rate of growth of Community GNP, would continue. Agreement on additional expenditure for external actions was also arrived at with relative ease. The member states were committed to enhancing the Union role in post-cold-war Europe and were willing to increase external spending.

There was no major change in the method of raising revenue. A radical proposal from Belgium for a new Community tax fell on fallow ground. Instead, there was some re-balancing between the third and fourth resources, aimed at strengthening the link between contributions to the budget and capacity to pay. The Commission had been a long-standing advocate of a progressive revenue base which would take account of the 'relative economic capacity' of the member states. It proposed that the fourth resource, based on GNP, should contribute more revenue than the third resource, based on VAT, which tends to be regressive. Despite strong objections from the Italian government in particular, which stood to lose most from the change, agreement was reached to give more weight gradually to the fourth resource. The existing 1.4 per cent of VAT was retained until 1995, but would be progressively reduced to 1 per cent between 1995 and 1999.

Overall, the Edinburgh agreement highlighted a commitment by the member states to relative budgetary peace in the Union and to a medium-term framework for the financing of the Community. Delors-2 built on the Brussels agreement of February 1988, emphasizing once more a pragmatic and incremental style of bargaining. Once again, the Commission played a crucial role in designing a package that found the agreement of the member governments and the European Parliament. However, there is good reason to doubt that the relative peace of recent years can be maintained. A host of difficult issues are jostling for attention and they are likely to make the budgetary politics of the late 1980s and early 1990s seem like a brief interregnum between two more troubled periods.

Can budgetary peace be maintained?

The 1988 and 1992 agreements on the Community's financial perspective were secured in large measure by the willingness of Chancellor Kohl and the German government to bear the burden of the increases as the budget's largest contributor. Since Edinburgh, there has been growing criticism within Germany of the level of its contribution to the EC budget, given the costs of

German unification. Chancellor Kohl has warned that Germany is at the limits of its budgetary contributions and that future payments must be linked more closely to per capita income. Put simply, equity considerations are back on the budgetary agenda. As a consequence of German unification, Germany has fallen from second to sixth place in the league table of per capita incomes in the Community, but remains the budget's main paymaster (see Fig. 3.2). Theo Waigel, the German finance minister, has said that Germany will seek an adjustment to its budgetary contribution before agreeing to new areas of spending. This is not the first time that senior German ministers have grumbled about the level of their contributions. However, the costs of reunification, heavy budget deficits, and an uncertain public opinion on integration suggest that the Germans are more serious this time. The era of German willingness to accept its position as the main net contributor to the budget may have come to an end.

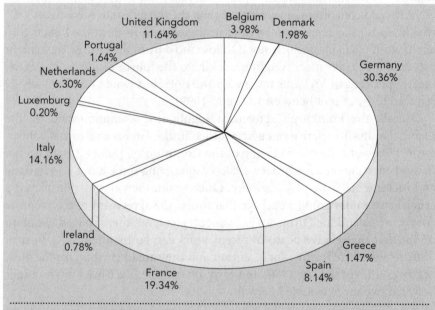

Fig. 3.2. Financing of expenditure from the general budget by Member State (1994)

Source: European Commission, *The Community Budget* (1994).

In addition, four more general problems face the Union and are likely to make it difficult to prevent the budgetary issues from becoming intertwined with wider policy concerns:

- inter-institutional relations, and in particular the respective roles of the Council and the Parliament in the budgetary process;
- the increasingly delicate question of budget management, highlighted by a succession of critical Court of Auditors reports, and the difficulty of ensuring that a larger budget is well spent;
- 'the ghost of fiscal federalism' and the implications for EU public finance of an evolving EMU; and
- the budgetary cost of enlargement, particularly to central and eastern Europe, and all that this means for the unwritten rules that have marked out budgetary policy up to now.

Inter-institutional relations

Community budgetary procedure poses a challenge of logic to any Cartesian mind, one which it is more sensible to bypass. Jean-Pierre Cot, MEP (Cot 1989)

In 1988 the decisions taken at the Brussels European Council left considerable scope for the Council and the Parliament, with the help of the Commission, to fill in the details of the financial perspective for spending up to 1992. In Edinburgh at the end of 1992 the member states agreed a much more detailed financial structure which specified spending by category for each year of the new financial perspective up to 1999. This left less room for manœuvre to negotiate on the figures and obliged the Council to consider more carefully the claims of the Parliament for wider powers within the budgetary process as part of the price for agreement on a new Inter-Institutional Agreement in October 1993.

One particular concern of the Parliament in these negotiations was the distinction between compulsory and non-compulsory expenditure. The Parliament has always sought to reduce the importance of this distinction and with some success. The volume of non-compulsory expenditure, over which the Parliament has the last word, has increased markedly over the last ten years and is now approaching 50 per cent of the total. In the negotiations on the new agreement the Council was willing to allow the financing linked to international agreements, previously considered compulsory, in future to be non-compulsory, thus giving the Parliament a greater say. The Council also agreed to a wider review of the distinction in the 1996 Intergovernmental Conference. Such a review can hardly fail to address the status of agricultural guarantee expenditure, which has up to now been untouchable as compulsory expenditure.

This stress on the inclusive nature of the Parliament's powers has also prompted it to insist that appropriations used to finance joint actions under the common foreign and security policy (CFSP) pillar of the Maastricht Treaty should be included in the budget of the Union and be considered non-compulsory, thus giving the Parliament a say in this area. Such an approach runs directly against the wish of the Council to keep the Parliament's involvement in the second pillar to the minimum laid down by Article J7 of the TEU. However, they have had to recognize the difficulty of financing such common actions by contributions from individual states. The risk is that money fails to materialize at the right moment and that policy falls subject to the same vagaries as affect the financing of UN operations. As a result, the two sides are obliged to try to seek an accommodation which will enable the Parliament to have a say, without being able to block actions which need to be sanctioned at speed.

These discussions pose the question of how far the member states are willing to go in granting further budgetary powers to the European Parliament. There will be strong resistance to widening those powers if they could seriously threaten core interests of certain member states, such as the CAP. At the same time, the Parliament has been encouraged by its relative successes in the Maastricht negotiations to push ahead in the search not simply for a larger budget, but for one that enshrines for itself a larger role in the budgetary procedure. This is certain to be a major issue long before the present financial perspective expires.

Managing a larger budget

The struggle for budgetary resources in the Community and the inter-institutional battles about budgetary power have tended to overshadow questions of 'value for money' and accountability. These may well become a major focus of budgetary politics in the coming years.

Increasingly questions of accountability and 'value for money' have impinged on the budgetary debate. As the budget has grown larger, so increasing attention has been paid to the Commission's management capacity and its control over expenditure. In the Santer Commission, Erkki Liikanen, the Finnish Commissioner, and his Swedish counterpart, Anita Gradin, are together spearheading a move to strengthen financial management within the Commission and to combat fraud against the budget. Unlike Delors, Santer is determined to address internal management problems in the Commission, an organization that for too long has paid more attention to policy innovation than to its management.

The implementation of the EU budget is characterized by a fragmentation of responsibility between the Commission and public authorities in the member states; 80 per cent of the budget is managed on behalf of the Union by the member states. The Commission relies to a large extent on the delivery and enforcement capacity of national authorities. At the same time, the complexity of European rules, particularly in agriculture and regional policy, creates loopholes which can be exploited by those intent on defrauding the EU budget. The sheer number of agricultural subsidy payments and export refunds creates considerable opportunities for abuse. Reports of fraud in olive oil, beef, wine, and fish, running in some cases to millions of pounds, undermine the credibility of the Community's policies. Press reports have highlighted various scams involving the forging of customs documents in order to claim export refunds, switching labels on various foodstuffs to claim higher refunds, claiming headage payments for non-existent animals, and putting non-existent food into intervention storage. No one knows with any degree of certainty the level of fraud against the EU budget: estimates of between 7 and 10 per cent of the budget are often cited but have never been convincingly proven. The issue is very much one of confidence in the system, where the Community has to meet higher standards than are sought at national level. That said, investigations of fraud suggest that some member states are dilatory in following up cases of fraud against the EU budget as this would involve devoting additional national resources to getting less money out of the Union budget. The absence of a common definition of what constitutes fraud among the member states and differences in national systems of criminal law exacerbate the enforcement problems.

The 1975 agreement to create the Court of Auditors (it began work in Luxemburg in 1977) enhanced the institutional commitment in the EC to systematic accountability. In its first report (1978) the Court raised the issue of fraud and it has continued to do so in all subsequent reports. The Court's reports have been consistently highly critical of the Commission's management of EC finances. Aware that continuing reports on fraud would undermine public confidence in European integration, the Commission, the Council, and the Court of Justice began to address the issue.[4] In 1988 the Commission established an anti-fraud unit, UCLAF, *Unité de Coordination de la Lutte Anti-Fraude*, and in 1989 produced a forty-five-point programme. The Court of Justice has also become involved, underlining the responsibilities of the member states in combating irregularities in the use of EU funds. For example, in a 1989 case of fraud concerning maize supposedly grown in Greece, but in fact imported from Yugoslavia, it held that the member states were legally obliged under the Treaty's 'loyalty clause' (Article 5 EEC) to deal with EC fraud cases in a manner similar to fraud against their national exchequers.

The European Council added its authority to the fight against fraud with a declaration at the Copenhagen meeting of June 1993, which underlined 'the importance of fully implementing the provisions of the Maastricht treaty according to which Member States are to take the same measures to counter fraud affecting the financial interests of the Community as they take to counter fraud against their own financial interests'.[5] The inclusion of this principle in the TEU (Article 209a) arose from a widespread feeling that national authorities probe cases of fraud involving national money with greater alacrity than those against the Community budget. The Commission is becoming more involved in monitoring and directing national enforcement of EU programmes. In December 1994 the Council (Justice and Home Affairs) reached an outline agreement on a system of 'penal sanctions' that allows the member states to impose fines on companies and individuals who have misappropriated EU monies. This makes fraud against the EC budget a criminal offence, in all but name. The Essen European Council (December 1994) supported further joint action to protect the Union's financial interests. The growing number of policy instruments financed by the budget raises issues not just of accountability but of performance. Are the Community taxpayers getting 'value for money' from EC expenditure on regional policy, training, education, research, and development? Views differ among politicians and academic experts about the rationale behind different types of public expenditure and the benefits accruing from direct intervention. The Commission has endeavoured to increase the effectiveness of the various budgetary funds, mindful that increases in its budgetary resources are partly dependent on assessments of the usefulness of current expenditure. In the past, the small size of the funds, the slender staff resources, the multi-level nature of the delivery system, and lack of information and expertise hampered the Commission in its evaluation of how EC monies were spent.

The 1989–92 expansion of the budget heightened the salience of such questions, as the Community embarked on a major expansion of its programme of grants and loans. The 1988 reform of the structural funds streamlined the procedures for allocating resources. The new framework strengthened the appraisal, monitoring, and *ex post facto* assessment of programmes. The emphasis on programmes released Commission officials from the tedious work of project evaluation and allowed them to build up a partnership with their national interlocutors. Pilot projects and innovatory programmes have contributed to an important learning process in the Commission and the member states. Moreover, a higher proportion of the budget is now devoted to evaluations of how Brussels money is spent. A myriad of sectoral and regional studies has been commissioned from outside experts. These studies feed back into Commission thinking about the future management of the funds, although they do not remove the 'grantsmanship' involved in the distribution of EU monies.

The Commission's commitment to evaluation has moved beyond the symbolic as the budget has expanded. The Commission is now much less timid in its criticisms of national programmes. That said, it tends to respond defensively to outside criticism of the structural funds. During the debate in the Ecofin Council on the Delors-2 package, a number of delegations questioned the value of structural fund spending, basing their criticisms on a special Court of Auditors' Report that was highly critical of the funds' organization and spending priorities. The beneficiaries of the funds and the Commission challenged the criticisms. President Delors argued forcefully that if the regulations were too complicated, it was because the ministers had wanted them that way. This defence, even if partially justified, does not overcome the lack of confidence in EU spending programmes. In turn this acts as a real constraint on the development of new spending, especially at a time when European integration has become more contested in public opinion.

The ghost of fiscal federalism

Discussion of the Delors-2 package did not begin from a coherent view on the role of public finance in European integration. The negotiations on EMU were remarkable for the fact that the budgetary implications of EMU were pushed to one side. As we have noted, before the 1980s it was widely assumed that there was an important link between the achievement of a single currency and an enhanced Community budget. The MacDougall Report (MacDougall *et al.* 1977) on the role of public finance in regional integration advocated a pre-federal budget for the Community amounting to between 2 and 2.5 per cent of GDP. According to MacDougall, the larger budget should be channelled to regional and counter-cyclical policies. The Padoa Schioppa Report (1987) downplayed the issue of fiscal federalism but still emphasized the need for those embarking on economic integration not to rely 'on simple beliefs about the benevolence of the invisible hands of the market'.

The assumption that EMU would require a substantial increase in the size of the budget and a reassessment of its policy instruments was effectively undermined in the run-up to the last IGC; DG II of the Commission went so far as to argue that cohesion had no place on the EMU agenda. It sought to sideline fiscal issues in an attempt to overcome British scepticism and German reluctance about EMU. This represented an important change in the intellectual and political climate. The previous consensus was that a monetary union required a strong fiscal capacity to aid adjustment problems and to deal with 'asymmetric shocks'. This was undermined by those who sought agreement to a treaty on EMU without confronting some of the more controversial issues surrounding the move to a single currency.

It will be increasingly difficult to avoid such issues in the coming years, as the prospect of EMU becomes a reality. Already the rules governing the Cohesion Fund establish a link between the economic policies of a member state and the budgetary benefits it draws from the Union; grants from the Cohesion Fund are conditional on the development of programmes leading to the fulfilment of the EMU convergence criteria. For the first time the Community budget is directly linked to the economic policies of the member states. The implication is that if states seek to meet the convergence criteria, they will benefit from Community structural support, thereby establishing the principle that EC finance can be used to reduce burdens on public finance. On the other hand, if they do not act to meet the criteria, then finance can be withheld. Either course of action implies a significant change in the philosophy of the Union.

Two possible routes forward can be imagined. One possibility is to promote EMU with a small hard core of members, because *inter alia* the budgetary implications of a wider group cannot be sustained. Alternatively, effective inter-regional stabilization would require a major increase in expenditure. The Commission has argued that a shock absorption scheme could be devised to provide a cushion against adverse economic change provoked by the move to EMU, and that it need not cost more than 0.2 per cent of Community GDP (Commission 1993b). This would, however, have to be accompanied by a fundamental change in the constitution and institutions of the Union, thereby raising the vexed question of the distribution of powers between the Union and the member states. 'The ghost of fiscal federalism' will not lie down!

Enlargement

Finally, there is the question of enlargement and its cost, something which may prove the most serious challenge to budgetary calm. Agreement on the Delors-2 package at Edinburgh allowed accession negotiations to open with Austria, Sweden, Finland, and later Norway. The prospect that the rich EFTA states would be net contributors to the EC budget was regarded as one of the main attractions of an EFTA enlargement. Already under the European Economic Area (EEA) Agreement, the EFTA states were willing to provide assistance, in the form of a EEA Cohesion Fund (not to be confused with the Cohesion Fund set up by the TEU), for structural adjustment in the Union's poorer regions. Towards the end of the enlargement negotiations, strong linkages developed between budget payments, agriculture, and regional policy. The Scandinavian governments pressed successfully for the development of a new budget line in the structural funds to deal with 'Arctic regions'. Sweden, potentially the largest new contributor to the budget, negotiated a generous package of budgetary compensation in order to phase in the consequences of membership. The prospect of referenda in all of the acceding states gave them added leverage in searching for additional compensation towards the temporary payments they are allowed to use under the agricultural agreement. In the end, the new members were granted a remarkably generous agri-budgetary package. The Commission estimates that there will be a net gain to the EU budget of 6,500 million ecus during the period 1995–9. (See Table 3.1.) This money is likely to be needed for any shortfall arising from lower GNP growth than anticipated at Edinburgh, though the amount of money is relatively small, given that the 1995 budget amounted to 79,000 million ecus. Paradoxically, the rich EFTA states won a far better budgetary deal than the Iberian states had in 1984. This partly explains Spain's tough negotiating stance towards the end of the enlargement negotiations. (See Table 3.1.)

However, a much more serious situation looms in the context of an eastern enlargement. At Copenhagen in June 1993 the member states accepted the principle of an eastern enlargement. The time-scale and the conditions

Table 3.1. Financial perspective for 'Europe 15': 1995–1999 (in million ecus, 1995 prices)

Budgetary categories	1995	1996	1997	1998	1999
1. Common agricultural policy	37,944	39,546	40,267	41,006	41,764
2. Structural operations	26,329	27,710	29,375	31,164	32,956
Structural funds	24,069	25,206	26,604	28,340	30,187
Cohesion fund	2,152	2,396	2,663	2,716	2,769
EEA financial mechanism	108	108	108	108	108
3. Internal policies	5,060	5,233	5,449	5,677	5,894
4. External action	4,895	5,162	5,468	5,865	6,340
5. Administration	4,022	4,110	4,232	4,295	4,359
6. Reserves	1,146	1,140	1,140	1,140	1,140
7. Compensation	1,547	701	212	99	—
Total: Commitments	80,943	83,602	86,143	89,246	92,453
Total: Payments	77,229	79,248	82,227	85,073	88,007
Appropriations for payments (% GNP)	1.20	1.21	1.22	1.24	1.25
Own resources (% GNP)	1.21	1.22	1.24	1.26	1.27

Source: *European Parliament*: PE 182.856 final, 13 Dec. 1994: 7.

under which this might take place have yet to be worked out in the second half of the 1990s. Such an enlargement will undermine existing budgetary bargains in the Union and will necessitate considerable changes in the policy *acquis*. It is difficult to envisage the Union being able to stretch the CAP and structural policies with ease to cover the countries of central and eastern Europe. The winners from the existing budgetary regime face difficult choices about budgetary reform. The prospective member states are poorer and more agricultural than the current member states. If the CAP, as presently structured, were to apply to low-cost producers like Poland and Hungary, huge surpluses, at a high cost to the Union budget and Union taxpayers, would be generated. Pressures will inevitably build up for further reform of the CAP beyond the MacSharry reforms, accentuated by the impact of the GATT agreement.

The countries of central and eastern Europe would also expect to benefit from Union 'solidarity' and policies on cohesion, because they are less prosperous than the poorer regions of the existing Union. Attempts to exclude these states from 'solidarity' would completely undermine the existing basis of the Union's policies on cohesion. If Greece, Portugal, and Ireland benefit from sizeable financial flows, why not the poorer east European states? Under the current arrangements, Europe Associates receive financial transfers that amount to less than 10 per cent of those allocated under the structural funds to the four cohesion countries. In 1992 Greece and Portugal received a per capita transfer of approximately 200 ecus from the funds, which is set to rise to 400 ecus under the terms agreed at Edinburgh. Equivalent transfers to the four Visegrad states would cost 26 billion ecus. The Commission has estimated that a Visegrad enlargement would cost 43 billion ecus gross, with a further 37 billion ecus required for Romania and Bulgaria (Commission 1993*b*). Together these amounts total as much as the entire present budget for twelve member states. It is also already clear that an eastern enlargement will not take place unless there is an enhanced and probably expensive policy for the Mediterranean countries of north Africa and the near east.

Are such enormous increases possible? The member states will be obliged to reconsider their order of priorities and their commitment to the existing *acquis*. Parts of the *acquis* may be jettisoned, particularly the commitment to cohesion. The CAP is set for further reform. An eastern enlargement will disturb internal bargains in the Union. Many of the existing members, including some of the poorer states, may find themselves as net contributors to the budget, which may have an adverse effect on their commitment to integration and on the state of public opinion. These are tremendously difficult issues and are likely to make the question of enlargement the scene for a budgetary battle far more severe than any the Union has known up to now.

Conclusions

The story of budgetary politics underlines the importance of taking a long-term view of the evolution of the Union. In a tortuous and conflict-ridden process, the Union succeeded with the Delors-1 package in establishing a coherent medium-term budgetary strategy. The policy process proved robust enough to transform a zero-sum conflict into something which enhanced the Union. This finding is at odds with the conclusion in the introductory chapter that instability pervades the Union's policy process. The shared objective of the 1992 programme, consolidated in the SEA and its cohesion provisions, was critical to the search for a stable budgetary framework. Community institutions proved rather effective in establishing a sound budgetary environment. The Commission was very instrumental in designing packages that were ultimately agreed to by the member states. Council presidencies, albeit for different reasons (Germany in 1988, UK in 1992), displayed considerable political skill in the final stages of very difficult negotiations on Delors-1 and 2. The structure of EU spending now leads to relatively broad satisfaction among the member states. Endemic competition on budgetary matters between the Council and the European Parliament has abated as the legislative process became the focus of EP attention. The focus has shifted, in the medium term, from the 'high politics' of budgetary reform to management issues, particularly fraud and 'value for money' considerations. The most significant development in terms of ideas and values underpinning the budgetary process was the emphasis on solidarity and cohesion. This marked a tentative move by the Union from being a problem-solving arena to becoming an incipient polity. That said, the challenge of enlargement and pressing demands from the Mediterranean underline the fact that a budgetary battle looms when Delors-2 runs its course. How the Union deals with the budgetary issues ahead will be important in determining whether the emerging polity can be consolidated.

Notes

1. For the most comprehensive and detailed treatment of the development of the EC budget and the rules that govern it, see Strasser 1992.
2. This chapter draws heavily on the analysis of Shackleton (1990; 1993a; 1993b).
3. For a detailed analysis of the UK issue see H. Wallace (1983a).
4. For a detailed analysis of fraud against the budget see Ruimschotel (1993).
5. European Council Copenhagen, 21–2 June 1993, Presidency Conclusions, point 16.

Further Reading

For an analysis of budgetary politics in the 1970s and early 1980s see H. Wallace (1980, 1983*a*). The Delors-1 package is thoroughly analysed in Shackleton (1990). For a recent overview of the budget and fiscal federalism see 'Stable Money, Sound Finances', special issue of *European Economy*, No. 53 (1993). For a provocative discussion of the future of EU finances see Brouwer *et al.* (1995).

Brouwer, H. J., *et al.* (1995), *Do We Need a New Budget Deal?* (Brussels: The Philip Morris Institute).

Shackleton, M. (1990), *Financing the European Community* (London: Pinter).

Wallace, H. (1980), *Budgetary Politics: The Finances of the European Community* (London: Allen and Unwin).

—— (1983*a*), 'Distributional Politics: Dividing up the Community Cake', in H. Wallace, W. Wallace, and C. Webb (eds.), *Policy-Making in the European Community*, 2nd edn. (Chichester: John Wiley and Sons, 1983).

CHAPTER FOUR

THE COMMON AGRICULTURAL POLICY: EXTERNAL AND INTERNAL DIMENSIONS

Elmar Rieger

The CAP in the 1960s represented a striking example of the evolution of European policy. Yet its establishment reflected defensive national strategies of economic modernization, attaching small farmers' loyalty to rebuilt democracies, with welfare state functions transferred to the European level and farming organizations as intermediaries. The CAP served to insulate agricultural policy both from competing domestic political constituencies and from American demands for trade liberalization, the latter weakened by the US GATT waiver. A highly segmented system of governance developed, operating through both national and supranational mechanisms. Changes in exchange rates and divergent national priorities led in the 1970s and 1980s to byzantine complexity and partial renationalization. Rising costs forced reform in the early 1990s; GATT negotiations paradoxically then became a lever to promote internal reforms.

Introduction

We have until recently conceived it as a moral duty to ruin the tillers of the soil and destroy the age-long human traditions attendant on husbandry if we could get a loaf of bread thereby a tenth of a penny cheaper. (Keynes 1933: 242)

I know of no illustration of our economic nationalism and our lack of internationalism more striking than this that . . . in almost all of the richer countries we have accepted it as a natural and perfectly normal thing to give the small minority of farmers higher and more stable prices . . . (Myrdal 1957: 46)

The two quotations aptly describe the dilemma of agriculture in modern industrial societies. On the one hand, there is the destruction of an economic sector with distinctive institutional and social features—family farming—through the introduction of industrial principles. The two typical examples are England and the former USSR. On the other hand, there is the costly and conflict-ridden preservation of economic and social features which are quite at odds with the usual social and economic structure of an industrial and urbanized society.[1] This policy choice has been made by the majority of the continental west European countries. In the first case, the agricultural sector is integrated into the overall economic and political system. In the second case, an elaborate and technically complex agricultural policy mediates between the agricultural sector, dominated by small-scale, broad-based family farming, and the encompassing economy by raising prices to a level which guarantees even small-scale farmers a lifestyle compatible with general standards, thereby compensating for inter-sectoral imbalances. In the first case the farming population becomes quite small, because of rapid concentration and specialization processes; in the second case the farming sector will remain both numerous and heterogeneous.

The common agricultural policy (CAP) of the European Union (EU) follows the rationale of the second option. This is the main reason for its distinctive features in the processes of transnational integration and supranational institution-building in western Europe. But this option of following explicitly non-economic objectives in agricultural policy-making is not only costly, but produces some additional problems, particularly to international agricultural markets. A shift to the first option is, therefore, a constant possibility and a basic threat employed by governments to civilize a strong farming lobby and to control public expenditures. Because of this peculiar constellation of institutional structures, social values, and cross-cutting economic criteria of legitimacy, the CAP has evolved into a highly ambivalent apparatus.

Between negative and positive integration

In the early years of the EU the particular features of the CAP were seen as the forerunners of more extensive European integration. Although the creation of the European common market was based on the elimination of national barriers, the integration of agriculture required the coordination and the fusion of six highly developed regulatory systems of state intervention (Pinder 1968).[2] Supranational organizations and agencies replaced major portions of national systems of agricultural support, thereby establishing direct links between the farming population of the member states and the European Community (EC). Slowly, but persistently, quite elaborate and increasingly complex forms of supranational political governance evolved around the CAP, making it a positive example for the evolution of a European polity.[3] Accordingly, the founding and the remarkable rise of COPA (Comité des Organisations Professionelles Agricoles) as the major organization of European farmers seemed to indicate a quantum leap towards a supranational political system. Many of the general principles of European law were formed in the context of agricultural disputes (Snyder 1985; Usher 1988). Finally, because of its size, sources, and format, the European Agricultural Guidance and Guarantee Fund (EAGGF) is closely intertwined with the Community budget as a whole.

These particular features of the CAP are mostly due to the fundamental asymmetry in the policy-making capacities of the EC. Only with regard to agriculture did the scale of political governance reach proportions resembling those of a federal government:

Here the Community institutions have the power to *legislate* for the Community as a whole, without being required to refer back to the national parliaments. The progress made in agriculture will be of definitive importance for the integrative potential of the EEC. It represents the Community's first effort to develop a common policy in a major economic sphere. Such common policies are central to the successful implementation of the broader goal of economic union as well as the efficient operation of the customs union. (Lindberg 1963: 219–20)

Contrary to the hopes of the first generation of Europeanists, the CAP 'as a proxy for European integration' remained 'an isolated relic for the ambitions of the founding fathers of the European Community' (Duchêne *et al.* 1985: 1). But one has to be careful with the notion of an institutional asymmetry. I wish to argue in this chapter that it is the very fact of being both a policy of central importance and, at the same time, an institutionally highly isolated one, that the major social and and political forces underlying *and sustaining* the CAP remain largely invisible (Stevens and Webb 1983: 321).

In this chapter I aim, among other things, to offer a concrete refutation of the view, quite common in the recent two decades, that the CAP is, in the words of *The Economist*, 'the single most idiotic system of economic mismanagement that the rich western countries have ever devised'.[4] I shall try to show that this mainstream view of the CAP is superficial and meaningless, because it is not attached to an analysis of the real forces shaping the agricultural policies of western Europe in the post-war era. In particular, the insistence on purely economic criteria of rationality and legitimacy in the analysis and evaluation of the CAP leads nowhere and misses the forest for the trees.

The first step in the right direction, therefore, is to see and understand the CAP as an entity in its own right, and not simply as the key to something else (be it integration theory, the dangers of supranationalism, a general theory of interest-group behaviour, of rent-seeking, or of joint-decision-making traps). The CAP is very much a self-contained world, following its own logic.[5] This is true in a more narrow sense, since 'all of the changes were implemented through partial adjustments and corrections of policies in place' (Anania *et al.* 1994a: 12). And it is true in a broader sense, since the CAP remained quite aloof from the shifts in the perception of Brussels as either a useful institutional setting or as menace, threatening national autonomy. Instead, the growing tensions in EC–US agricultural trade relations served the member states as a steady reminder of the uses of a truly 'common' policy. Relieving member states from the 'ugly work' of both satisfying the farm lobby and managing agricultural trade relations in an international environment dominated by liberal multilateralism is one of the most important factors sustaining the CAP (Vaubel 1994: 174).

The key to a better understanding of this particular policy is to see the CAP as an integral part of the west European welfare state and its particular moral economy.[6] Unlike welfare-state institutions, however, agricultural policies typically fuse production, that is output-increasing, with income-related goals. As a consequence, political and economic markets are not separable. The negative side of this fusion has been deplored by agricultural economists since the first inception of modern, welfare-oriented agricultural policies (Schultz 1943).

The positive—or functional—effects of the fusion can be summed up in the following way:

- It facilitates the administration of an income-securing and income-raising policy vis-à-vis a large and quite heterogeneous group, since it requires no direct and administratively costly contacts with individual farmers.
- It releases governments from having to decide on the needs of different farming groups or of individual farmers.
- It provides the farming sector with an overwhelming interest in a policy of this sort, since it improves the situation of all farmers, albeit the larger more so than the smaller.
- Concentrating on comprehensive and product-specific prices enables governments to reach the compromises necessary to balance the heterogeneous needs of a highly differentiated farming sector.

These factors provided not only the governments of the member states but national interest groups as well with the shared purpose necessary for sustaining a supranational policy.

If the CAP is a defensive strategy to modernize west European agriculture against both the internal threat of the industrial society and the external threat of American efforts to open agricultural markets, this helps to explain three points. First, it clarifies why the governance of the CAP emerged as so complex a form of multilateral decision-making instead of either pure intergovernmentalism or supranationalism. The CAP is sustainable because it represents an *addition* to the national political systems of the member states. Secondly, it explains why the CAP, on first sight, distributed benefits quite perversely despite its explicit welfare orientation. The creation of a single agricultural price-support system from what had been several produced an increase in social and economic inequalities. This development went unchecked, since big farmers, with considerable political clout, profited most from it. Thirdly, it gives some clues why, and, more important, on which terms, agriculture was quite surprisingly included in the last round of GATT negotiations. The main point here concerns the political control of the CAP. It will be argued that the CAP reform of 1992 and the GATT compromise on agriculture reached in 1993 were mutually reinforcing policies. The GATT negotiations could provide the *external* binding constraint to facilitate *and to sustain* the transition to a new institutional structure which may regain political control over CAP expenditure.

Defensive modernization: the CAP as a welfare state institution

What is perhaps most surprising in the history of State support is the fact that little of it has drawn its inspiration from the farming community itself. This does not mean that these policies were arbitrarily forced on the farmers, but it does show that agriculturalists in general seem to have had little idea about the specific manner in which they wished State aid to operate. If there is any moral in this story, it is that the pattern of State support for agriculture is only a facet of the national attitude towards the growth of the whole of the economy. The approach to agricultural affairs is a very good indicator of the national economic philosophy at any given time, since it reflects relatively quickly the changes in that philosophy. Far from being the last, farming is among the first in the queue for government action. This, however, has by no means always been for farmers an unmitigated advantage. (Attwood 1963: 148)

Like welfare state institutions with regard to labour, the CAP is essentially an economic and political mechanism which integrates the national farming population into both the *national* society and the *national* polity. Agricultural protectionism, reflected in domestic food prices that are often more than twice as high as international prices, is the most basic instrument employed to achieve this end. This feature is by no means unique to the CAP, which in this respect simply continues the agricultural policies of the founder member states. Agricultural protectionism is a common feature of economic modernization:

cross-sectional evidence suggests that as economies grow they tend to change from taxing to assisting or protecting agriculture relative to other sectors, and this change occurs at an earlier stage of economic growth the weaker the country's comparative advantage in agriculture. (K. Anderson and Hayami 1986: 1)

It is quite revealing that in the period 1955–9 agricultural prices in Korea were 15 per cent and in Taiwan 21 per cent *below*, in the period 1980–2 166 per cent (Korea) and 55 per cent (Taiwan) *over*, international prices (K. Anderson and Hayami 1986: Table 2.3).

What was at stake in the immediate post-war years was the political and social participation of severely disadvantaged and politically isolated social groups. The CAP, like its national forerunners in the six founder members, represents a compromise to this end. It was not a deliberate policy to increase food production in the prospective Community.[7] The much-repeated assertion that there was a political bargain between the industrial interests of Germany and the agricultural interests of France should also be laid to rest. The record indicates no such reasoning, nor does it make much sense economically (Willgerodt 1984: 111–4; Milward 1992: 283–4; Vaubel 1994: 174). Much more relevantly, the bigger countries in particular had sizeable farming populations which they did not want to exclude from the European Economic Community (EEC), and Dutch agriculture was an integral and important part of the national export economy (Lindberg 1963: 220–5).

The *national* integration of agriculture, though in a supranational policy, had two consequences. First of all, farmers did not block European integration. This was no small achievement, given the weight of the agricultural sector both as a proportion of the labour force and as a share of the economy at the end of the 1950s.[8] The second relates to the welfare state function of the CAP: the CAP provided political protection *vis-à-vis* the trade interests of overseas agricultural exporters, most notably the United States. Without the CAP it seems unlikely that EC members would have been able to sustain such protection under pressure from the US. In sum, the CAP is not only a striking illustration of major political responsibilities being Europeanized, but also a structure designed to strengthen and to pressure the national features of the political governance of agriculture.

To analyse and interpret the CAP in this sense goes well beyond the stress on the 'non-economic objectives' of the articles of the Treaty of Rome dealing with agriculture (see Box 4.1).

Box 4.1. Objectives of the CAP

Art. 39 of the EEC Treaty sets out five objectives of the common agricultural policy:

- to increase agricultural production by promoting technical progress and by ensuring the rational development of agricultural production and optimum utilization of the factors of production, in particular labour;

- to ensure a fair standard of living for the agricultural community, in particular by increasing the individual earnings of persons engaged in agriculture;

- to stabilize markets;

- to assure availability of supplies; and

- to ensure that supplies reach consumers at reasonable prices.

The history and features of the CAP are much more the product of how west European societies were being transformed after the Second World War than of the formulations of the Treaty of Rome. Since the rise of the welfare state is such a major element of this transformation, the CAP's part in this transformation had far-reaching implications: for the functions of the CAP's main instruments; for the meaning of the common tariff and the role of external trade relations; for the position and the role of interest groups, and for the 'moral economy' of the conflicts accompanying the governance of the CAP.

Income maintenance and income security for farmers

The Treaty of Rome was remarkably brief on agriculture. The essence of Articles 38–47 was basically an 'agreement to agree'; they gave little indication of the terms for integrating agriculture within the common market (Dam 1967: 219). This imprecision was an early indication that agriculture promised to become a contentious issue.

Given that national policies were well established, wide-ranging, and highly protective, it was no surprise that the basic instruments of the CAP were wide-ranging and equally protective market organizations for agricultural products (Lindberg 1963: 223–5). For several reasons they were judged the most appropriate means to provide public goods to the agricultural sector. First of all, market organizations with fixed prices reduce the risk and the uncertainty associated with large variations in commodity prices and in the volume of production. Secondly, market organizations redistribute incomes. Since the Great Depression of the early 1930s, which hit agriculture even harder than industry, 'equity' and 'parity' concerns have been central to the farm policies of all developed societies. Income redistribution to financially stressed or low-income farmers became a major impetus to farm policies (Schiller 1939; Schultz 1943). The European market organizations were, at least at the outset, expressly designed to foster small-scale and medium-size family farms. Thirdly, market organizations, with the institutional apparatus to fix and to guarantee prices, are a rather efficient way to provide farmers with higher incomes. This was of particular importance when, as in the 1950s and 1960s in west Europe, the farming populations consisted of millions of smallholders quite remote from the normal apparatus of the state bureaucracy.

The democratic political machinery, with a little help from agrarian interest groups, has been continuously used to bring agriculture closer to the ideals and principles of the welfare state. This explains why the CAP also evolved into something which could be called a welfare state for farmers; this explains, among other things, the dominant role of agrarian interest groups. Without the engagement of interest groups the political governance of the CAP and the administration of the commodity regimes and various structural programmes would be an impossible task; the complexities of these are breathtaking—but for all farmers they are of vital importance.

The role of the European Parliament remains limited and cannot be compared to national legislatures. Parties are absent as mediators of interests at the European level. Both national and transnational farm groups therefore evolved as the basic channels of mediating between the farmers and their needs on the one hand, and the supranational political system on the other hand. One has thus to see the power of the agrarian interest groups as a function of the CAP—or of agricultural policies in the developed countries in general. The institutional structure of these agricultural policies is not so much the result of the power and the strategies of agrarian interest groups; its character derives more from its functional equivalence to the basic institutions of national welfare states.

The CAP as a safe haven

Perhaps the most basic feature of the welfare state is its systematic disregard for the international consequences of its institutions and their peculiar mode of operation. For this reason, in the post-war years national borders were transformed into social and therefore highly exclusive borders. Citizenship rights now extended to the lower classes and took on a social dimension, thereby transforming the moral economy as well as the basic structure of the post-war developed world.[9] However, with regard to the highly protectionist, welfare-oriented agricultural policies of western Europe, this revolutionary turn produced some extra problems.

It was noted above that the social-policy functions of agricultural policies were not an invention of the CAP; they came into being in the early 1950s in all west European countries, not only in the countries that founded the EEC. The need to increase agricultural production was widespread, and there was a special need to integrate farmers into the newly founded democracies. This basic constellation of quite heterogeneous driving forces resulted in heavily protectionist agricultural policies, high institutional prices for nearly all products, and other means of raising agricultural incomes. In the interwar period major sections of the agrarian population turned to radical right-wing parties to protest against governments that tolerated the collapse of agricultural prices. This experience proved to be of crucial importance. It helped to create a dramatic change in both the meaning and the institutional structure of agricultural policies (Milward 1992). For this reason Gunnar Myrdal, in a lecture in 1957, called the agricultural policies of the developed world the most striking illustration of the importance of the welfare state (1957).

Nearly all west European countries ran into serious trouble with the

United States because of its protectionist agricultural policies. The USA had a highly efficient agricultural sector which was, due to the Second World War, directed towards exports. The welfare of American farmers therefore depended on access to foreign markets. In the 1950s the USA tried to use the GATT, first, to get a waiver for their own programmes of agricultural subsidies—which they obtained—and then to force west European countries to open their agricultural markets.[10]

The protectionist agricultural policies of Germany, Belgium, and Luxemburg were explicitly denied a legitimate position in the GATT, while France, Italy, and most other west European countries had already been singled out by the US Tariff Commission because of the protectionist intention of their agricultural import restrictions (Dam 1970: 263).[11] It was this situation which made the inclusion of agriculture in the EEC a last resort for the agricultural protectionism of the original six member countries.[12] It was not by accident that in 1962, when the market organization for cereals was introduced as the first step in building the CAP, after heavy conflicts with the USA in particular about agricultural policies, *The Economist* called the CAP a '*deus ex machina* which looks to most agricultural exporters in the GATT to be pretty diabolic'. The CAP was, 'as a system of protection . . . about as watertight a system as could have been devised' (15 Dec. 1962). In particular the system of variable levies guaranteed that no imports would disturb the workings of the CAP's market organizations. Consequently, agricultural protectionism was not much of a topic in the following GATT negotiation rounds.

The normal rules of the GATT do not apply to custom unions and other forms of regional economic integration, which are exempted from the mostfavoured-nation principle. And since the USA supported European integration for other, mostly non-economic, reasons, they had to swallow the closing of the EEC's agricultural markets.[13]

Despite the problems in agricultural world markets created by the systemic overproduction and the patterns of export subsidies in the EC, the exclusion of agriculture from the GATT regime and from the various rounds of linear tariff-cutting survived. In every GATT round there was heavy criticism of the CAP from the USA, but even American lawyers and economists have dismissed this criticism as hypocrisy, in that the USA had its own systems of deficiency payments and export subsidies (Dam 1970: 260–1).

The political and social dynamics of the CAP

An examination of its history shows that the CAP was not merely set up to complete the common market, promote an economic union and to pave the way for a politically united Europe, although all of these were certainly motives of the founding fathers of the Community. It was also designed to finesse a number of problems arising out of national agricultural policies at the time of its inception in the late 1950s. A particular aim was to postpone—if possible to avoid altogether—placing limits on farm production, even though national surpluses were already beginning to emerge. This and other evasions were built into the CAP and hobbled it ever since. In many ways, the problems it has constantly faced are merely the logical consequence of the ambiguities wished on it at birth. (Duchêne *et al.* 1985: 1)

Adjustments and piecemeal reforms of the CAP became increasingly necessary because of developments which the CAP itself triggered and shaped. The international environment, changing world markets, and political pressure applied from outside, most notably by the USA, were of secondary importance. After all, insulating the farming sector was the major goal of the CAP, for which a supranational policy was a precondition. But this achievement should not be viewed as a high level of supranationalism at the expense of the autonomy of the member states. Formally it can be claimed that the implementation of collective market organizations in the 1960s seriously reduced the freedom of the member states to pursue an agricultural policy devised in purely national terms. The formal structure of decision-making also suggests strong elements of supranationalism. But this picture is wrong. A peculiar system of governance with an integration of national and supranational elements emerged to secure national interests. On the other hand, however, this unique and thus highly segmented system of governance developed a dynamic of its own. Consequently, the same features which helped the member states to maintain the national integrity of their agricultural communities created a situation in which it became increasingly difficult to achieve a balance between supranationalism and national control, *and* to keep the CAP functioning.

Political control of supranationalism

In the early negotiations over the regulations establishing common agricultural market organizations, the member states took pains to retain a number of powers, particularly the power to fix prices for the various agricultural commodities to be included.[14] Secondly, the main power to agree a substantive legislation was retained by the Council of Ministers. Thirdly, although the responsibilities and powers of the European Parliament increased substantially over time, its actual role as regards agriculture has been particularly limited. A fourth element was the creation of the Special Committee on Agriculture (SCA), which helped to preserve the pre-eminence of national interests in the legislative process on agricultural matters. The SCA is composed of senior national civil servants, who prepare the meetings of the Council of Agriculture Ministers.[15]

A fifth major element of the policy process provides for a vertical integration of national and supranational decision-making through the management committee system. The Council delegated most powers to implement the rules of the market organizations to the Commission, but required the Commission to consult a committee representing national interests. Each market organization has its own management committee. The committees are composed of representatives from member states, with votes in accordance with the qualified majority procedures laid down in Article 148 (2) of the EEC Treaty.[16] A representative of the Commission serves as chairman. Under this procedure the Commission submits proposed measures to the committee for endorsement. If the measure receives a 'favourable opinion' by a qualified majority, the Commission can put it into operation.[17] If there is a qualified majority against, the issue goes to the Council. The same procedure is applied to the Guidance Section of the EAGGF.[18]

To sum up then, a basic thread in the development of a vertical integration of policy through national and Community mechanisms has been to limit the power of the Commission and to retain key functions in the Council. The thick web of committees ensures the prominence of national interests in supranational decision-making. In addition, with the establishment of the European Council in 1974 the function of the Commission as the sole initiator of Community legislation has in practice been circumscribed. Therefore, at both the preparatory and the executive stages the national elements prevail.

The renationalization of west European agriculture

This *juxtaposition* of national and supranational agricultural policies was further strengthened and perfected by the introduction of monetary compensatory amounts (MCAs), which in effect renationalized the common market with regard to agriculture. Monetary compensations have an effect equivalent to a customs duty and became necessary when the system of fixed exchange rates collapsed in 1971. MCAs emerged as a measure to maintain the form of a *common* price structure for the agricultural products while insulating *national* agricultural markets from the repercussions of highly volatile exchange rates. The major function of MCAs, therefore, is the elimination of trade movements *not* originating from shifts in agricultural markets.[19] Over the years the agrimonetary system developed a 'byzantine complexity that almost defeats rational exposition even by those expert in the system' (Neville Brown 1981: 509). Worse, however, from the point of a common market, is the fact that member states used MCAs as side-payments to offset compromises in the Council, thereby sustaining the CAP.

MCAs have the merit of allowing countries considerable freedom in determining the level of their domestic farm-product prices and in general bring about inter-country transfers which are politically acceptable and economically reasonable. But unlike purely national agricultural pricing, the MCA system places limits on price divergencies... This allows the retention of common financing, which would not be possible if countries were allowed complete freedom to choose their own farm-product price level (Heidhues *et al.* 1978: 48–9).[20]

It comes as no surprise, then, that one of the consequences of the CAP is an increase in national and regional differences in the farming sector. The CAP was increasingly redesigned to allow member states—and sometimes the regions too, when they have the autonomy and the political and financial resources to do so—to make use of the CAP's instruments in their own way and for their own purposes. Two examples suffice to describe this development. First, some member states tailor the CAP's structural programmes to discriminate against part-time farmers, particularly in southern Europe. Other member states, however, use the same structural programmes to make part-time farming more viable, for example in Germany.[21] A second example is the introduction of milk quotas. One reason for the quota regime was to protect small milk producers. The UK, however, allowed the selling of quotas and the evolution of a market for quotas, which has resulted in considerable concentration in the dairy sector. Thus, the member governments can use the CAP for quite different purposes; they can structure and influence their farming sector with the help of supranational instruments, thereby eliminating or at least weakening the role of national parliaments. The example of the CAP shows that the EC became a genuine *arcanum imperi*.

The continuous strengthening of the national elements in the CAP's institutional structure, however, poses a dilemma. There is, on the one hand, the desire to make use of the CAP in a thoroughly national style. On the other hand, member states are in need of a CAP with enough autonomy to fulfil its basic functions, that is to insulate west European agriculture both from its international environment and from encroachment on the welfare function of agricultural policy as well. The institutional development of the CAP is clearly marked by this dilemma.

Permutations of the CAP

The CAP, because of its peculiar institutional format, evolved into the most important agent for transforming west European agriculture.[22] It changed the original socio-structural foundations from which it emerged and which had given it legitimacy and rationality. This development alone presents a formidable challenge to the CAP. At the same time, however, the system of agricultural decision-making became more and more hampered by the *de facto* requirement of unanimous decisions by the Council of Ministers.[23] The substantial increases in the heterogeneity of the agricultural sector as a consequence of successive enlargements provided an extra source of problems, causing major difficulties in making the decisions necessary to govern the CAP.

Three factors contribute to the problems inherent in European agricultural policy-making. First, the CAP's institutional arrangements require annual decisions on the prices of the various market organizations. Decisions *must* be reached, otherwise the existing level of prices will continue or the Commission, endowed with emergency powers, will step in.[24] Secondly, the relevant Council of Ministers consists only of ministers of agriculture from the member states, which gives this forum for decision a distinctive and homogeneous character.[25] Thirdly, since the EC had no administration of its own and since the CAP, because of its central importance to the economic situation of all farmers, has an extra need for legitimacy, the agrarian interest groups early on became an integral part of the work of the Commission.[26] Thus there are powerful tendencies in the system to generate price agreements which ignore some of the pressing budgetary and distributional demands on the CAP.

All in all it has proved extremely difficult to control CAP expenditure or to limit its more perverse distributional effects. It should be noted that the EC has not acquired its own powers of direct taxation, nor can its budget go into deficit. Instead, the budget depends on import levies and, much more importantly, upon revenue contributions from the member states. Originally the CAP was expected to be self-financing; since the EC at the time of its inception was still dependent on agricultural imports, levies on these imports, so the argument ran, would easily cover the costs of the CAP. But after the rapid increase in agricultural productivity in western Europe, induced by high institutional prices, the EC soon emerged as a major exporter. From the second half of the 1950s to the beginning of the 1970s the degree of self-sufficiency of the Community of the Six rose from 90 to 111 per cent in the case of wheat, and from 101 to 116 per cent in the case of butter.[27]

Despite early warnings by agricultural economists regarding the costly consequences of an output-geared system of price support, until the end of the 1970s the CAP could be characterized as a textbook case of an income-oriented policy.[28] Market organizations for basic products with Community-wide institutional prices were at the centre of the CAP's institutional framework. Following the 'objective method', agricultural prices increased as the costs of agricultural production increased. In that period, the growth of average agricultural income per capita corresponded closely to general income growth. The years between 1979 and the mid-1980s saw the beginning of a more cautious price policy and the introduction of 'producer co-responsibility', with farmers bearing part of the costs of disposing of surplus production. In particular, the Commission acknowledged, through its price proposals, the need for a change in the CAP, but these bore little fruit.

The years between 1984 and 1987 brought more serious changes: the introduction of milk quotas, and restrictive price policies for most other agricultural products, to reduce surpluses. High budgetary costs for this started the push for a policy change; this reduced agricultural incomes and forced some small producers out of business, but all in all the price reductions had little effect on production growth. In 1988 budget ceilings and stabilizers were introduced to control expenditure: first, by binding it to the level of the general budget of the Community—no price increases to be funded beyond the 'agricultural guideline'; and, secondly, by introducing production thresholds, which automatically trigger price cuts. (See also Ch. 3) This meant a further reduction in real agricultural prices.

The agreement on CAP reform in summer 1992 and its implementation in 1993 marked the beginning of a new phase. Its basic feature is the attempt at decoupling the income problem of west European agriculture from price policy: price policy is to be more oriented towards the efficient functioning of agricultural markets, and the introduction of direct income payments will help to improve the income situation of farmers. In addition, these will be accompanied by a system of mandatory fallowing and other means of production control at the level of the individual farm.

This is a very short account which gives only a limited view of the real complexities of the issues and conflicts accompanying the development of the CAP. It should provide, however, some idea as to why the CAP proved so resistant to major changes. Two factors, which have been already mentioned, are of particular importance. First, the institutional apparatus of supranational policy-making brings a considerable inertia into the governing of the CAP. Secondly, the distinctive 'moral economy' of the CAP, as primarily welfare-oriented and inward-looking, makes it quite difficult to legitimate and rationalize policy changes by using purely economic and budgetary criteria.

Not surprisingly, then, the most extraordinary development characterizing the CAP, particularly in the last decade, is the explosion of its costs, despite a shrinking farming population and a diminished economic role of agriculture (see Table 4.1). The most revealing indicator of this development is CAP expenditure per head of the population. It rose from 64.2 ecus in 1987 to

Table 4.1. The declining position of farming in selected member states

	Percentage of total working population in agriculture		Percentage of GDP derived from agriculture	
	1959	1992	1959	1992
Germany	17.9	3.3	8.4	1.5
France	26.6	5.8	16.0	3.3
Italy	39.8	8.5	21.2	3.6
Netherlands	12.4	4.5	10.9	4.0

106.5 ecus in 1994.[29] More important than the absolute figures, however, is the manner in which CAP money is spent. Most of CAP expenditure goes into storage and export subsidization. This means that the money is benefiting the farmers in an indirect way. In addition, such a system of price support benefits the bigger farms much more than smaller and medium-size farms. The Commission estimates that 80 per cent of CAP expenditure goes to 20 per cent of farmers. Therefore, the more pressing problem is the distribution of CAP expenditure. At the same time, however, for the reasons already mentioned, it is extremely difficult to change the CAP, particularly the undifferentiated system of price support, which survived the introduction of quotas, stabilizers, and budget ceilings.

The politics of including agriculture in the GATT

Domestic politics and international relations are often somehow entangled, but our theories have not yet sorted out the puzzling tangle. It is fruitless to debate whether domestic politics really determine international relations, or the reverse. The answer to that question is clearly 'Both, sometimes.' The more interesting questions are 'When?' and 'How?' (Putnam 1988: 427)

In this section I argue that it was *neither* the state of international agricultural markets *nor* the pressure applied by the USA that convinced the European policy-makers to take the inclusion of agriculture in the Uruguay Round more seriously. There is no point in denying the grave international problems resulting from the CAP because of its substantial export subsidization. But this development alone, despite trade wars between the EU and the USA, did not result in more resolute attempts to change the institutional format of the

CAP, nor in the decision to include agriculture in the new GATT Round. Of crucial importance were the difficulties of maintaining the social-welfare function of the CAP.

The primacy of domestic welfare

The basic similarity between the general welfare-state institutions of west European countries and the institutional structure of the CAP is most evident in its strongly autarchic character, that is the more or less total disregard for its international consequences, despite the obvious interdependencies of agricultural markets. External relations therefore took a low priority in the governance of the CAP:

> In international negotiations, the Community has insisted that the CAP is an internal policy: exports are a means of disposing of internal production when it exceeds internal requirements, imports serve to satisfy internal requirements when they exceed internal production . . . The Community declines to admit any responsibility for the instability of world trade but uses it to justify protecting the internal market. (Pearce 1983: 148–9)

Until the second half of the 1980s this was still true, but then the constellation changed quite dramatically. Structural policies for agriculture have achieved a new prominence and a new meaning by being removed from the closely hedged agricultural policy-making process and made into a much more integrated general approach to structural problems.[30] But the real change in the agricultural policy-making arena was brought about by the decision to make *domestic* agricultural support systems—and not just agricultural tariffs and export subsidization—a topic of high priority in the Uruguay Round of the GATT. This change came to the fore in a way which is characteristic of the EU—slowly, and with surprising twists (for a different account, see Ch. 12).

Much more important than the international consequences of systemic overproduction or neo-liberal pressures were the mounting budget costs and, even more so, the farm income problem. More and more the CAP resembled a bargaining system with the national ministers in the driver's seat. At the same time, despite the huge increases, CAP expenditure produced fewer and fewer benefits for the farming sector itself.[31] Big proportions of CAP expenditure were eaten up by the costs of storage and export subsidization. For this reason, the GATT Round opened a 'window of opportunity' for CAP reform. It provided a context for overcoming internal disagreements over the future course of the CAP. These disagreements have sharpened national profiles. Only by deliberately upgrading the external threat—and not pushing it aside, as in former Rounds—could the member states convince themselves that they were better off with a reformed CAP than with an unreformed one.

The move to open this 'window' was quite simple: by blaming the abandonment of the laws of competition in agriculture, and by openly acknowledging that the CAP was distorting trade, European policy-makers tried to gain an extra lever to deal with both their internal disagreement and the agricultural interest groups and national parliaments as well. In reconstructing the terms in which agricultural policies are discussed, a new level of agricultural policy-making and a different and more circumscribed negotiation forum may bring a solution to the long-term problems of the CAP. However, the politics of making agriculture an item of high priority in the Uruguay Round had consequences not intended by those who devised them in the first place.

The irony lies in the fact that initially the EU, and most notably the French, tried to block the inclusion of agriculture in the new GATT Round, whereas the US administration explicitly tried to internationalize its domestic farm problem in order to overcome a political stalemate at home (Paarlberg 1992: 30). Five years after the opening of the GATT Round, and one year after the Round should have been concluded, the EU suddenly found it a useful device to overcome its political inertia. Only then did the peculiar intermeshing of internal problems, finding acceptable terms of a CAP reform, and external problems, the GATT Round, convince the European policy-makers that something could be won by moving from the deadlock of the exclusively European level to the international level. The US government was not able to exploit the new situation, although it had explicitly tried to shift its domestic agricultural problems to the international level precisely to circumvent the strong position of agricultural interests in the Congress. On the contrary, these interests found new means to use the EU–US conflict over agriculture to secure the continuation of the American system of agricultural subsidization (Paarlberg 1992, 1995).

Changing course: the role of the budget

When the Uruguay Round opened in 1986, it was not particularly difficult for the EC to agree to include agriculture as a major topic. After all, the experience from other GATT Rounds had been that a last-minute exclusion of agriculture was the only way to complete them. Why should it be different this time? In both the Kennedy Round and the Tokyo Round the EC succeeded in defending the CAP. Since the USA started the new Round with an extreme negotiating position, by demanding an elimination of *all* agricultural subsidies that tended to distort production of trade over a ten-year transition period, the EC saw no reason to take the agricultural negotiations very seriously (Paarlberg 1992: 35; Anania *et al.* 1994a: 6–11). The EC proposed only to reduce the overall level of support and refused to negotiate specific policy instruments (Anania *et al.* 1994a: 25). Moreover, at the beginning of the Uruguay Round it was assumed that the introduction of stabilizers and budget ceilings in 1988 had succeeded in attempts to control CAP expenditure. Whereas from 1987 to 1988 CAP expenditure increased by roughly 20 per cent, from 1988 to 1989, for the first time in the history of the CAP, expenditure decreased by 5.5 per cent. From 1989 to 1990 CAP expenditure began to rise again, but slowly, and stayed, in absolute figures, below the level of 1988.[32]

According to the Punta del Este declaration of 1986, the Round was to be concluded by 1990. Between 1988 and 1990 the budgetary situation of the EC with regard to the CAP was relatively relaxed, leaving both the Commission and the Council of Ministers not much impressed by the demands of the USA. But the situation changed when CAP expenditures again began to rise. From 1990 to 1991 CAP expenditures increased by 22 per cent. The budget forecasts for 1992 and 1993 promised even higher outlays.[33]

Box 4.2. The parameters of CAP reform in summer 1992

....

- Reduction of grain prices by 30 per cent within three years and beef prices by 15 per cent.

- Introduction of compensatory direct income payments, which are, however, not production-neutral: farmers can obtain the payments only if they grow eligible produce (grain, pulses, oil-seed, feed maize).

- Introduction of a compulsory set-aside scheme: all farmers with farms above a certain size are required to set aside 15 per cent of their arable land in order to be eligible for transfer payments.

- So-called small producers are exempt from the set-aside requirement.

The return of budgetary problems was the main reason for the Commission formulating, at the end of 1991, those concepts which became the corner-stone of the CAP reform in summer 1992: a reduction of the intervention price for wheat by one-third, and the introduction of direct income payments to all farmers to compensate for price cuts. To receive these, farmers would have to take part in a system of mandatory fallowing, thus introducing new and hopefully more efficient elements of production control.

The move to direct income payments

The experience of the 1980s showed that limits on the budget alone, in the form of stabilizers and thresholds triggering price cuts, would not help to reduce expenditure in the long term. The main problem was the direct connection between the institutional prices of the market organizations and the income of farmers, which turned price policy into an instrument which increasingly bolstered the incomes of trading and exporting companies, but not the incomes of farmers. The annual price settlements, required for over twenty different systems of market organizations, had become a highly unsatisfactory means of guaranteeing adequate income levels. Therefore, the basic idea of the Commission was to decouple prices from incomes.[34] The argument of the Commission was straightforward and focused directly on the welfare issue: eighty per cent of the expenditure of the CAP went to 20 per cent of farmers. The introduction of direct income payments promised more equitable means of distributing CAP money.[35]

Box 4.3. Summary of the final agreement on agriculture in the Uruguay Round

....

- A reduction of domestic interventions in agriculture, measured by an aggregate degree of support over a six-year implementation period, starting in 1995 and taking 1986–8 as base period.

- Direct payments to farmers under production-limiting programmes are not to be subject to the commitment to reduce support, as long as they are based on fixed area and yields or on livestock numbers.

- All non-tariff barriers (quotas and other import restraints) are to be subject to tarification.

- A reduction of the average tariff by 36% over the implementation period; each tariff line will be reduced by at least 15%.

- A reduction of export subsidy expenditures by 36% and a reduction of the volume of subsidized exports by 24% over the implementation period.

- An introduction of safeguard clauses specifying the circumstances under which countries are allowed to impose additional duties to prevent undesired market and price distortions due to imports.

- A guarantee of minimum market access equal to 3% of average domestic consumption in the base period.

Once the US relaxed its negotiation position in agriculture after 1989 to keep the Uruguay Round going, the EC saw a possible solution to both problems: to accept the proposal of the Commission as a long-term solution to the budget problems of the CAP and thus change the CAP's institutional structure; and to offer this to the US as a compromise in the GATT negotiations. To combine the domestic solution with the external solution helped to overcome the internal difficulties in the Council of Ministers.[36] This was no easy task since the Council of Ministers was once described as a '13–headed dragon with 26 arms and 26 legs' and the job of the Commission as 'to slay the 13–headed dragon and to replace it with one beast with all 26 legs moving in the same direction' (Hardy-Bass 1994: 240).

The role of agrarian interest groups: policy-makers or policy-takers?

Although the Commission did not try to introduce social criteria into the calculation of compensation payments, thus deliberately *not* discriminating between the bigger—and supposedly more competitive—and the smaller farms, the reaction of the agrarian interest groups was unanimously negative. The direct income payments foreshadowed a change in the institutional structure which would, in the eyes of the farming lobby, enable the EC to rationalize the CAP, in ways not previously possible: 'Since they are financed from national budgets, they are more transparent and thus can be scrutinized more easily than price supports, which are hidden in higher consumer prices' (Legg 1993/4: 26). Unlike support through market prices, direct payments open up the questions of 'who, where, what, and how much?'. The difficult issue of eligibility could easily fragment the interest base of agriculture, particularly if direct income payments were structured to function as a minimum income scheme, in which eligibility was determined in conformity with criteria established for other schemes of social welfare. Moreover, the farming lobby was strongly opposed to direct income payments, because they would result in increased economic insecurity and a sharpened political dependence. Budgetary constraints, agrarian interest groups argued, would have a much more direct impact on agricultural policy-making because of the higher visibility of the new transfer payments.[37]

The agrarian interest groups, at both national and European levels, did not have the power or the means to influence the terms of the CAP reform significantly. They shape agricultural policy-making by pointing out the likely economic and social consequences of decisions.[38] But they are not able to influence the parameters of agricultural policy-making—which derive for the most part from the imperatives of welfare statism—nor are they in a position to formulate *positive* statements and proposals. The role and the power of COPA was a function of the institutional structure of the CAP—but not the other way around. COPA was of crucial importance in securing the acceptance of the CAP in the sixties. In particular the Commission used its close relationships with COPA to bolster its—supranational—position *vis-à-vis* the Council of Ministers (Averyt 1977). Once the CAP had turned into a weight-bearing test for European integration, and once the Commission had reversed gears and tried to put brakes on the development of the CAP, the politics of COPA became heavily constrained (Buksti 1983). Therefore, the much more important problem for the EC was the disagreement over CAP reform in the Council of Ministers.

Thus it was mainly the conflict over the distribution of the costs and benefits of the CAP reform among member states which made the integration of the reform in the GATT Round a promising strategy. The GATT negotiations provided a binding *external* constraint needed to bring about the necessary price cuts and the transition to a more transparent and a more liberal economic order in the agricultural support system. In short, the EC was in dire need of a constellation of pressures to sustain substantial CAP reform beyond summer 1992 and to secure its implementation. Without the heightened political pressure from the Uruguay Round it seems unlikely that the Council of Agriculture Ministers would have been able to agree. The inclusion of the CAP reform in the compromise on agriculture in the Uruguay Round helped to ensure that the reform would survive, and that making the social and economic effects of the new CAP on individual farmers more transparent would eventually strengthen political control of CAP expenditure. In addition, it was the only feasible option for the USA. Not to give the EC credit for its own reform, as the so-called Dunkel Text proposed, would have meant that, first, there would be no agreement in the agricultural part of the Uruguay Round, and, secondly, the CAP reform would have been jeopardized, faced with the strong resistance of the French government.[39]

Conclusion

The task is to determine the claims of social justice as these reflect and contribute to the general interest. It confronts us with the whole matter of values in modern society. There is, it would appear, sufficient agreement to make some of the criteria sufficiently objective to serve policy making . . . Social welfare thus conceived is positive and forward-looking and not a compound made up of the values inherent in the income structure of some historical period. . . . (Schultz 1943: 34)

There are several reasons for arguing that the supranational politics of the CAP and the Uruguay Round of the GATT have strengthened social and economic nationalism in the member states of the European Union. Both the CAP reform and the Uruguay Round pushed west European agriculture in an even more nationalistic direction, which implies policies which try to block or to avert the potentially disturbing influences of international trade and of a highly interdependent, global economy on the economic and social situation at home. But it is the supranational character of the policy which has helped to achieve this goal, increasing the autonomy of governments not only *vis-à-vis* national parliaments and interest organizations, but *vis-à-vis* the interests of agricultural exporting countries as well. The CAP sits in a kind of a halfway-house between the nation-state and something else. Since it is still unique in the experience of European integration it contains its own limitations and is fundamentally contradictory. But contrary to all expectations, both member governments and Commission have been able to exploit this contradiction. In a certain sense, then, the paradoxes of the CAP are in fact a source of stability.

The CAP was, and still is, an important means for member governments to defend national agricultural policies with their highly protectionist, welfare-oriented institutional structure. In addition, it helped to take agricultural policies out of domestic distributional conflicts, since the decision-making process at the level of the Community is, to say the least, highly complicated and quite cumbersome. For this reason, the CAP in general and the Uruguay Round in particular have to be seen as political instruments used by member states in defence of their own agricultural sectors, with their distinct structures. This sort of supranational politics means an increase in the autonomy of member states rather than a decrease. Contrary to the expectations of the early theories of European integration, the CAP is not about a transnational integration of farmers or about a basic shift in the loyalties of farmers from the national to the supranational level. The CAP is much more about the internal problems of the member countries and about the political control of agrarian interest groups.

Notes

1. 'Family farming' is a particular system of agriculture which evolved on the European continent in the last century in parallel to the industrial revolution. Farms tend to be owner-operated and rely on family labour. There is no division of labour between home and work; ownership is widespread and production is diversified to reduce dependence on a single market for income and to ensure year-round use of labour. In contrast, an industrial agri-business is based on a sharp differentiation between economic roles: 'Some people work for wages, others invest for profits, and yet others manage the affairs of both workers and owners' (M. Strange 1988: 36). It should be noted, however, that these two models of farming systems are ideal-types. They represent the opposite ends of a continuum. (For a general—non-specialist—discussion, see Cochrane 1965.)

2. Because of the agreement on the basic institutions of the CAP in 1961, Walter Hallstein, then President of the Commission of the EEC, did not see the Treaty of Rome as a step back from the supranationalism of the ECSC to a much more restricted intergovernmentalism (1962: 19).

3. In 1970 two leading American scholars on European integration looked upon the CAP as 'a story of action and success' (Lindberg and Scheingold 1970: 41).

4. *The Economist*, 29 Sept. 1990. Agricultural economists tend to agree: 'Save for that of Japan, the Common Agricultural Policy of the EC is perhaps the most distortionary agricultural intervention in the world' (Rausser and Irwin 1988: 355).

5. See *inter alia* Rosenblatt *et al.* (1988: 19): 'The issues facing policy-makers and the arguments put forward in the discussion of policy have remained largely unchanged since the inception of the CAP. Surplus production and its budgetary and international implications were actually discussed at the Stresa conference—that is even before the formation of the CAP'.

6. To avoid possible misunderstandings I stress that this is intended to be a positive, and not a normative statement. (For a fuller description of this approach, see Rieger 1995*b*.)

7. For a similiar view see the important contribution of the economic historian Alan S. Milward (1981).

8. In 1958 Ernst B. Haas, a leading scholar on European integration, listed agriculture as the first among several other groups 'finding it difficult if not impossible to reconcile their aims with the supranational economic and political organization of Europe. Spokesmen for agricultural associations . . . stress that a common market for agricultural commodities would ruin the tariff-protected and nationally subsidized peasantry in most ECSC countries' (1958: 296).

9. The best description of this institutional revolution is still T. H. Marshall (1963).

10. See the telling comment by an American legal scholar: 'The breadth of this waiver, coupled with the fact that the waiver was granted to the contracting party that was at the same time the world's largest trading nation and the most vocal proponent of freer international trade, constituted a grave blow to the GATT's prestige' (Dam 1970: 260).

11. The position of the USA in the GATT Round at the time pressed the west European countries to consider extreme measures: 'The Austrians reportedly said that if the GATT gave them too much trouble on agriculture, they were indeed ready to put all agricultural products under state trading' (Richter 1964: 14).

12. Ernst Haas saw in the international dimension a major impetus for the supranational integration of the EEC: 'The practical need for co-operation in other international economic organizations is especially striking. The six countries had to act in unison in being recognized as a single contracting party in the GATT and in being exempted from extending liberalization requirements in OEEC. Had GATT permissions been denied, the six countries would automatically have been compelled to pass their tariff relaxations to third countries under the most-favoured-nation clause, thus possibly eliminating any special benefit to the integrated sectors' (1958: 297–8).

13. For the details of the capitulation of the USA in the 1961–2 GATT game see J. H. Richter (1964: 14–15).

14. Art. 43 of the EEC Treaty provided the Commission with powers to make pro-
    posals on the organization of the CAP and its implementation. In its proposals
    the Commission stressed that 'it is easy to see the limitations of price policies as
    a means of under-pinning incomes' because 'extensive support for prices and
    markets easily leads to overproduction'. Accordingly it asked for close links
    between structural policy, market policy, and social policy for agriculture
    (Lindberg 1963: 238). Such an approach never materialized.

15. The segmented character of agricultural policy-making is also characterized by
    the fact that outside agriculture most meetings of the Council of Ministers are
    prepared by the Committee of Permanent Representatives.

16. This means that the number of votes and the majority required has been changed
    every time Article 148 (2) has been amended with the successive enlargements.

17. For details see Usher (1988: 147–9). Since originally the management system had
    no legal base in the EEC Treaty, there were some doubts about its status. In a rul-
    ing issued in 1970 the European Court of Justice held that the committees were
    an admissible method of permanent consultation between the Council and the
    Commission. Since then management committees have mushroomed in other
    areas of legislation.

18. For details see Weatherhill and Beaumont (1993: 55–6).

19. Take as an example the floating of the German currency upwards, which initially
    caused the problem: 'If . . . an intervention price had been fixed in German marks
    at the official parity, then, during the course of the marketing year, the value of
    the mark went upwards in relation to other currencies, but the number of marks
    a producer received per metric ton of cereals still remained the same, then there
    would be a very considerable inducement, for example, to French cereal grow-
    ers, to sell all their cereals to the German intervention boards and, in terms of
    French francs, get considerably more than they would from the French inter-
    vention agencies' (Usher 1988: 108).

20. The completion of the internal market programme and the abolition of border
    controls were accompanied by a decision to phase out MCAs. This has not been
    achieved because of the still very high volatility of exchange rates among mem-
    ber states. Unless there is a monetary union—*ceteris paribus*—MCAs play a
    significant role in the governance of the CAP.

21. In Germany nearly half the farmers are officially counted as part-time farmers.
    The importance of this steadily increased over the years. For a good analysis of
    this general transformation of west European agriculture see the research report
    of the Arkleton Trust (1992). The inclusion of the eastern Länder with their
    larger land-holdings may change the domestic politics in Germany.

22. For a detailed discussion of the social change of west European agriculture under
    the CAP see Bowler (1985) and Rieger (1995b).

23. It should be noted that it was for the CAP that the transition to qualified majority
    voting envisaged by the EEC Treaty was blocked and thus a 'national veto' intro-
    duced. In the 1980s, the option of member states to invoke 'vital national interests',
    as stipulated by the Luxemburg Compromise, was severely restricted (Vasey 1988).

24. This happened in 1985 and again in 1988 (Vasey 1988: 731).

25. This does *not* imply that there are no cross-cutting restraints on supranational policy-making. The restraints, however, are to be found not at the European level, but at the national level. See the remark by Peter Walker, former British minister of agriculture: 'One of the great myths in articles which are written about the CAP is this myth that there is a group of Agriculture Ministers who meet together with a great and deep desire to push up farm prices and, to the horror of their Finance Ministers, pile up enormous bills. The reality is that every Minister of the Agricultural Council negotiates from a brief agreed by his Cabinet and agreed by his Finance Minister' (Swinbank 1989: 304).

26. One of the best studies on the origins and early history of COPA is Averyt (1977). The recent revival of interest in European interest groups does not extend to agriculture. In one major study it is mentioned only in passing (Mazey and Richardson 1993*a*).

27. At the beginning of the 1960s, when the Community of the Six devised the first European market organizations, it was already clear that the Community would have substantial surplus production. In fact, this was one of the main arguments for 'financial solidarity'. As a French negotiator stated: 'It is true that in the future the Common Market may no longer have such large import deficits; over-all surpluses may even appear for sugar and milk. But the Community, whose agricultural policy would by then be defined by common decision, would quite naturally have to bear the consequences of that common definition; a common policy calls for common financing' (quoted by Tracy 1989: 258).

28. The following is based on Rosenblatt *et al.* (1988), Tracy (1989), and Anania *et al.* (1994*b*).

29. Commission of the EC, *The Agricultural Situation in the Community*, Report for 1990, Report for 1994.

30. For a discussion of the changed internal and external environments in which the CAP operates, see Villain and Arnold (1990).

31. In particular, the smaller farms are able to survive because of two elements more important than the CAP: the first is the rapid increase in other activities, thereby opening up new sources of income for agricultural households; the second is the availability of non-agricultural welfare-state payments, mostly child allowances and pension payments. For a detailed analysis of the income of farm households in different member states see Arkleton Trust (1992).

32. Calculations based on the statistical annex to the *The Agricultural Situation in the Community, Annual Reports* by the Commission, for the years 1990, 1991, and 1992.

33. See the *Annual Report for 1992*, statistical annex T84.

34. It should be noted that the coupling of price and income functions is an old problem in the agricultural politics of developed countries. For an early discussion see Schultz (1943: 6–19.

35. There is no evidence that the 'new' ideas of the Commission about a more serious reform of the CAP were influenced by the agricultural negotiations in the

Uruguay Round. The American Trade Representative Carla Hill stated in late September 1991 that 'we have not (yet) had one day of political negotiations on agriculture' (Paarlberg 1992: 38–9).

36. For a detailed discussion of the different implications of the CAP reform for northern and southern member states see Sarris (1993). He points out that, on the one hand, the farming sectors of the northern member states are much more vulnerable to price cuts, whereas on the other, the introduction of direct income payments presents enormous administrative problems to the southern member states.

37. A crucial factor determining the political security and reliability of transfer payments is the willingness of the courts to interpret direct income payments as property rights of individual farmers, thus limiting the discretion of policy-makers. Proponents of direct income payments argue that 'the amount of payments should . . . remain outside the farmer's control' (Legg 1993–4: 27). If this attitude prevails, farmers indeed would become beneficiaries of 'handouts'.

38. Consumer groups concerned with high food prices and the perplexing paradox of surplus within the EU at a time of severe food shortages in other parts of the world, and conservation groups concerned with the quality of the countryside receive increased media coverage, but influence agricultural policy-making in an indirect way at best.

39. In December 1991 the then Director General of the GATT presented a Final Draft for the GATT Agreement in Agriculture. In May 1992 the CAP reform was finally agreed. In November 1992 the so-called Blair House agreement settled the differences between the Dunkel Text and the CAP reform: 'All major E. C. objections to the Dunkel Text appear to have been augmented in the Blair House agreement' (Anania *et al.* 1994*a*: 30). In addition, in autumn 1993 the French government was first able to force the EC to reopen negotiations on the Blair House agreement, and succeeded in getting even more concessions from the USA.

Further Reading

For a more detailed political history of the CAP see Tracy (1989). Milward (1992, ch. 5) provides the best account of the pre-history. For the social consequences see Bowler (1985). Those interested in current developments should consult *The Agricultural Situation in the Community*, published annually by the European Commission.

Bowler, I. R. (1985), *Agriculture under the Common Agricultural Policy: A Geography* (Manchester: Manchester University Press).

Milward, A. S. (1992), *The European Rescue of the Nation-State* (Berkeley: University of California Press).

Tracy, M. (1989), *Government and Agriculture in Western Europe* (New York: Harvester Wheatsheaf).

CHAPTER FIVE

THE SINGLE MARKET: A NEW APPROACH TO POLICY

Helen Wallace and Alasdair R. Young

The single market and the SEA marked a turning-point in European integration, the roots of which, however, stretch back well before 1985. Heavy harmonization had proved a frustrating approach to common standards, especially as the pressures of external competition bore down on European industry. New ideas about market regulation permeated the EC policy process and, supported by ECJ judgements and Commission entrepreneurship, facilitated institutional reforms and the 1992 programme. These led to remarkable legislative activism and important changes in the policy-influencing and policy-implementing processes. Their longer-term impact is harder to assess, and the task of 'completing' the internal market remains unfinished. None the less the single market has drawn other European countries towards EU membership and changed the context in which many other policies are shaped.

Introduction

The plans to complete the single market induced an explosion of academic interest in the European Community (EC).[1] Before 1985 the theoretical debate on political integration was stalled, studies of EC policy-making were sparse, and few mainstream economists devoted themselves to the analysis of European economic integration. In the late 1980s that all changed as competing analyses proliferated and the nooks and crannies of the new legislative programme and its economic consequences were examined. Indeed new theoretical approaches to the study of European integration have taken the single market as their main point of reference, just as the earlier theorists had taken agricultural policy as their stimulus. The single market has been elevated so much that for many it is taken to constitute the critical turning point between stagnation and dynamism, between the 'old' politics of European integration and the 'new' politics of European regulation.

Our task in this chapter is to re-examine the renewal of the single market as a major turning-point in European policy-making. We draw on the study by Alan Dashwood in previous editions of this volume, a salutary reminder that the single-market programme had roots that were overlooked in much of the commentary that focused on developments in the late 1980s. Essentially we argue that many of the analyses that proliferated in response to the Single European Act (SEA) and the 1992 programme overstated their novelty and understated some of the surrounding factors that helped to induce their 'success'. We also suggest that some elements of the policy process around the single market contributed to the subsequent public disquiet about European integration.

None the less we also believe that the embedding of the 1992 programme represents a very significant redefinition of the ends and means of policy. It enabled the European integration process to adapt to new constellations of ideas and interests and to produce a different policy mode of regulation that has permeated many other areas of policy (Majone 1994b). Other chapters in this volume illustrate the consequences, both direct and indirect, of giving so definite an emphasis to market liberalization and different forms of policy regulation, as Giandomenico Majone (1993) argues. Hence we situate these developments in the broader context of structural shifts in the (west) European political economy, in the expectations and behaviour of entrepreneurs, and in the debate about adapting the European welfare state.

These developments are therefore as important for their impacts on the European public policy model *within* the member states as they are for their repercussions at the transnational level. We can observe market regulation, heavily based on a transnational level of European governance, jostling, often uneasily, with other issues on the political and economic agendas evident within the EC member states. We can also see the bifurcation between transnational regulation for transnational markets, engaging transnational regulators and large market operators, and encapsulated intra-national politics, engaging those charged with and dependent on the reduced domestic political space, smaller-scale entrepreneurs, local regulators, and national or regional politicians.

Nor have these ricochets been confined to the member states that accepted the SEA and '1992'. The extraterritorial impact on neighbours, partners, and competitors has been powerful. The alignment of the EFTA countries to the single market, first through the Luxemburg process, then through the European Economic Area (EEA),[2] and some eventually by full accession reveals the soft boundaries of a European economy that never did coincide with the political boundary of the EC. But the costs, social and political as well as economic, of adjustment within the single market have also generated rearguard action, sometimes focused on other intra-EC policies that might provide compensation, and sometimes by displacement to external competitors.

Several themes thus run through the story of the single market:

- the impact of new ideas, as views about the European 'welfare state' altered and Keynesianism was forced to compete with neo-liberalism as an alternative and potentially predominant paradigm in economic policy;
- the mobilization of industrial opinion and pressure in novel ways as a transnational phenomenon and a stimulus to policy change;
- the critical conjunction of changes to EC decision-rules with alterations in the relationships between the business community and policy-makers and in business responses to global markets;
- evidence of policy 'entrepreneurship', especially by the Commission, backed by a new coalition of supporters of change and the recasting of the old argument about 'Community preference';
- the impact of 'statecraft' by and 'collusion' between top policy-makers from key member states;

- the pervasive impact of European law and rulings from the European Court of Justice (ECJ) on the ways in which policy options were defined;
- the external dynamic of third-country competition and technological innovation; and
- the external projection of EC policy.

Un peu d'histoire

The aim to establish a single market started with the Treaty of Rome. This set targets for creating a customs union and the progressive approximation of legislation, as well as for establishing the 'four freedoms' of movement for goods, services, capital, and labour, all within a single regime of competition rules. In this it followed Bela Balassa's steps towards full economic union (see Box 5.1), though the path was more clearly defined for the customs union than for the single market (Balassa 1975; Pelkmans 1984). The policy-makers of the 1950s were more concerned about tariffs than non-tariff barriers to trade (NTBs), a preoccupation and 'set of ideas' also reflected in the General Agreement on Tariffs and Trade (GATT).

Box 5.1. Stages of economic integration

....

free trade area	no visible trade restrictions between members
customs union	free trade area plus common external trade regime
internal commodity market	customs union plus free movement of goods (no invisible trade restrictions)
common market	internal commodity market plus free movement of services, capital, and labour
monetary union	common market plus a common currency
economic union	monetary union plus a common economic policy

But entrepreneurial ingenuity to segment markets combined with the activism of governments, under pressure from domestic firms, to circumscribe production, sales, and consumption by product, safety, and process standards. Thus, as tariffs came down, other barriers were revealed, even reinforced. With the new technologies and new products of the 1960s and

1970s came new standards, which, whether so intended or not, were a frequent source of protection. Local market preferences, as well as national policy and industrial cultures, were divisive. Market fragmentation was often buttressed by operating rules, such as those for public procurement, that promoted local suppliers.

Harmonization and its increasing frustration

The harmonization of national legislation, especially for standards and market management, was one important policy instrument for moving towards the common market goal. We do not argue that standards as such were the be-all and end-all of policy. But we do argue that the impact of the initial efforts at harmonizing standards played an important part in testing policy methods that proved inadequate during the 1960s and 1970s. Frustrated, Commission officials, with some allies from the member states, sought a new regulatory approach, which was then more broadly applied.

The principal EEC instrument for advancing the four freedoms was the directive, in principle setting the essential framework of policy at the European level and leaving the 'scope and method' to the member states. In the case of NTBs harmonization was based on Articles 30 and 100 (EEC). Other articles provided the legal foundation for the freedom of movement for services, capital, and labour and for aligning many other national regulations. The Commission began to tackle the negative impact on trade of divergent national standards and differing national legislation in the early 1960s. These efforts gathered pace after the complete elimination of customs duties between member states on 1 July 1968 (Dashwood 1977: 278–89). Initially the Commission tended to regard uniform or 'total' harmonization as a means of driving forward the general process of integration. After the first enlargement, however, the Commission adopted a more pragmatic approach and pursued harmonization only where it could be specifically justified. It insisted on uniform rules only when an overriding interest, such as the protection of consumers or the environment, demanded it, often using 'optional' rather than 'total' harmonization.

Harmonization measures were drafted by the Commission in cooperation with working groups, one for each industrial sector, composed of experts nominated by member governments. Advice from independent specialists supplemented the Commission's resources and provided a depth and range of expertise comparable to that of the much larger national bureaucracies. The Commission also regularly invited comments on their drafts from

European-level pressure groups (Dashwood 1977: 291–2). Beginning in 1973 with the 'low voltage directive', the Commission, where possible, incorporated the work of private standard-making bodies into Community measures by 'reference to standards' (Schreiber 1991: 99). The two principal European-level standards bodies—the European Committee for Standards (CEN—from French title) and the European Committee for Electrical Standards (CENELEC)—did not, however, provide adequate technical assistance (Dashwood 1977: 292). Thus the complex and highly technical process produced very uneven results.

Progress was also greatly impeded by the need for unanimity in the Council of Ministers. Different national approaches to regulation and the pressures on governments from domestic groups with an interest in preserving the *status quo* made delays and obstruction frequent (Dashwood 1977: 296). The Commission exacerbated this problem by over-emphasizing the details and paying too little attention to the genuine attachment of people to familiar ways (Dashwood 1977: 297). Technicians and special interests often further constrained the opportunities for decision. As a result, only 270 directives were adopted between 1969 and 1985 (Schreiber 1991: 98).

ECJ jurisprudence, however, began to bite at the heels of the policy-makers. In 1974 the *Dashonville* ruling established a legal basis for challenging the validity of national legislation that introduced new NTBs. The famous *Cassis de Dijon* judgement of 1979 insisted that under certain specified conditions member states should accept in their own markets products approved for sale by other member states (Alter and Meunier-Aitsahalia 1994: 540–1; Dashwood 1983: 186). None the less there was cumulative frustration in the Commission and in the business community at the slow pace of progress and the uncertainties of reliance on the ECJ, since its impact depends on application to cases lodged. European firms kept encountering other countries' regulatory barriers in the knowledge that the international regime offered by the International Standards Organization (ISO) was weak, as was its US affiliate (Woolcock 1991). Stronger European standards would have provided a basis for negotiating more effectively for multilateral standards.

Pressures for reform

The governments of western Europe confronted an economic crisis in the early 1980s. The poor competitiveness of European firms relative to those of their main trading partners in the United States and, particularly, Japan contributed to large trade deficits (Pelkmans and Winters 1988: 6; and see Fig.

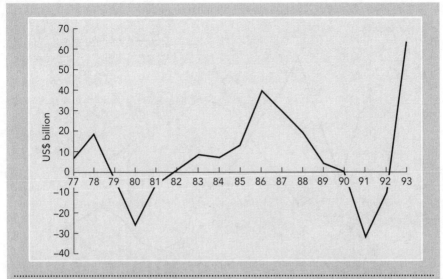

Fig. 5.1. EU trade balance with the rest of the world, 1977–1993
Note: EC9 1977–80; EC10 1981–5
Source: OECD (1994)

5.1). Transnational companies proliferated, producing and selling in multiple markets, and often squeezed the profit margins and markets of firms confined to national markets. The sharp increase in oil prices following the Iranian Revolution in 1979 contributed to the trade deficit and helped to push the west European economies into recession. Inflation and unemployment both soared during the early years of the 1980s (see Fig. 5.2). Business confidence was low and international corporations began to turn away from the Community (Pelkmans and Winters 1988: 6). American direct investment began to flow out of the Community and European companies sought destinations outside the Community for their investments and production facilities.

During the late 1970s and early 1980s the member states increasingly used economic regulations as NTBs to protect their industries (Dashwood 1983; Commission 1985; Geroski and Jacquemin 1985). This undid some of the earlier progress in harmonization, contributed to a decline of intra-EC imports relative to total imports (Buigues and Sheehy 1994: 18), and sharply increased the number of ECJ cases concerning the free movement of goods (see Fig. 5.3). The high level of economic interdependence within the EC made these NTBs costly and visible (Pelkmans 1984; Cecchini *et al.* 1988).

While the crisis was clear, the response was not (see, for example, Tugendhat 1985). Large trade deficits and high inflation constrained the

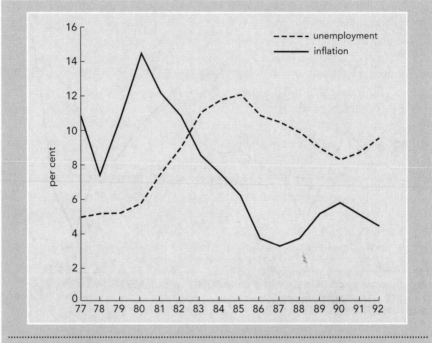

Fig. 5.2. Unemployment and inflation rates in the EU, 1977–1992

Note: EC10 for 1977–82.
Source: Eurostat (1986, 1988, 1993).

ability of member governments to use expansionary economic policies to bring down unemployment. Economic interdependence further reduced the efficacy of national responses to the crisis and provided an incentive for a coordinated response to the region's economic problems. The scope for a coordinated response was enhanced by changes within the member states. These are widely described in the political integration literature as a convergence of national policy preferences during the early 1980s (Sandholtz and Zysman 1989: 111; Moravcsik 1991: 21; Cameron 1992: 56).

But a note of caution should be added here: new government policies certainly emerged, but they differed between countries. The British government was radically neo-liberal, the French government switched policy after a factional contest within the Socialist majority, the German government's policy was the product of cross-party coalition and European market strength, while the Spanish government sought to link Socialist modernization at home to transnational market disciplines, and so on. Convergence is thus something of a misnomer: European market liberalization served quite different purposes for each government and for different economic actors.

Parties that advocated neo-liberal economic policies came to power in the UK, Belgium, the Netherlands, and Denmark, in part due to a rejection of the parties that had overseen the economic decline of the late 1970s (Hall 1986: 100). The rejection was less marked in Germany, where the underlying strength of its economy preserved an attachment to the established 'social market' framework. Elsewhere the Keynesian policies of the past attracted much of the blame—in France the 'policy learning' was explicit. Expansionary fiscal policies had led to increased inflation and unemployment, exacerbated the trade deficit, and swelled the public debt (Hall 1986: 199). By 1983 the French government had started to look for European solutions, reversing its threat of autumn 1982 to obstruct the Community market. The threat had been prompted by the trade deficit with Germany, attributed by some ministers to the impact of German product standards (H. Wallace 1984; Woolcock 1994).

New ideas about markets and competition thus started to be floated in

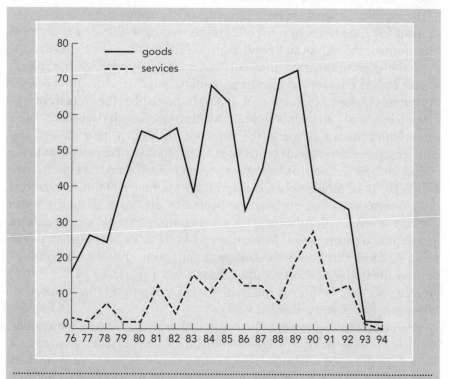

Fig. 5.3. Actions brought before the European Court of Justice regarding single-market matters

Source: General Report on the Activities of the European Communities, 10th–27th and 1994 reports.

response to the problems of the European economy, as the label of Eurosclerosis started to stick. Some transnational firms started to voice criticisms. The shape of an emerging policy consensus was influenced by the wave of deregulation in the United States in the late 1970s and early 1980s (Hancher and Moran 1989: 133; Sandholtz and Zysman 1989: 112; Majone 1991: 81). The ECJ's 1979 *Cassis de Dijon* judgement, although not deregulatory, advanced the concept of mutual recognition of national standards. This provided the Commission with a lever with which to pursue greater market integration (Dashwood 1983).

From the early 1980s European Council communiqués reflected a concern about the poor state of the single market. The European Council in December 1982 discussed a Commission communication that recommended the removal of NTBs, simplification of frontier formalities, liberalization of public procurement, and closer alignment of taxes (*Bulletin EC* 12-1982). The European Council responded by creating an Internal Market Council to meet regularly to consider such issues.

During 1983 support for revitalizing the single market continued to grow. In April 1983 the heads of some of Europe's leading multinational corporations formed the European Round Table of Industrialists (ERT) to advocate the completion of the single market (Cowles 1994). In July 1983 the Spinelli Report linked the costs of conflicting national regulations to the need for institutional reforms (Spinelli *et al.* 1983). In September the French government circulated a memorandum advocating the development of a Community industrial 'space', the reduction of NTBs within the EC, and compensating external trade protection. The proposals were a response to the realization that France (or any other member state) could not on its own redress the basic problems of industry and that reinforced EC measures were needed (Pearce and Sutton 1985: 68). A month later UNICE, the European confederation of national employers' associations, added its voice to calls for greater market integration. In February 1984, with its adoption of a draft treaty on European Union, the European Parliament sought to focus attention on institutional reform, calling *inter alia* for increased parliamentary powers and greater use of qualified majority voting in the Council of Ministers (European Parliament 1984).

Meanwhile the Commission also began to sharpen its focus on these issues. Karlheinz Narjes, the responsible Commissioner, and his staff started to look for ways of attacking market barriers, both by systematically identifying them and by exploring ways of relaxing the constraints on policy change. They suggested the 'new approach' to standards harmonization, which encouraged 'mutual recognition' of validated national rules, subject to a core of agreed essential requirements. Towards the end of 1983 Commission officials were able privately to persuade key officials from Britain, France, and Germany to accept the new approach, which was endorsed in July 1984, but not formally adopted until May 1985 (*Bulletin EC* 5-1985). This built on earlier British efforts to argue the deregulation case and on bilateral exchanges between the French and Germans to coordinate the activities of their standard-setting bodies, AFNOR and DIN (H. Wallace 1984). Also in this period concern to mitigate the impact of border controls led the French and Germans in 1984 to agree the Saarbrücken Agreement, later converted at the insistence of the Benelux governments into the first Schengen Agreement of 1985.

The new approach to harmonization developed the principle of reference to standards and built on the jurisprudence of the ECJ, notably the definition in *Cassis de Dijon* of essential safety requirements (Schreiber 1991). It was to be paralleled by 'home country control' for financial services. The new approach limits legislative harmonization to minimum essential require-ments and explicitly leaves scope for variations in national legislation (sub-ject to mutual recognition). It delegates the maximum possible responsibility for detailed technical standards to CEN and CENELEC, the private European standard-setting bodies, subject to Commission mandates, with deadlines and financial provisions. We can see here three important developments: first, a greater reliance on national definitions of acceptable standards, albeit bounded by some collective requirements; second, a devolution of greater responsibility to the private sector and to external agencies for taking policy forwards; and third, the involvement of standards bodies from the EFTA countries.[3]

The European Council's Fontainebleau meeting in June 1984 marked a renewed commitment to accelerate European integration. It resolved the question of Britain's budget rebate and the outstanding issues of the Iberian enlargement, thereby clearing the way for serious consideration of revision of the treaties. At this meeting, Commissioner Narjes presented his plan to con-solidate the single market, and the British government tabled a memoran-dum that called *inter alia* for the creation of a 'genuine common market' in goods and services (Thatcher 1984). The meeting also established the Ad Hoc Committee on Institutional Reform (Dooge Committee) to consider reforms to the Community's decision-making procedures, with a worrying southern enlargement in prospect.

The remaining piece of the puzzle was put in place in January 1985 with the arrival of the new Commission with Jacques Delors at its head and Lord Cockfield as Commissioner for the single market (Cockfield 1994). Delors's preliminary discussions in national capitals convinced him that a drive to 'complete the single market' was perhaps the only strategic policy objective that would find a consensus. In his inaugural speech to the European

Parliament Delors committed himself to completing the single market by 1992. The Milan European Council in June 1985 endorsed the White Paper (Commission 1985) drawn up by Lord Cockfield, containing 300 (later reduced to 282) measures that would complete the single market by 1992. (For the main features of the programme, see Table 5.1.)

By December 1985 a remarkably tight Intergovernmental Conference (IGC) had completed the political relay by agreeing the terms of treaty reform which became the SEA. Apart from its important focus on accommodating enlargement, it specifically endorsed the single market and altered the main decision-rule for single-market measures (taxation excepted) from unanimity to qualified majority voting in the Council. Thus a strategic policy change and institutional reform were linked symbiotically and symbolically.

Three points should be emphasized about the SEA. First, it locked together institutional change and substantive policy goals. Second, the agreement to proceed with the single market was embedded in a set of wider agreements, in particular the accommodation of new members and budgetary redistribution. Third, it met relatively little resistance at the point of ratification in the member states, except in Ireland, for special reasons to do with neutrality, and in Denmark, where the Schlüter government escaped domestic censure only by calling a consultative referendum on the SEA.[4]

The theoretical debate about how the single-market programme and the SEA came about ranges between two main approaches, one emphasizing the role of supranational actors, the other stressing the importance of the member governments. Comparisons of the two views are complicated by the fact that some observers focus on the '1992' programme, while others concentrate on the SEA. It is quite possible that different actors exerted different levels of influence in processes shaping the two linked, but different, policy areas (Cowles 1994; J. Peterson 1995a). Maria Green Cowles (1994) stresses the importance of supranational business interests in shaping the EC agenda in favour of completing the single market. Wayne Sandholtz and John Zysman (1989) also give pride of place to supranational actors, though they cast the Commission in the leading role, with big business lending its support. Andrew Moravcsik (1991), on the other hand, argues that the SEA was the product of interstate bargaining between the British, French, and German governments in particular, and that traditional tools of international statecraft, such as threats of exclusion and side-payments explain the final composition of the '1992' programme and the SEA. Geoffrey Garrett (1992) and David Cameron (1992) also stress the role of the member governments. Garrett argues that the member states were willing to constrain their sovereignty because they were engaged in an iterated prisoner's dilemma and wanted to avoid the high transaction costs of monitoring compliance with agreements. Cameron concludes that ultimately the member governments, particularly in the context of the European Council, were the crucial actors, although he concedes that supranational actors, such as the Commission, ECJ, and big business, may have influenced their preferences.

As the theoretical debate implies and our history shows, a wide array of influences came to bear on the redefinition of market regulation: the impacts of the international economy, the inadequacies of national policies during the 1970s, the redefinition of interests, and the emergence of new ideas, helped by 'policy learning'. The story also shows the involvement of a plurality of public and private actors in the redefinition and the channelling of their activities within the EC institutional process over a period of years before 1985, as well as afterwards. We are therefore reluctant to endorse any interpretation of events in 1985 that seeks to offer monocausal explanation—the striking picture is of a clustering of factors (Scharpf 1994a).

The oddity of what happened is that an array of individually dull, technical, and everyday items were combined into an overarching programme that attracted such high-profile attention. The congruence of preferences of governments in power around the instrumentality of European market liberalization for both domestic and external purposes partly explains this. That these preferences could be expressed as embodying new ideas as well as satisfying specific interests was, in our view, crucial. The EC institutions, having experimented with a different and heavier approach to policy cooperation and having failed to produce results in the 1970s, were able to engineer an alternative and to fashion it into a convincing joint programme. But that programme engaged some political and economic actors more intensely than others, an imbalance for which a price was to be paid subsequently as the immediate excitement of 1992 gave way to more sober assessments, compounded by the pressures of economic recession in the early 1990s.

The 1992 programme and the ratchets of institutional change

With the formulation of the 1992 programme, drafted by Narjes and crafted by Cockfield, the EC institutions moved into top gear to drive forward an extraordinarily ambitious programme of legislation. The Commission set to producing draft directives speedily and the Internal Market Council, meeting at ministerial and official levels, kept up a remarkable rate of legislative endorsement. Qualified majority voting became an established procedure, though more by implication than by observance, in that small minorities

Table 5.1. The White Paper on the single market: a taxonomy

Markets measures	Products	Services	Persons and labour	Capital

Market access	• Abolition of intra-EC frontier controls
	• Approximation of:
	  * technical regulations
	  * VAT rates and excises
	• Unspecified implications for trade policy
		• Mutual recognition and 'home country control', removal of licensing restrictions (in banking and insurance)
		• Dismantling of quotas and freedom of cabotage (road haulage)
		• Access to inter-regional air travel markets
		• Multiple designation in bilaterals (air transport)
			• Abolition of intra-EC frontier checks on persons
			• Relaxation of residence requirements for EC persons
			• Right of establishment for various highly educated workers
				• Abolition of exchange controls
				• Admission of securities listed in one member state to another
				• Measures to facilitate industrial cooperation and migration of firms

Competitive conditions	• Promise of special paper on state aid to industry
	• Liberalization of public procurement
	• Merger control
		• Introduction of competition policy in air transport
		• Approximation of fiscal and/or regulatory aspects in various services markets
			• European 'vocational training card'	• Proposals on takeovers and holdings
				• Fiscal approximation of:
				  — double taxation
				  — security taxes
				  — parent–subsidiary

Market functioning	• Specific proposals on R&D in telecoms and IT • Proposals on standards, trade marks, corporate law, etc.	• Approximation of: — market & firm regulation in banking — consumer protection in insurance • EC system of permits for road haulage • EC standard for payment cards	• Approximation of: — income tax provisions for migrants — various training provisions • mutual recognition of diplomas	• European economic interest grouping • European company statute • Harmonization of industrial and commercial property laws • Common bankruptcy provisions
Sectoral policy	• CAP proposals: — abolition of frontiers — approximation and mutual recognition in veterinary and phyto-sanitary policies • Steel: call to reduce subsidies	• Common crisis regime in road transport • Common air transport policy on access, capacity, and prices • Common rules on mass risks insurance	• Largely silent on labour-market provisions	• Call to strengthen EMS

Source: Pelkmans and Winters (1988: 12). Reproduced with kind permission of the publisher.

often tolerated decisions that they could not obstruct rather than pressing for formal votes. The rather few decisions, ninety-one out of 233 during 1989–93 (*Financial Times*, 13 Sept. 1994), adopted by qualified majority perversely sometimes isolated member states that had a substantive interest in the outcome. The German government, for example, was out-voted on a directive that permitted road hauliers from one member state to transport loads entirely within another (cabotage).

The SEA also increased the European Parliament's role in policies concerning the single market, among others, by giving it the power, under the cooperation procedure, to reject or amend proposals. This power is, however, significantly constrained. The Parliament must vote to amend or reject a proposal by an absolute majority of its members; the Commission can choose not to integrate parliamentary amendments into its revised proposal to the Council; and the Council can overturn the Parliament's amendments or rejection by a unanimous vote. Consequently, the Parliament only very rarely rejects proposals under the cooperation procedure and only about 40 per cent of its amendments, many of which are only minor changes to the substance of the text, end up in directives (European Parliament 1993).

The introduction of the co-decision procedure under the (Maastricht) Treaty on European Union (TEU) further augmented the Parliament's importance in single-market matters, particularly strengthening its ability to reject proposals. It is already clear that the Commission must pay more attention to the Parliament's concerns when drafting proposals that are subject to the co-decision procedure, just as the Council has to weigh them carefully at the decision stage.

Evidence was compiled to demonstrate to the doubters the benefits of agreeing the measures in the White Paper. The 'Cecchini studies' of the 'cost of non-Europe' produced chapter and verse of justification for sectoral and horizontal policy change (Cecchini *et al.* 1988; Commission 1988a). The overall static gains from full integration were estimated at 4.3–6.4 per cent of Community GDP (Emerson *et al.* 1988: 203), deliberately not differentiated too explicitly by region. The argument focused on the overall, long-term welfare gains, thus downplaying the estimated costs of adjustment. In the wake of the Cecchini studies, the interest of economists in the single market was aroused. The spate of expert economic studies that followed, such as those by the Centre for Economic Policy Research (CEPR), helped to shift the debate cross-nationally on to the gains from market liberalization.

The institutional process gained huge new credibility as the transmission belt for delivering policy effectiveness, perhaps most sharply demonstrated by reactions from outside the EC. The neuralgic debate in the USA and Japan about 'fortress Europe' and the increased urgency of the EC/EFTA dialogue, started in Luxemburg in 1984, bore witness to the policy impact of the EC. Statistics were accumulated to show the strike rate of achievement in legislating and to maintain momentum and later to demonstrate the relative rates of implementation by member states. (See Table 5.2.) The exercise, piloted by the Commission, was to carry governments, business, and wider opinion along with the notion that legislative success would breed economic gains all round.

Table 5.2. The transposition of White Paper measures (situation as of 1 June 1995)

Member state	%
Denmark	98.6
Luxemburg	96.8
France	95.0
Netherlands	95.0
Spain	95.0
Sweden	95.0
United Kingdom	94.1
Portugal	93.2
EU (14)*	**92.6**
Italy	92.2
Ireland	90.9
Belgium	90.4
Germany	90.0
Greece	86.3
Finland	84.5

* Austria is not included because it had not forwarded complete transposition data to the Commission by 1 June 1995.
Source: DG XV News, July 1995.

National institutions were conscripted as additional endorsers of the programme. National parliaments had to implement the directives, a necessary but not very visible process, since in most cases they are enacted by subordinate legislation and not much debated. Here it should be noted that subsequent criticisms of 'Brussels bureaucracy' often relate to rules that had been transposed into national law without debate and with little attention from national parliamentarians, but then 'Brussels' is always an easy scapegoat for unpopular changes. In most member states there were public-relations campaigns around the 1992 theme—in the UK the Department of Trade and Industry set up a new 1992 unit and hotline. In France Edith Cresson spearheaded large public colloquia. It was the first time in the history of the EC that its policy process had stimulated so wide a span of attention and engagement, remarkably so, given the obscure and technical character of most of the legislation.

The hyperactivism of public policy institutions was matched by an extraordinary involvement of private-sector bodies. Indeed the change in business attitudes and business behaviour, much of it anticipating legislation, is probably the most important direct 'output' of the 1992 programme (Jacquemin and Wright 1993a). 'Brussels' had for a long while attracted pressure groups and lobbyists from the *cognoscenti* among the would-be influencers of Community legislation. 1992 took this phenomenon to unforeseen levels and forms of activism. In part this was the simple result of the range and quantity of sectors and products affected by the single-market programme and the speed with which they were being addressed. Organizations (pressure groups, firms, local and regional governments, and non-governmental organizations (NGOs)) which had previously relied on occasional trips to Brussels, reasonably so given the slow pace of earlier harmonization, started to prefer to establish their own offices or to hire lobbyists on retainers. The Commission, pressed for staff and expertise, opened its doors readily and even took some of the outsiders on to the inside on consultancy or expert contracts.

But other factors began to alter the character of the policy-influencing process. The previous quasi-monopoly of conventional peak and trade associations in the formal consultative processes was challenged by more direct lobbying from individual firms and by the emergence of the ERT. This is perhaps not so startling a phenomenon as some have argued. Firms had been players within member countries, some more than others, and some like Fiat and Philips had long been established in Brussels. As early as 1979 the Davignon Round Table had established a direct partnership between key IT firms and the Commission.

Another change in style came with greater reliance on consultancy (an import from the USA), which started to erode the old distinctions between public policy-making and private-interest representation. The Commission, member governments, and firms found themselves relying increasingly on consultants to inject interpretative and facilitating 'expertise' through both public and private contracts. It is an interesting question just what kind of groupings were developing around the single-market programme, whether or not they warrant description as an epistemic community (P. Haas 1992), and whether the clusters of policy activists formed policy networks (Atkinson and Coleman, 1989) or 'advocacy coalitions' (Sabatier 1988). Overlapping groups of consultants worked for the Commission, member governments, and firms. Some were providing expertise on the economy, markets, and business opportunities; others advised on procedures, legislation, and litigation. Thus the conventional delineation of public policy-makers and private influencers became blurred.

Consumer and other 'civic interest' groups also put their feet in the door, though they found it much harder to exercise effective political muscle. The consumer and the purchaser had been intended beneficiaries of the 1992 programme, and the 'minimum essential requirements' of harmonizing and liberalizing directives were often intended to help them or their assumed interests. But it is easier to discern consumers as objects of policy than as partners in the process. The point is important when we compare the favourable attitude of business opinion across the EC towards 1992 with the waning of wider public expectations of benefits from 1990 onwards (Reif 1994, Franklin *et al.* 1994) or with the subsequent emergence of public antipathies to the apparent efforts of 'Brussels' to remove differences of local taste—food standards being a particularly emotive issue.

The bulk of the activity to achieve the goals of 1992 was focused on the legislative programme rather than on the follow through, on designing rather than implementing policy rules. Its impact on economic sectors also differed: the early breakthrough on freeing capital movements was crucial in opening up opportunities for cross-border investment; the repeated obstacles to progress on food, veterinary, and phytosanitary standards were a barely heard signal of problems to come; the stubborn resistance to tax harmonization was more predictable.

As 'flanking' policies were developed, especially the cohesion programmes for social and regional spending, other layers of governance became engaged, especially at the regional level. Traditions of local political territory jostled with adjustment to transnational markets. EC institutional links to regional and local authorities and cohesion programmes to some extent filled the policy gaps at the regional level (Marks 1993). But, as Andy Smith (1995) points out so convincingly, the policy gaps were different in different member states. These issues are discussed further by David Allen in Chapter 8 of this volume.

In the social arena a big push was made by the Commission, in coalition with the old social-market protagonists, to buttress the single market with the reassurance that it would also deliver social progress for the labour force. This effort was a logical extension of the analysis of earlier experience of European integration and national policy adjustments, but it was to prove much harder to embed, as Leibfried and Pierson show in Chapter 7. The shared new doctrine of market liberalization and the new regulation had a narrower political base than the old social-market doctrine used to have. Indeed it began to emerge that regulatory instruments and competition, the mode for liberalizing movements of products and services, were beginning to define the scope for collective social policy (Majone 1993).

Policy implementation

The single market was not complete two years after the supposed deadline, and EU business leaders contend that national standards persist (Coutu *et al.* 1993; *Financial Times*, 17 Nov. 1994). Transposition of directives into national law is also a problem that is particularly severe in insurance and public procurement (Commission 1995*d*). The Commission has undertaken a study of priority sectors to identify the obstacles to the proper functioning of single-market rules. It simply does not have the resources to police the transposition, let alone implementation, of directives across sectors and across countries. To a significant extent it relies on complaints filed by non-governmental organizations or injured firms to bring infringements to its attention.

In March 1992 the Commission established the High Level Group on the Operation of the Internal Market, under the chairmanship of Peter Sutherland, a former Commissioner, to assess ways of realizing the full potential of the single market. The Sutherland Report (Sutherland *et al.* 1992) focused on only those aspects of the single market concerned with goods and services. It called for improved consultation with the actors affected by new regulations, greater accessibility of the Community's laws, and better cooperation between the Commission and the member governments to ensure that uneven implementation of directives did not create trade barriers. The Commission (1992*c*) largely accepted these proposals, though it made clear that the onus for implementation lay with the member states. It also argued that additional funding would be required in order to oversee and coordinate the member states' implementation of directives.

In spite of the manifest importance of effective implementation of the 1992 programme, the Sutherland Report has received a relatively low-key response. Or at least it has been difficult to give the hard grind of implementation an exciting public face and even to rally the many relevant national agencies in the member states. Architects (Cockfield 1994) of the single market had often signalled that the real test of success would lie in thorough implementation, and critics (Metcalfe 1992) had warned of the obstacles. The Commission has taken some steps to improve implementation. An Internal Market Advisory Committee, chaired by the Director General responsible for the single market, was formalized in 1992, and networks of national and Commission officials dealing with individual areas of regulation are being established. In addition, the first subject on the agenda of an informal meeting of the Internal Market Council in March 1995 was the effective application of Community law and the issue of reducing the disparity between and increasing the transparency of national sanctions for violations of Community law. The French presidency also asked the Commission for proposals for a model clause on sanctions that could be introduced into all future European directives (Lamassoure 1995).

We note, however, the enormity of the task of implementation. In one sense results are simply to be sought in changes in business behaviour. However, these necessarily engage some entrepreneurs more than others. In France, for example, only 140,000 firms export out of a total of a million and a half businesses, and only 30,000 of those export on a regular basis (Lamassoure 1995). Small and medium-sized enterprises (SMEs) are less engaged than the large transnational conglomerates or highly effective niche players. It is thus not surprising that SMEs have been caught up in a defensive criticism of the 'burdens on business' that flow from 'Brussels'. Such criticisms contributed to the establishment of the Molitor Group in 1994 to work on the simplification and lightening of EC and national implementing legislation. Results also depend on national implementation, both by formal transposition of directives and by systematic enforcement, as necessary through the national courts. The transaction costs of implementation are not trivial and account in part for the failure as yet of a process of 'regulatory competition' to gather pace (Sun and Pelkmans 1995; Woolcock 1994) or of really open markets in, for example, public procurement (Cox 1992) and insurance (Woolcock et al. 1991).

Meanwhile the standards bodies have plugged away at their part of the task of adding voluntary industry standards to the legislative rules. Although the output of European standards bodies has increased substantially (Schreiber 1991), the number of standards that the member governments notify to the Commission has remained high and was higher in 1994 than in 1989.

Winners and losers

Who has gained and who has lost from the single-market programme is a politically charged issue. As yet there have been no serious empirical studies that attempt to provide answers. The series of studies set in train by the Commission in 1994 to assess the effectiveness of measures taken in creating the single market may shed some light on this question.

A 1990 Commission study (Buigues et al. 1990: 4), which was targeted at member governments, predicted that completion of the single market should 'neither upset the mix of sectoral specializations across member states nor lead to massive transfers of economic activities between geographic zones'.

The study acknowledged that the processes in question were too complex to allow predictions of which regions or sectors would gain or lose from the single market.

While such analysis is relevant for governments and political parties wishing to draw up balance sheets, it is not satisfying in that 'national economies' may not be helpful units of analysis. Indeed, as the Cecchini studies had tried to argue, the single market was designed to improve the competitiveness of firms, not of countries as such.

Although all businesses are supposed to benefit from greater market access, large corporations are better placed than SMEs to take advantage of new market opportunities and economies of scale. None the less, even some large corporations do not view the single market with enthusiasm (Coutu *et al.* 1993). The less competitive ones, in particular, fear increased competition.

Consumers were also supposed to benefit from a wider choice of goods and services. As yet there are no systematic data on the impact of the single market on consumer welfare. Theoretically, the single-market programme should lead to increased competition, which brings greater choice and lower prices, without compromising consumer protection. In practice, however, there are some indications that, at least in countries with the highest standards, there has been some erosion of consumer protection (Millstone 1991).

In addition to improving the competitiveness of firms and increasing consumer choice, the single-market programme was also expected to increase employment in the long run (Emerson *et al.* 1988: 213–17). In the short term, however, jobs would likely be lost as firms fail or restructure in the face of increased competition. In addition, the absence of internal frontiers and the guarantee of the right of establishment would increase the likelihood that workers in different locations might be forced to compete with each other to attract investment.

It is possible that the overselling of the benefits of the single-market programme has contributed to business and public disappointment with its achievements.

Policy-making results

What have been the main results of SEA/1992 as conceived for the single market? Our concern here is with the results in terms of the policy process rather than with the economic impact of liberalizing the single market. None the less we note that policy substance and policy process interact. For the Community level of governance, with its fragile sources of direct legitimation and dependence on support from within the member states, evidence of policy effectiveness and tangible gains is of particular relevance in establishing policy and political credibility (Schmitter 1992; H. Wallace 1993; Reif 1994;).

Several yardsticks can be proffered of the impact of the single market programme:

Legislative output

By the middle of 1994, of the White Paper's 282 measures 262 had been adopted by the EC institutions. Sixteen others, of which eleven had priority, were still awaiting adoption by the Council and one (not priority) measure was awaiting the Parliament's opinion.

Judicial support

Cases lodged with the ECJ related to the single market increased during the late 1980s, but tapered off in the early 1990s. The *Frankovitch* ruling and the addition of Article 171 to the TEU (which introduced fines for member governments that fail to implement Community legislation) suggest a readiness by both the ECJ and the Council to tighten judicial enforcement. However, some cases have also limited the definition of those national variations of rules or practice that can be successfully contested.[5]

Policy development by the Commission

Under the two Delors Commissions there was an emphasis on policy innovation rather than implementation. Other policy areas claimed attention, and responsibilities for different aspects of the single market have been uneasily divided between different DGs: III for industry; XIII for high-technology sectors; XV for financial services; and in parallel DG IV for competition. Recently DG XV became the main co-ordinator. There are indications that under Jacques Santer the Commission will concentrate more on increasing the effectiveness of Commission activity. The Commission's Work Programme for 1995 states that 'the consolidation, completion and enhanced visibility of the single market . . . must be the cornerstone of the Commission's activity', and includes the strict enforcement of existing single-market rules among the Commission's priorities (Commission 1995a: 7). This approach meshes well with the concerns of the European business community, which wants to see the gains of the single market consolidated before the EC launches any new endeavours (Coutu *et al.* 1993).

Policy 'performance' and implementation by Commission and member states

The Commission has belatedly begun to address the deficiencies in implementation. Transposition of EC directives into national laws is still incomplete. Two years after the '1992' deadline the member governments had transposed an average of only 90 per cent of EC directives (Commission 1995c). However, as of the middle of 1994 only just over half of the measures requiring national implementation had been implemented in all twelve member states (Buiges and Sheehy 1994). Differences in the pace of national compliance with EC legislation pose serious problems for business, as do persisting barriers to trade, particularly in the form of state subsidies and public procurement procedures, and new barriers, such as environmental regulations (Coutu *et al.* 1993). In addition, some firms encounter difficulties in understanding the new rules of the game. Not only is the regulatory environment in which they operate in flux, but they sometimes have to cope with conflicting national and EC regulations—hence the sub-theme of deregulation as a corrective.

Industrial behaviour

Despite its shortcomings, the single-market programme has had a significant impact on business behaviour. Following the programme's adoption, there was a dramatic increase in transnational investment within and into the EC (see Fig. 5.4). There has also been a substantial degree of consolidation of business activities, as businesses seek to capitalize on economies of scale (Jacquemin and Wright 1993a). This trend has progressed faster in retailing than manufacturing (Bayliss *et al.* 1994), and some industries, such as retail banking and automobile manufacturing, have not been able to realize the gains they had anticipated from rationalizing production (Coutu *et al.* 1993). In addition, firms in various industries have collectively adopted voluntary codes of conduct as a means of implementing or deferring new regulations (Matthews and Mayes 1995).

Public reception

Public support for European integration increased following the adoption of the single-market programme and the SEA, but began to decline during 1991 (*Eurobarometer* 1994). Popular expectations regarding the single market began to change at about the same time, with increasing evidence of fears of

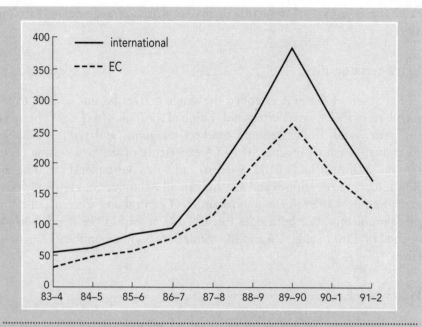

Fig. 5.4. EU and international acquisitions of majority holdings *(including mergers)* in the EU

Note: In 1994 the Commission (1994*d*) began reporting acquisition data that been gathered in a different fashion from those used in this figure. Those data cover the period only from 1987–8 and so do not clearly illustrate the change that took place in the wake of adoption of the single-market programme. The data gathered under the two methods are not directly comparable, but for the period of overlap (1987–8 to 1991–2) reveal similar trends. The most significant difference is that the new data-collection method captures relatively more international activity. The data indicate that in 1992–3 both international and intra-EC cross-border investment stabilized at roughly the 1991–2 level.

Source: Commission (1988*b*: 230; 1993*g*: 497).

becoming worse off (*Eurobarometer* 1994). In addition, the momentum of the late 1980s has already proved difficult to maintain (as the problems experienced during the Maastricht Treaty ratification process demonstrated), as popular support for integration is undermined by poor economic conditions (Eichenberg and Dalton 1993).

External impacts

Through promoting policy collaboration among the member governments, the single-market programme has enhanced the EC's effectiveness in international trade negotiations (see Ch. 12). European industries, such as chemicals, that are affected by other countries' regulations recognize the advantage that common EC policies provide in international negotiations regarding the mutual recognition of standards or market-access arrangements (Paterson 1991).

Extra-territoriality

The creation of the EEA extended the single market beyond the EC to the members of EFTA, save Switzerland. This had the logic of a closer fit between the patterns of production and markets for goods and services and the emerging regulatory regimes. The EEA experience made the accession negotiations with Austria, Finland, Sweden, and Norway simpler. Extending the single market to developed and relatively small economies was relatively easy; extending it to the reforming economies of central and eastern Europe is more problematic for both sides (see Ch. 14). In early 1995 a policy debate opened on how to align US and EU regulation, building on the EEA experience.

Policy blockages

Shortcomings in national transposition and enforcement of EC legislation persist. Stubborn 'differences of taste' inhibit uniform implementation and generate controversy. Under the label of 'subsidiarity' some member governments have sought to enhance their freedom of action in 'flanking policies', especially social and environmental. Differences in both social and environmental standards allow different regions or countries to compete with each other to attract investment. These factors highlight the issue of whether a single market implies similar standards for production processes, as well as for products, and the consequences for who sets which rules.

Policy linkages

The drive to complete the single market has pervaded the policy agenda of the EC. The single market and its associated thousands of pages of *acquis communautaire* took the central place in defining the EC that the common agricultural policy had once occupied. Its scope was much greater, in that it included not only vertical measures for particular products or sectors, but horizontal measures, such as the right of establishment, that impacted across the economy. Its impact depended on the corollary availability of a tough competition policy (see Ch. 6). Its political acceptability was buttressed by the development of cohesion programmes and was argued to require an active social dimension. As products came to move more easily attention also focused on the processes and conditions under which goods and services were produced and provided. Irrespective of other arguments for European policies on environmental and social issues (see Chs. 9 and 7 respectively), the preoccupation of entrepreneurs with operating on a 'level playing field' turned attention to the relevance of these other factors for costs, competition, and profitability. Necessarily the change to a single policy also had to be followed through into the projection of external policy, both on trade issues in general and as regards particular economic partners (see Chs. 12 and 13). The abolition of border controls within the EC meant that national trade regimes with third countries in products such as cars, bananas, and textiles had to be replaced by EC trade agreements.

In addition, the single market was invoked in aid of the two big policy initiatives that followed it: the plans for economic and monetary union (EMU) and the goal of removing internal borders for people. In the case of EMU it became commonly argued that a single market logically implied a single currency. In the case of borders between member states two arguments were made: first, that the increase in cross-border economic transactions would facilitate illicit and illegal transactions, drugs, fraud, arms trade, and the like, which would in turn require European-level responses; and, second, it was asserted that border controls on persons would have to be removed lest they be used as substitute controls for economic purposes. These subjects are covered in Chapters 11 and 15.

A new approach to policy

The new policy ideas of liberalization and regulatory governance that underpinned the single-market policy continue to resonate in Community parlance. The White Paper of December 1993 on competitiveness (Commission 1993g) and the pre-accession strategy for the (eastern) Europe associates (Commission 1995b—this being the most up-to-date summary by the Commission of the state of the single market) both reveal the embedding of these ideas, albeit with persistent echoes of other policy perspectives. The Santer Commission, installed in January 1995, was quick to signal its commitment to continue in the same broad line (Santer 1995).

A new policy mode pervades much of the work of the EU institutions and their outputs of policy and legislation. Majone's general line of argument on this (1994b) seems well substantiated, even though some of the consequences for policy management remain ambiguous. It remains to be seen, for instance, whether a choice will be made to go the route of establishing independent regulatory agencies, a direction not particularly welcome to the Commission, whose powers would then risk being diffused. It is harder to tell how far the management of public policy within the member states has also changed or whether such changes as have occurred were the consequence of European policy or a parallel response to the same factors that stimulated a new European policy. Competition among national rules had been envisaged as one possible outcome, but several key factors militate against this kind of competition. Woolcock (1994) argues that some of the basic pre-conditions—notably transparency of national regulations and certainty about the effects of specific regulations—for regulatory competition to occur are absent. Sun and Pelkmans (1995) contend that, even if those conditions existed, the impact of regulatory competition would be restricted. First, some factors of production are less than completely mobile, and so producers are unable to take full advantage of differences between regulatory regimes. Second, when capital investment decisions are made, the regulatory environment is just one among many considerations. In addition, all note that local market characteristics, particularly 'differences of taste', might make adhering to foreign regulations an unattractive option for firms.

New relationships have been established by the business community with the policy process, as is argued by Cowles (1994) and McLaughlin et al. (1993). Although it is widely accepted that the process is pluralist rather than corporatist (Streeck and Schmitter 1991), there are significant opportunities for policy capture at the European level (Cawson 1992). The continued segmentation of EU policy-making contributes to skewed access to influence. Bigger firms have and take more opportunities than SMEs. Other kinds of organized interests can less easily gain a foothold than the producers or the providers of services (Young 1995).

The dependency of many economic actors on national policy has also been reduced. The scope for national policy-makers to control economic transactions on their territories has become more limited and will stay so as long as the transnational legal regime of the EU holds together. But this is not to say that the political turf has been won by EU-level policy-makers, since the new regulatory mode involves a diffusion of policy authority rather than its concentration elsewhere. Though the Commission has been heavily engaged in promoting the single market, its own net gain in authority is open to debate, not least since it has also become the butt of residual criticism about the downside effects of market liberalization. Moreover, the member governments—as the enforcers of most EC legislation, the guardians of 'home country controls', and the proponents of subsidiarity—retain important footholds in the regulatory process. In particular, member governments are forced into the position of defending the losers from the single market against the incursions of European regulation. Hence the single-market programme has to be seen as an important element in the legitimacy test faced by the EU since the early 1990s. It is moreover a paradox that this test has been most severe in member states that have had governments strongly committed to market liberalization—the UK is a clear case in point.

The discussion continues on how far the European regulatory model can be applied more broadly. Other Europeans and immediate neighbours have become increasingly drawn within the regulatory scope of the EU by direct membership or close association, or as free-riders. Indeed the closeness of the association seems increasingly to depend on how much of the single-market *acquis* can be absorbed by the third country. The creation of the single market also limited the scope of derogations that can be granted to new members to ease their transitions. In this sense the threshold for full membership has been raised much higher by the single market, though the dependence on EU markets may in any event force unilateral alignment with many EU rules—at least as far as products are concerned. Issues—such as the impact of competition, environmental, and social policies on trade—that became increasingly important in the EC are pushing their way into the wider international discussions in the World Trade Organization (WTO) and Organization for Economic Cooperation and Development (OECD) (Devos *et al.* 1993).

The adoption of the single market thus in many senses marks a critical change in the European policy process as well as a different choice of policy content. But its prominence and predominance in the late 1980s must also be seen in proportion, given that the follow-through to implementing the programme remains patchy and it now has to jostle with several other big subjects for the prime attention of strategic policy-makers at both European and national levels.

'1992' probably would not have become such a relative 'success story' had not policy and industrial entrepreneurs been able to talk up the importance of what they were seeking to do and thus to give political sex appeal to what otherwise consisted of a rather dreary list of separate and very technical proposals. Politicians found it convenient for a variety—we stress a variety—of reasons to use the single market and the constraints from 'Brussels' as cover for changes in domestic policies and as explanation for both inaction and action at home. Commission officials were delighted to have found a theme that had such wide resonance and to play it for all it was worth in developing the symbolism of European integration and its impacts on citizens as well as on firms. Sustaining political integration on the back of a programme of market liberalization has, however, proved elusive. Modernizing or adapting the European welfare state to the exigencies of external competition and the pressures of a changing industrial society at home is a much taller order.

Notes

1. This chapter draws heavily on research funded by Phase II of the Economic and Social Research Council's Single European Market Programme (Award * L113251029).
2. The Luxemburg process well preceded the official recommitment of the EC to the single market. After preliminary discussion it was launched with the Luxemburg Declaration of April 1984. It was when the limits of this were reached that the moves began which led to the EEA debate from January 1989 onwards.
3. CEN and CENELEC then took decisions by qualified majority votes, weighted as in the Council, allowing assenting EFTA members to implement them, but not allowing EFTA dissent to prevent an EC majority from agreeing a standard.
4. The issues raised were specifically focused on security policy but also revealed the split between the bourgeois parties and the 'alternative majority' (N. Petersen 1993).
5. The two most prominent examples of this are the cases concerning *Danish Bottles* (302/86) and *Keck* (267 & 268/91). In the former the ECJ ruled that the Danish government could require the recycling of beverage containers even though it might impede imports. In the latter, the court held that national laws restricting or prohibiting certain selling arrangements do not infringe the Treaty of Rome's rules on the free movement of goods, provided that the laws are not aimed at imports and that they have the same effect on commercial freedom to market domestic products as on imports.

Further Reading

On the original development of the 1992 programme see Cecchini (1988), Cockfield (1994), and Pelkmans and Winters (1988). For an economic evaluation see Jacquemin and Sapir (1991) and Siebert (1990*b*). The introduction to Commission (1995*b*) summarizes the programme and its development (in identifying a 'pre-accession strategy' for the Europe associates to adapt to the single market). For the theoretical debate see Cowles (1993), Majone, in particular (1992) and (1994*b*), Moravcsik (1991), and Sandholtz and Zysman (1989).

Cecchini, P., with Catinat, M., and Jacquemin, A. (1988), *The European Challenge 1992: The Benefits of a Single Market* (Aldershot: Wildwood House).
Cockfield, Lord (1994), *The European Union: Creating the Single Market* (London: Wiley Chancery Law).

Commission (1995*b*), 'Preparation of the Associated Countries of Central and Eastern Europe for Integration into the Internal Market of the Union', COM (95) 163 final, 3 May.

Cowles, M. Green (1994), 'The Politics of Big Business in the European Community: Setting the Agenda for a New Europe', Ph.D. thesis (American University, Washington, DC).

Jacquemin, A., and Sapir, A. (1991) (eds.), *The European Internal Market: Trade and Competition* (Oxford: Oxford University Press).

Majone, G. (1989), *Evidence, Argument and Persuasion in the Policy Process* (New Haven: Yale University Press).

—— (1991), 'Cross-National Sources of Regulatory Policymaking in Europe and the United States', *Journal of Public Policy*, 2/1: 79–106.

—— (1992), 'Regulatory Federalism in the European Community', *Environment and Planning C: Government and Policy*, 10/3: 299–316.

—— (1993), 'The European Community Between Social Policy and Social Regulation', *Journal of Common Market Studies*, 31/2: 153–70.

—— (1994*a*), 'Independence vs. Accountability? Non-Majoritarian Institutions and Democratic Government in Europe', in J. Hesse (ed.), *European Yearbook of Public Administration and Comparative Government* (Oxford: Oxford University Press, 1994).

—— (1994*b*), 'The Rise of the Regulatory State in Europe', *West European Politics*, 17/3: 77–101.

—— (1996), 'Public Policy: Ideas, Interests and Institutions', in R. E. Goodin and H.-D. Klingemann (eds.), *New Handbook of Political Science* (Oxford: Oxford University Press, 1996).

Moravcsik, A. (1991), 'Negotiating the Single European Act: National Interests and Conventional Statecraft in the European Community', *International Organization*, 45/1: 19–56.

Pelkmans, J., and Winters, L. A. (1988), *Europe's Domestic Market* (London: Royal Institute of International Affairs).

Sandholtz, W., and Zysman, J. (1989), '1992: Recasting the European Bargain', *World Politics*, 42/1: 95–128.

Siebert, H. (1990*b*) (ed.), *The Completion of the Internal Market* (Tübingen: J. C. B. Mohr).

CHAPTER SIX

COMPETITION POLICY: POLICING THE SINGLE MARKET

David Allen

The legal base for competition policy in the Treaties gave the Commission direct powers, which the Court has consistently supported. In this policy sector the Council and the European Parliament have thus played minor roles. The Commission's use of its discretion in selecting and deciding cases limited conflicts with national governments, at the cost of criticisms of secrecy and inconsistency. The Commission succeeded in gaining the authority it had long sought to approve mergers under a Regulation of 1989, reluctantly and conditionally approved by the Council. Efforts to limit state aids, hampered by national reluctance to provide information, also gained ground in the late 1980s. Unease among member governments about a policy sector in which the Commission may act as policeman, prosecutor, judge, and jury has, however, lent rising support to proposals for a separate European Cartel Office.

Introduction

Competition policy has been a major concern since the original European Coal and Steel Community (ECSC) was agreed in the early 1950s. After all, the political *raison d'être* of the ECSC was to ensure a thriving industry, subject to the workings of the free market rather than manipulated by the government of any one state. From early on in the evolution of the European Community (EC) the Commission has enjoyed unique powers to use, with the full support of the Court of Justice, judicial procedures to develop competition policy. Competition is one of the few areas where the Commission is given the power not only to make policy, but also to implement and to enforce it, with little formal involvement of the governments of the member states. Directorate General IV (DG IV), responsible for competition, has acquired considerable prestige over the years as the operator of one of the more successful areas of Commission competence. By playing the role of 'policeman, prosecutor, judge and jury' (Wilks and McGowan 1995b) DG IV has also attracted the suspicious hostility and envy of both private and public organizations within the member states. It is a measure of the importance of its role that some now seek to 'hive off' many of its activities into an independent agency.

Informally, of course, the long arm of national power and interest reaches into the Commission itself and plays a major role in influencing the decisions on which competition policy has come to be based. Decisions are reached in the college by simple majority votes, not by qualified majority voting, because only rarely is the Council called upon to act. The complex legislative powers of the European Parliament (EP) are of little relevance either. The competition policy that the EU operates today owes nothing to subsequent treaty amendments (with one or two minor exceptions); it is based on powers that existed in the original Treaties of Paris and Rome. Neither the Single European Act (SEA) nor the Treaty on European Union (TEU) had much new to say about competition policy, although both led to extremely important subsequent developments within that policy by reinforcing the market philosophy of the Community. Competition matters may preoccupy the negotiators at the 1996 Intergovernmental Conference (IGC), if the German government succeeds in forcing a decision to establish a European Cartel Office (ECO).

Competition policy has developed incrementally in response to both internal and external stimuli. Internally, the zeal of three successive commissioners in the 1980s (Andriessen, Sutherland, and Brittan) drove policy in a new direction, even though they could be said to be responding to the prevailing economic and political climate. The difficulties experienced by the subsequent Commissioner, Karel van Miert, clearly reflected concerns about the uneven development of the European economy and the questioning of the European integration process.

During the golden years of integration in the 1960s competition policy developed relatively uncontroversially, because policy development was protected from the intergovernmentalist backlash that occurred in the Council of Ministers in the mid-1960s. It was also relatively unambitious, concentrating as it did on the area of restrictive practices—so-called negative integration. An emphasis on restrictive agreements could be seen as part of the process of creating the common market by removing non-tariff barriers to internal trade. As economic conditions worsened in the 1970s and as people wondered if the Treaty of Rome was not just a 'fair weather' treaty, so the principles of competition policy were challenged. What few restrictions there were against state aids became enveloped in the bid to reduce unemployment in Europe. Competition policy became concerned with the regulation, rather than the prohibition, of 'crisis' cartels. For the first time competitiveness, or the lack of it, rather than competition was the name of the game.

Attempts by the Commission to develop a policy on merger control, based on its existing powers to control monopolies (or the abuse of a dominant position, as the Treaty put it) were inhibited by the desire of many member states to participate in the rescue and restructuring of their national industrial bases. Their aim was to challenge the power of US and Japanese multinationals. During the late 1970s the member governments of the Community basically sought national solutions and attempted to create national champions in their search for economic salvation. At the European level they were mainly interested in mitigating their inability to compete internationally by collective protection (Tsoukalis and da Silva Ferreira 1980). They were interested in giving state aids, rather than outlawing them, and they cared little about the inflationary impact of restrictive agreements between firms, when inflation rates, for other reasons, were running at such high levels.

The transformation came in the 1980s, with the swing away from Keynesian policies towards a new enthusiasm for the workings of the market. This time it was to be the European market, non-tariff barriers were to be swept away, and competition policy was seen as an essential adjunct to the market so as to ensure that it did not deviate too much from the perfect competition model. In some member states, most obviously Britain, full-blown Atlantic capitalism, as opposed to modified Frankfurt capitalism, led to calls not just for more than the free workings of the market *vis-à-vis* private enterprise: it also provoked restraints on state aid and policies to privatize state-owned companies, as well as to liberalize restricted markets.

At the European level, this new atmosphere generated the single-market programme and a new demand for an enlarged and developed competition policy. In the new European market the dangers of monopoly control were that much greater. The Commission's long search (the first proposal had come in 1973) for enhanced powers to control mergers was eventually satisfied with the passing of the merger regulation in 1989. Furthermore, the new circumstances and the effective drive of Peter Sutherland and Leon Brittan as commissioners led to a renewed interest in dealing with restrictive practices, this time supported by a willingness to impose serious fines on miscreants. As the spirit of liberalization spread, the Commission sought to make greater use of its powers to ensure competition in previously restricted markets. All this, combined with a renewed assault on state aids, marked a period of great activity on the competition front in the Community.

More recently, in response to the set-backs heralded by the Danish referendum and to the increased enthusiasm for the concept of subsidiarity that followed from it, we find a slackening of the pace of competition policy development. This has been accompanied by demands for a reconsideration of the powers enjoyed and exercised by the European Commission. For those concerned above all with competitiveness, the dogmatic pursuit of a 'pure' competition policy can be seen, at times, as an unnecessary hurdle. Some within the Union would argue that European business needs aid from the state to encourage investment in research and development and that some restrictive and monopolistic practices may be necessary in order to compete effectively on the world market. Others, including those who work in DG IV, retain their commitment to the free market and the essential role of competition policy within it. These competing views are argued over within the Council of Ministers and are reflected also within the Commission. So it seems likely that this debate will mark just one more stage in the story of the development of competition policy within the EU.

The legal base and the philosophy of competition

A key objective of the original European Economic Community (EEC) was, first, to establish a common market, and then to remove and to prevent distortions of free trade within that market. Article 3(f) of the Treaty of Rome referred to 'the institution of a system ensuring that competition in the *common market* is not distorted' (my italics). This objective is preserved in Article 3(g) of the TEU; this states that the activities of the Community shall include 'a system ensuring that competition in the *internal market* is not distorted' (my italics again). Taken at face value, this language implies that any deviation from the idealized model of perfect competition is to be outlawed. However, as we shall see, in practice the principle of pure competition has found itself in conflict with rival notions of industrial protection and restructuring, as well as with ideas concerned with the international or global competitiveness of European industry. This has been particularly marked in periods of low growth.

Even when prevailing economic conditions are benign, most economists have come to accept that perfect competition cannot be achieved in the real world. Instead they have developed the notion of 'workable competition', in which firms and governments should be encouraged to work towards the most competitive structure which can be achieved. The Commission has regularly invoked this notion; it accepts that neither is perfect competition achievable, nor is it a state of affairs that will naturally develop in the marketplace. Workable competition has to be sought, hence the need for competition policy. A further, more ambiguous, notion of 'contestable competition' has recently been associated with moves, under Article 90, to liberalize restricted or regulated markets. This idea accepts that a market does not have to be perfectly competitive, but argues that firms must be free to enter and to leave that market without significant cost (Kent 1992).

Competition policy clearly has both a negative and a positive role to play in the evolution of the Community. On the one hand, it is concerned with policing the market by the application of sanctions against those who would seek to abuse or to restrict its freedoms. Here it might be argued that there is no need for policy as such, merely a requirement that the laws of perfect competition, as enshrined in the treaties, should be upheld, using the procedures established by them. On the other hand, competition policy can also be seen as having the more positive role of contributing towards the creation of the free market and, some would argue, of the industrial structures and behaviour that characterize it. To the extent that the latter role requires choices to be made, there is a need for policies to guide them.

In practice, competition policy finds a sort of equilibrium between the negative and positive roles suggested above. Guided by the original treaties and supported by the Court of Justice in its rulings, DG IV of the European Commission has been inspired mainly by a desire to police the European market in the interests of creating and preserving a system of undisturbed or perfect competition. Nevertheless political and economic realities, as well as the practical problems of implementation, have often required DG IV either to turn a blind eye to uncompetitive practices, by both companies and member states, or to rationalize them in terms of other objectives being pursued by the Commission. Thus, in its management of all aspects of competition covered by the treaties, as well as in seeking to extend its competition competences, the Commission has had to develop policy guidelines.

As internal and external circumstances have altered, so DG IV has had to decide which aspects of competition covered by the treaties it should concentrate on, as well as which aspects it should seek to develop further. As noted above, the formal treaty base for competition policy has hardly altered over the years; everything that the Commission has achieved to date stems fundamentally from its interpretation and implementation of the powers that it was originally given. Only twice (1962 and 1989) has the Commission felt the need to seek significant further powers (in both cases via regulations) from the Council of Ministers.

The competition rules are directed against both private companies and the governments of the member states. In the early days the Commission showed more enthusiasm for taking on companies than governments. Policy towards private undertakings is covered, in the main, by Articles 85 and 86 of the Treaty of Rome (EEC).[1] Article 85 refers to agreements and 'concerted practices', either vertical or horizontal, between undertakings that, by their restrictive nature, are liable 'to affect trade between member states and which have as their object or effect the prevention, restriction or distortion of competition within the common market'. Article 86 refers to the prohibition of abuse by 'one or more undertakings which have a dominant position within the common market'.

Both Articles 85 and 86 have served as a springboard for the Commission in its search for the power that it was given in the ECSC treaty, but not the EEC treaty—the power to control mergers. Under Article 87, the Council is required to adopt 'any appropriate regulations or directives to give effect to the principles set out in Articles 85 and 86'. In particular it is to 'define the respective functions of the Commission and the Court of Justice'. The Council issued the key implementing Regulation 17 in 1962.[2] This Regulation, along with Article 89, establishes that it is the responsibility of the Commission to ensure that Articles 85 and 86 are applied. It is the Commission that is called upon to investigate cases, either at the instigation of member governments or on its own initiative; it is the Commission that determines whether an infringement has taken place; it is the Commission that is required to propose 'appropriate measures' (which can include a substantial fine) to bring any infringement to an end; and it is the Commission that is charged with determining further measures to 'remedy the situation', if any infringement is not brought to an end. Any party can of course challenge the legality of the Commission's actions before the Court of Justice.

Competition policy, as it affects the governments of the member states, is provided for in Article 37 (state monopolies), Article 90 (public undertakings and undertakings to which member states grant special or exclusive rights), and, most significantly, Articles 92–94 (state aids). Under all these provisions, actions by the member states which distort competition and impact adversely on free trade within the common market are in principle outlawed. As with Articles 85 and 86, the drafters of the treaties were careful to establish the ground for permissible exemptions and to give some preliminary guidance as to when these might be allowed.

Here too it is the Commission, supported by the treaty articles and implementing regulations from the Council (Article 94), that is charged with ensuring that member states do not take actions that challenge the principle of competition. The Commission must be kept informed by the member states of any system of aids or privileges that they are operating. It is up to the Commission to decide whether or not any practice is 'compatible with the common market'; if it judges a practice to be incompatible, the Commission may propose a remedy, which can include both abolishing an aid or privilege and the repayment of aid previously granted. Commission decisions can be challenged in the Court of Justice, and, in the event of member states not complying with Commission decisions, the matter can be referred to the Court. In an amendment introduced into the TEU, the Court, acting under Article 171, can now impose a fine on a member state which it finds to be in breach of any of its rulings.

The following sections examine specific aspects of competition policy and associated policy processes. As we shall see, the treaties have clearly made the Commission and the Court of Justice the major actors. The only power given to the European Parliament with regard to competition was the right to be consulted by the Council before it laid down (Article 87) the implementing regulations for Articles 85 and 86. The TEU extended this minimal right to implementing regulations made by the Council under Articles 92 and 93. The Council of Ministers, whilst instrumental in the early days in agreeing the necessary implementing regulations, has had little formal role to play since then. From 1973 until 1989 it failed to agree on a merger control regulation, though in the mean time the Commission sought to develop merger control powers by other means. In the end the Council agreed the 1989 Merger Regulation because it feared that the Commission might obtain even greater powers by making a success of one of the alternative routes.

This does not mean that the member governments are unable to exert any significant influence on the evolution of competition policy. As we shall see, their influence has been keenly felt within the Commission because competition policy decisions have never been left to DG IV alone. Commission decisions are made by and in the name of the entire college of commissioners, many of whom, by dint either of their nationality or their functional responsibilities within the Commission, have had cause to oppose the policies advocated by the Commissioner for competition and by DG IV.

First steps: restrictive practices and monopolistic behaviour

As we have seen above, the Council complied with Article 87 in 1962 when it issued Regulation 17. This established the procedures for giving practical effect to Articles 85 and 86, and, in combination with later regulations, gave the Commission independent powers of interpretation and implementation. Regulation 17 covered all spheres of economic activity with the exception of transport, since rail, road, and inland waterways were covered by a separate regulation. Air and sea transport were to prove more contentious and were not made subject to the competition rules until the late 1980s, despite the palpable lack of competition in both these sectors and its adverse impact upon the consumer. However, air and sea transport apart, the Commission was able to build up a policy on restrictive practices and monopolistic abuse. It did so incrementally and independently of the Council and thus without the need to achieve a consensus among the member states. In the early years of establishing these policies, the Commission was greatly assisted by the actions of the Court of Justice. These appeared consistently to interpret the competition provisions so as to enlarge the Commission's scope for intervention.

Under Regulation 17, proceedings to enforce Articles 85 and 86 can be triggered in three ways: by notification directly from firms; on the Commission's own initiative; and by complaints from third parties. The Commission then works through a process of investigations and hearings in three stages. At the end of each stage the Commission can take an administrative decision: to drop the case altogether; to issue a 'comfort or discomfort letter', whereby, without taking a formal decision, the Commission can indicate its intentions and thus hope informally to influence the behaviour of firms; or to proceed formally to either of the next two stages (see Fig. 6.1).

The Commission procedures can be laborious, partly because of the need to hear opinions of, and defences by, the firms concerned, as well as the views of other interested parties, and partly because they must be decisions of the whole college, not just of DG IV. Although many European firms have criticized the Commission, both for its lengthy procedures and for its fondness for informal 'short cuts', others have been more positive. Joel Davidow (an American official from the US Department of Justice who was seconded for a time to the Commission) praised the Commission for its 'development of one

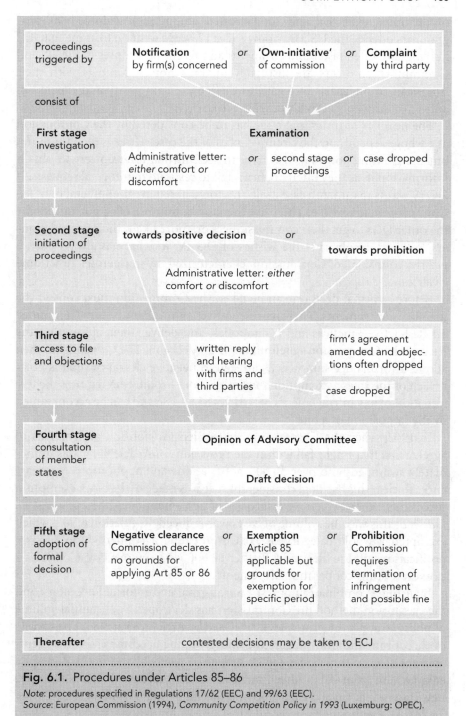

Fig. 6.1. Procedures under Articles 85–86

Note: procedures specified in Regulations 17/62 (EEC) and 99/63 (EEC).
Source: European Commission (1994), *Community Competition Policy in 1993* (Luxemburg: OPEC).

of the most extensive and sophisticated competition enforcement pro-grammes outside the US' (Davidow 1977: 175). He went on to argue that 'EC hearings are often faster and cheaper than present procedures in the US and Community enforcers are justifiably anxious not to compete with their American counterparts in regard to the length, complexity and expense of anti-trust litigation' (Davidow 1977: 176).

The number of times that a file has to be considered by the Commission as a whole depends on the circumstances of the case. Three is the minimum, but the Commission may also have to seek formal legal powers to obtain information if, as they often do, firms fail to respond to informal requests. If DG IV wants to initiate a case itself, this too will require the agreement of the Commission as a whole. There are therefore a number of opportunities for the other DGs to get their commissioners or, more likely, their commission-ers' *cabinets*, to obstruct the wishes of DG IV, in what has always been a poorly coordinated Commission characterized by numerous functional rivalries.

Before a formal decision can be taken the Commission must consult the member governments. Their representatives sit on the Advisory Committee on Restrictive Practices and Monopolies, and deliver their opinion on the basis of the Commission's preliminary draft decision. The Commission then adopts a formal decision which can lead to negative clearance (that is not a breach of the rules), exemption, or prohibition. Prohibition requires the ter-mination of any infringements of Articles 85 and 86 and can be accompanied by the imposition of a fine.

In 1962, when the Commission first called for notification by firms of agreements that might fall within the provisions of Article 85, it was imme-diately swamped with over 35,000 notifications and applications for clear-ance. To clear this backlog, considerable use was made of the block exemption provisions, whereby certain categories of agreement could be exempted *en masse*. In this way the volume of cases was reduced. Regulation 67/67, which exempted a whole category of exclusive-dealing agreements, took 13,000 notifications off the files and a further 12,000 were removed by a 1968 deci-sion exempting 'cooperation agreements'.

By using a combination of test cases, formal and informal decisions, and the exemption method the Commission has developed a substantial volume of policy to apply the principles established by Articles 85 and 86. The success of the Commission in categorizing and grouping acceptable and unaccept-able practices in these areas can be measured partly by the fact that, despite enlargement from six to fifteen member states, there are now less than 500 new cases a year under Articles 85 and 86 (EEC) and 65 and 66 (ECSC) (see Table 6.1). Furthermore, in recent times, the Commission has also succeeded

Table 6.1. New cases before the Commission under Articles 85 and 86 (EEC) and Articles 65 and 66 (ECSC)

Year	Notifications	Complaints	Own-initiative	Total
1980	190	58	51	299
1985	213	66	25	304
1994	281	151	23	455

Source: European Commission, *Community Competition Policy in 1993*, and European Commission, *General Report on the Activities of the European Union in 1994*.

in reducing the backlog of cases in this area from around 5,000 at the start of the 1980s to around 1,000 at the start of 1995.

Those concerned with the transparency of Commission procedures have, however, commented adversely that policy has been built on relatively few *formal* Commission decisions. The Commission would argue that the cases that it pursues through to a formal decision are deliberately selected for their strategic importance and are designed to establish policy precedents. None the less, others have expressed concern about the considerable use of discretion and of informal procedures. The Commission makes full use of its discretion to select cases, including the power to impose fines. It may thus be vulnerable to charges that decisions which should be taken solely on competition grounds are influenced by political considerations. Its informal procedures, such as the issuing of comfort or discomfort letters, are less than transparent. As a result the Commission is open to charges of secrecy and inconsistency.

As we shall see in the last section, concerns such as this have led some to call for the establishment of an independent European Cartel Office (ECO) to take over much of the work of the Commission in this and other areas of competition policy. In 1993, although the Commission closed 865 cases, only twenty-one were terminated by a formal decision that was published in the *Official Journal* and which could be challenged in the Court of Justice. The other cases were terminated by the sending of comfort or discomfort letters, or by firms voluntarily giving up practices which the Commission deemed to be anti-competitive rather than waiting for the formal Commission decision. Whilst some see this as part of a drive for greater efficiency, others fear the lack of transparency and the danger of unrestrained Commission discretion.

The Commission's activities under Articles 85 and 86 developed apace in the 1960s and early 1970s. But as integration slowed along with economic growth in the late 1970s and early 1980s, so did the Commission's enthusiasm for pursuing these cases, even though they did not have to run the gauntlet of

the Council. The Commission certainly exercised a great deal of political discretion in recognizing that the member governments would not take kindly
to excessive zeal in applying competition policy during harsh economic
times. At this time, in addition to being relaxed about certain state aids to
industries in crisis, DG IV also made full use of its powers of exemption
under Article 85(3).

On occasion, where this was not possible, the Commission simply ignored
the treaty obligations and permitted the creation of crisis cartels. For
instance, in the synthetic fibres sector, which was characterized at the end of
the 1970s by decreasing demand and surplus capacity, Étienne Davignon, the
Commissioner for industry, persuaded the eleven European firms which
dominated the sector to cooperate with one another. They signed an agreement which involved a number of restrictive practices; these included capacity reduction and market-sharing outlawed by the established competition
rules. This agreement led to a conflict between Davignon and Raymond
Vouel, the Commissioner then responsible for competition, who argued that
the agreement could not be exempted under Article 85(3). This episode
graphically illustrated the recurrent potential for conflict between the objectives of competition policy and other Community objectives—in this case a
quasi-industrial policy. In this instance, partly as a consequence of economic
circumstance and partly because he had the stronger personality, Davignon
had his way, and a compromise had to be devised (Allen 1983). The
Commission found against the agreement, but decided not to serve formal
notice on the participating firms, leading *Agence Europe* (10 Nov. 1978) to
comment that 'considerations of economic and social expedience have thus
prevailed over Community law'.

However, after the relative gloom of the late 1970s and early 1980s, the
revival of EC dynamism in the form of the single-market programme led to
a renewed enthusiasm for all aspects of competition policy, including those
covered by Articles 85 and 86. The Cockfield White Paper (Commission
1985) did not say much explicitly about competition policy—it was after all
produced by the DG responsible for industrial policy, a traditional functional
rival of DG IV. None the less, as Swann (1992) notes, Peter Sutherland, as
Commissioner responsible for competition, took full advantage of the single-
market programme to enforce the competition rules with increased vigour.
This was reflected in the use of greatly increased fines. In 1986 the
Commission used Article 85 to impose a fine of 57 million ecus on the fifteen
petrochemical companies which it had found guilty of price-fixing and
market-sharing—two practices which it had been forced to sanction in a
number of earlier crisis cartels. This use of spectacularly high fines has been
maintained to the present day. In 1994 seventeen steel companies were fined

a total of 100 million ecus, nineteen carton-board producers were fined 132 million ecus, and thirty-three cement manufacturers found themselves facing a fine of 248 million ecus, all for attempting to fix prices and to share markets.

Competition policy based on Articles 85 and 86 thus continues to be a major area of activity, although, with the development of the merger regulation and a renewed interest in policy towards state aids, it is less central to the overall policy than in earlier years. Activity in this area covers an enormous range of sectors,[3] with a large recent growth in the service sectors. As noted above, there is a growing concern about the Commission's multi-faceted powers and its Article 85 and 86 activities are clearly in the sights of those who advocate the establishment of an independent ECO.

Furthermore, it should be noted that several member states possess their own national legislation covering restrictive practices and monopolistic abuse. The existence of parallel legislation in the past presented an administrative problem rather than one of particular political sensitivity. Regular contacts took place between Commission and national officials on the basis of a general acceptance of both Community competence and precedence. Before the days of the single market the ability to distinguish between national markets and the European market avoided serious problems, other than those of interpretation between national courts and the Court of Justice. These persist and it is theoretically possible for a firm to be fined twice, by the Commission and again by a national court. It is possible to envisage the Commission's powers in this area being transferred to an independent agency. However, the logic of the single market makes it harder to invoke the principle of subsidiarity to justify a greater reliance on national authorities.

The control of mergers

The Treaty of Rome did not make any specific reference to the control of mergers, although Article 66 of the ECSC Treaty had given the High Authority the right to declare a merger in the coal or steel industry 'unlawful' and to prohibit it, if it so chose. At the time this was a considerable power to give to a supranational institution, especially since no such legislation then existed in any of the six member states. This had been agreed specifically to 'control' any potential German domination of these two industries. The absence of national legislation meant that there was also no institutionalized opposition to the transfer or sharing of powers. Under the ECSC coal and steel companies were faced by what has been called a 'one-stop shop', with only the supranational authority having the power to act.

From the beginning of the EEC the Commission regarded the absence of comparable merger-control powers as a serious omission that needed to be rectified in the interests of ensuring competition in the market. Article 86 of the EEC Treaty applied only to the case of firms which could be shown to be *abusing* a dominant position. This seemed to imply that to hold a dominant position was not *per se* unacceptable. Coal and steel apart, the Commission did not have the right of prior approval of mergers, even if they were likely to lead to the creation of a dominant position for the firms involved. One possibility might have been to amend the EEC treaty at the time that the institutions were merged and to extend the ECSC merger powers to the common market as a whole. In 1967, however, this was not a feasible option as the member states were in no mood so obviously to extend supranational competence. The Commission was thus forced to pursue its merger-control aspirations by interpreting the treaty as it stood and by exploiting Regulation 17.

As Bulmer (1994*b*) notes, the Commission therefore attempted to create a merger regime using Articles 85 and 86. Initially, it turned to Article 86, and in 1972 issued the Continental Can decision (Marston 1973); this sought to establish that a dominant position could, by itself, be regarded as an abuse and that a firm in a dominant position which took over a rival was also guilty of an abuse. In 1973, following an appeal against this decision, the Court of Justice supported the Commission's interpretation of Article 86 by linking it with the general competition objectives laid down in Article 3(f). This was a classic example of the Court liberally interpreting the treaty in the interests of advancing supranational integration. However, even this ruling, which many considered to be an outrageously bold leap, did not give the Commission the legal base for a *prior* control of mergers, since it provided for only a retrospective intervention. Although the Court agreed with the Commission's legal reasoning, it nevertheless found in favour of the company on the grounds that the Commission had got the facts of the case wrong by failing properly to define the market in which it claimed that Continental Can was dominant. This was by no means the only case of the Commission establishing its formal legal competence, yet also demonstrating a degree of practical incompetence.

Even before the Court's Continental Can verdict was announced, the Commission had decided that it needed a stronger legal base for merger control than Article 86. It had been encouraged by statements made at the Paris Summit of 1972 and by the Council of Ministers' pronouncements on inflation to propose a new regulation in 1973. It had to wait until 1989 for the Council to act, helped by the push of the single market. The proposed merger regulation was relaunched in 1987, despite the considerable reservations expressed by Britain, France, and Germany. Once again it was the Court that provided the necessary stimulus, albeit by accident rather than design. In 1987 in the Philip Morris case (Swann 1992; Bulmer 1994b; Cini 1994) the Court ruled that Article 85 could apply to situations in which a concentration was created by one company reaching an agreement with a competitor so as to acquire commercial control over it. The attraction to the Commission of using Article 85 to control mergers in this way was that there was no need to establish the existence of a dominant position, as Article 86 required.

The Continental Can ruling gave the Commission a legal base which it could threaten to use, if the Council was not prepared to consider positively its proposed merger regulation. The Commission indicated its intent by suc-cessfully forcing British Airways to give up some of its routes to its rivals after it had taken over British Caledonian, despite the fact that the UK authorities had approved the takeover. It used Article 85 similarly to block a proposed takeover of Irish Distillers by GC and C Brands. The result was that both companies complained to their national governments about the Commission's activities and sought to cover themselves for the future by notifying the Commission of any proposed mergers.

From the Commission's perspective the great advantage of this merger regime, using Article 85, was the uncertainty that it generated. This served to put pressure on the doubting member states to settle for a better worked-out and potentially more limited merger regulation. By a combination of luck and skill the Commission had managed to create a problem which the Council felt could be eased only by passing the legislation it had previously refused to consider.

The merger-control regulation as finally agreed[4] was an inevitable com-promise between the Commission and those member states which already had their own national merger controls. On the sidelines were the represen-tatives of industry who wanted the least restrictive rules possible. They also, for the sake of clarity and efficiency, were interested in not being caught, as they sometimes are under Articles 85 and 86, between national and suprana-tional regulatory authorities. The business community had a preference for a 'one-stop shop' as under the ECSC merger regime.

The Commission and the member states instead fought their competence battles over the thresholds beyond which Community powers could be exer-cised. The Commission would have preferred a threshold of 1 billion ecus, so that any merger that produced a concentration with a combined turnover of more than that figure would become subject to its authority. In the Council the governments from the smaller countries, which mostly did not have their own merger authorities, were inclined to support the Commission's bid for a low threshold; but the governments from Britain, France, and Germany all proposed thresholds of over 10 million ecus. The compromise was a principal threshold of 5 million ecus, to be subject to a review in 1993. When this review fell due, the Commission was keen to see the threshold lowered, but it recognized that, with integration under challenge and subsidiarity so much in fashion, it was not an auspicious time to push this issue. The German government subsequently made it clear that it would consider lowering the threshold in future only if its proposal for an ECO were taken seriously.

A number of other restrictions on Commission competence were added to the Regulation, of which the most significant was the so-called German clause. This enables a member state to ask the Commission to allow a national investigation of a merger if it is likely to have a significant effect on its domestic market, even if it goes beyond the threshold. The reverse image of the 'German clause' is to be found in the 'Dutch clause': this permits a member state also to ask the Commission to investigate a merger that does not reach the threshold if it might seriously affect the competitive situation in that member state (presumably a member state which does not have merger controls of its own).

Much controversy has arisen over the criteria that the Commission should use for considering the desirability of mergers; should the criteria be restricted only to those relating to competition, or should the Commission allow other considerations to be taken into account? The Regulation refers to the need to preserve effective competition, but it also says that 'other factors' should be taken into account, including the development of technology and economic progress. It is the new Merger Task Force within DG IV that implements the Regulation, but the Commission as a whole that takes the final decisions. Although those who work in DG IV may be interested only in competition criteria, other interests come into play within the Commission as a whole.

The policy procedure laid down by the Regulation requires that the Commission be informed of any merger with a Community dimension (i.e. above the threshold). Such mergers cannot be pursued either before or for three weeks after notification. The Commission must decide within one month of notification whether to initiate proceedings. Then within four months, after consulting the Advisory Committee of the Member States, the Commission must decide whether to approve the merger, to prohibit it, or to approve it subject to certain conditions (see Fig. 6.2). In exceptional cases, at the request of a member state, the Commission may decide that a case, although it meets the criteria for consideration at Community level, can instead be considered under national procedures. Between 1990, when the Regulation first came into effect, and the end of 1994 the Commission had

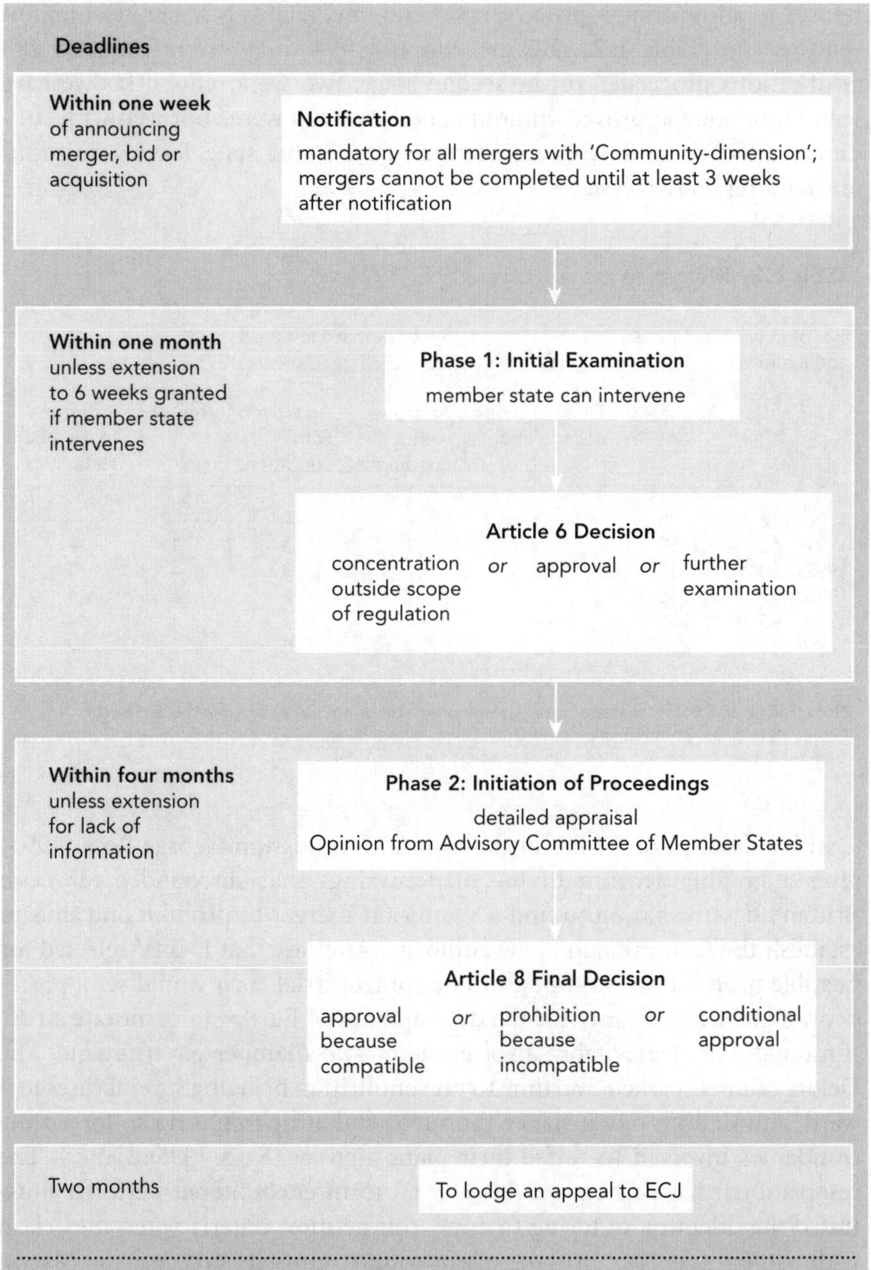

Deadlines

Within one week
of announcing
merger, bid or
acquisition

Notification

mandatory for all mergers with 'Community-dimension';
mergers cannot be completed until at least 3 weeks
after notification

Within one month
unless extension
to 6 weeks granted
if member state
intervenes

Phase 1: Initial Examination

member state can intervene

Article 6 Decision

concentration   *or*   approval   *or*   further
outside scope   examination
of regulation

Within four months
unless extension
for lack of
information

Phase 2: Initiation of Proceedings
detailed appraisal
Opinion from Advisory Committee of Member States

Article 8 Final Decision

approval   *or*   prohibition   *or*   conditional
because   because   approval
compatible   incompatible

Two months

To lodge an appeal to ECJ

Fig. 6.2. Procedures for controlling mergers

Notes: Procedures specified in Regulation 4064/89 (EEC). Criteria for 'Community-dimension' based
on aggregate turnover worldwide (5 billion ecus+) and in Community (250 million ecus), unless 2/3 of
latter in only one member state.
Source: European Commission (1994), Community Competition Policy in 1993 (Luxemburg: OPEC).

refused to allow one proposed merger and one, relatively insignificant, joint venture (see Table 6.2). By the end of 1994, only twenty out of 288 notifications proceeded to the second stage: two were refused (as we have seen); four were approved without conditions; ten were approved with conditions attached; and two were referred to a member state, leaving decisions outstanding on two cases.

Table 6.2. Merger regulation cases, 1990–1994

No. of new notifications	First phase		Second phase initiated			
	Approved	Outside scope of regulation	Approved without conditions	Approved with conditions	Refused	Referred to member state
1990: 12	5	2	—	—	—	—
1991: 63	50	5	1	3	1	—
1992: 60	47	9	1	3	—	1
1993: 58	48	4	1	2	—	1
1994: 95	105	—	1	2	1	—
Total: 288	256	20	4	10	2	2

Note: The figures do not necessarily tally; some proceedings are not decided in the same year.
Source: European Commission, Community Competition Policy in 1993, and European Commission, General Report on the Activities of the European Union in 1994.

In his description of life in the Delors Commission George Ross (1995) gives a graphic account of the manœuvrings that surrounded Sir Leon Brittan's determination to find a significant merger to prohibit and thus to establish the Commission's new authority. Any case that DG IV selected for possible prohibition was bound to be controversial, as it would set a precedent of great significance for the development of European corporate structures and would inevitably involve one or two member governments. The Delors cabinet worked overtime to prevent Brittan bringing several cases forward, although in one instance the rumour that he might do so forced the companies involved to revise their plans anyway (Ross 1995: 132–5). The reservations that Brittan and his DG IV team encountered perfectly illustrated the dilemma of trying to apply competition criteria when they clash with other criteria. Brittan often found himself up against Martin Bangemann, arguing from his functional perspective as Commissioner for industry, and he also found himself obstructed by commissioners from the member states from which companies were likely to be affected.

In June 1993 Brittan found the case that he needed when he sought successfully to prohibit the acquisition of de Havilland, a Canadian aircraft manufacturer, by ATR, a French-Italian consortium (Hawkes 1992). Even then he won the crucial vote in the Commission by only nine votes to seven (with Delors abstaining once he saw that he would lose). Brittan found himself challenged by an alliance of DG III with French and Italian interests. More fundamentally, as the only Commissioner in a position to stand up to Delors, Brittan found himself defending his free-market principles against the more interventionist bent of the President. As Ross put it 'if Delors wanted to "organize" a European industrial space, Brittan wanted quite as much to "open" that space' (Ross 1995: 176).

Two distinct features of the merger policy process are evident. First of all, it is clear that, though the Council has no formal role to play in the control of mergers, the member governments both can and do seek to influence the decisions of the college via the commissioners of their nationality. Secondly, the arguments about mergers, and indeed competition in general, are not just technical and legal, but are fundamental. There are thus constant debates about how the narrow competition objectives of DG IV can or should be reconciled with the broader and sometimes contradictory objectives of the Commission as whole. The merger saga reveals how much politics permeates this apparently technical area. The Commission has succeeded, for the time being at any rate, in adding to its supranational armoury by extracting the merger-control regulation from the Council. It has not, however, insulated itself from national, sectoral, or political influences and disagreements, nor are the Commission's new powers likely to remain unchallenged in the future.

State aids

The other area of competition policy that had been neglected for some time, but which received a significant boost after 1985, was state aids, covered by Articles 92–4 (EEC). In the early years of the Community, the Commission adopted an essentially pragmatic approach to its powers to take on national governments. Article 92 states that any aid that is likely to distort competition and thus affect interstate trade is incompatible with the common market. It then goes on to list, in much the same way as do Articles 85 and 86, those types of aid which are regarded as compatible. Article 93 requires the member states to notify the Commission of all existing aids and proposals for new aid. If the Commission finds any aids incompatible with the common market, it can require the member state to abolish them and in some cases it can require the beneficiary to repay the relevant money (see Fig. 6.3).

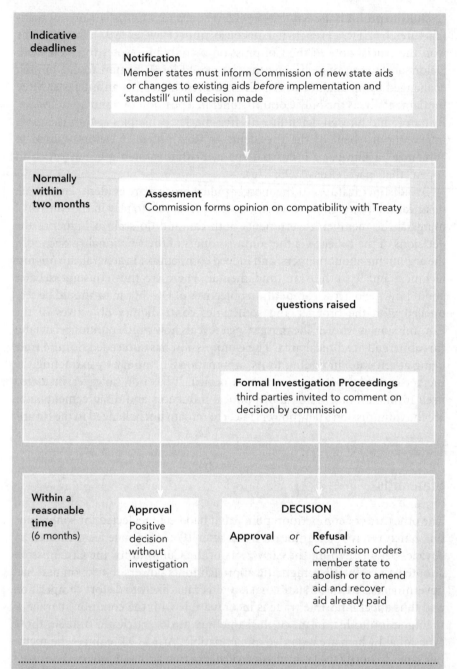

Indicative deadlines

Notification
Member states must inform Commission of new state aids or changes to existing aids *before* implementation and 'standstill' until decision made

Normally within two months

Assessment
Commission forms opinion on compatibility with Treaty

questions raised

Formal Investigation Proceedings
third parties invited to comment on decision by commission

Within a reasonable time (6 months)

Approval
Positive decision without investigation

DECISION

Approval *or* **Refusal**
Commission orders member state to abolish or to amend aid and recover aid already paid

Fig. 6.3. Procedures for assessing new state aids

Notes: Legal bases Art.4 (ECSC) and Arts 92–4 (EEC). Decisions may be appealed in ECJ.
Source: European Commission (1994), *Community Competition Policy in 1993* (Luxemburg: OPEC).

The member states have a rather poor record of notification, especially of general aids.[5] Several state-aid investigations have been initiated on the basis of a complaint by an aggrieved third party in the private sector. In the event of a state refusing to comply with a Commission decision, it can be referred to the Court. This can in turn, under the TEU and the revised Article 171, impose a fine at a level recommended by the Commission.

Although the Treaty of Rome provided a basis for a policy on state aids and the means for the Commission to implement that policy, little real progess was made until the 1980s. Before then, the Commission was hard pressed to keep abreast of the aids of which it was notified and its usual response was to approve them. During periods of recession pressures from member governments and those within the Commission who favoured industrial intervention limited the scope for a restriction on state aids and pushed rather for a policy of 'coordination'. Such decisions as the Commission issued in this early period were more often than not ignored by the member governments. Although the Commission had some success in constraining national regional aids, the large industries in decline—shipbuilding, textiles, and steel in particular—all received substantial national subsidies.

Many observers would argue that a sea change in the evolution of the Commission's state-aids policy came with the Court of Justice's ruling in the Philip Morris case in 1981 (Cini 1994). Although it had not been intended as a test case, it became one, in that the Commission's right to exercise discretion 'in the European interest' was firmly upheld by the Court. Although the judgment was clearly helpful to the Commission, the Court seemed also to be reminding the Commission of its obligations and responsibilities in this area. The case revolved around an appeal by Philip Morris against a Commission ruling that the subsidies that it had received from the Dutch government were incompatible with the state-aids provisions of the treaty. In finding for the Commission, the Court appeared to be firmly stating that, even if the national interests of a member state could be seen to be furthered by the provision of certain types of aid to a company within that state, this could not be advanced as a basis for the Commission exercising its powers of discretion and thereby allowing the aid. In effect the Court was saying to the Commission that, if it wanted to adopt a firmer line than before, it would be supported by the Court.

Emboldened by the Philip Morris ruling, the Commission began to prohibit more of the aids about which it was notified. However, it is hard to see this as anything other than scratching the surface of the problem, in that the Commission had little detailed knowledge about either the extent or the effect of state aids in the European market. What it did know about, it sought to categorize, to clarify, and, where possible, to coordinate. Thus DG IV endeavoured to develop general rules to govern regional, general, and sectoral aids (Swann 1992; Cini 1994). It sought to use its powers to ensure that aid to industries in distress took account of the need to progress, rather than simply to preserve the *status quo*; it tried to increase transparency across the Community in an area where even the member governments themselves had limited knowledge of the exact extent of the aids that they were granting; and it sought to provide a Community-wide framework to prevent governments competing to provide attractive regional aid.

A further and more significant change came with the single-market programme. Although DG IV shared the general Commission enthusiasm for the dynamism that this provided, it did feel threatened by the new prestige of its 'rival' DG III. As Cini (1994) points out, DG III seemed to be encroaching on DG IV's territory with its espousal of the competitive advantages to be had from the single market. On the other hand, it was also clear that the single-market programme, by endorsing the competitive philosophy that D IV had for so long advocated alone, presented an enormous opportunity, as well as a challenge. The initiative was seized enthusiastically by Peter Sutherland as Commissioner and from 1989 by his successor Sir Leon Brittan.

Sutherland's most significant move was to pursue the objective of greater transparency and efficiency by initiating a major review of state aids. A Task Force was set up and it produced its First Survey of State Aids in 1989.[6] This important survey indicated that state aids within the EC averaged just over 100 billion ecus per annum between 1981 and 1986 and that within each member state they accounted for between 10 and 20 per cent of public expenditure or between 3 and 5 per cent of GDP. Subsequent surveys in 1990 and 1993 confirmed these levels and suggest that, despite the renewed efforts of DG IV in the 1980s and early 1990s, there has been very little change in the scale of aid granted during this period, even though the UK, Denmark, and Italy had substantially reduced their own contributions to the overall total.

These surveys were accompanied in 1989 by a major review of policy prompted by Sir Leon Brittan. He, like Peter Sutherland, was a firm believer in market forces and in the steady application of the competition provisions. However, they were both also keen to concentrate on those state aids which were most obviously and publicly distorting the market, those which had an 'appreciable effect'. Two spectacular examples, which served to raise the profile of the Commission in this area, came with the Commission's insistence that first Renault and then Rover (both historically national flagships) pay back subsidies that they had received in the mid to late 1980s.

The policy review at the end of the 1980s also drew attention to the Commission's interest in using the treaty provisions both to police and further to encourage the processes of liberalization and privatization. Here the

Commission has demonstrated a 'light touch' in its use of Article 90, even though it has been keen to ensure that liberalization does not lead either to a creaming off of profitable activities to the detriment of the public interest or to abuse from member governments, which might give illegal subsidies to public enterprises which are about to be privatized. As far as liberalization is concerned, the Commission has preferred to use Article 100a (which provides for the approximation of national legislation) to develop a consensus, rather to make policy by test cases under Article 90. The Commission has effectively devised long-term framework programmes for the orderly liberalization of four key European markets: telecommunications, postal services, energy, and transport.

As in other areas of competition policy, the Commission's procedures for examining state aids have received some criticism. Once an investigation has been formally opened, the Commission deals almost exclusively with the member state which has given the aid or committed the apparent infringement. Companies affected by the aid in question do have a right to present their case to the Commission, but without an oral hearing. Companies can also go to the Court of Justice, but it is claimed that companies are generally unaware of their rights in this area, and in particular, unaware of the fact that damages can be claimed against member governments which have granted illegal aids without informing the Commission. Nevertheless a steady flow of cases continues (see Table 6.3); the Commission continues to approve most aids of which it is notified, while doubling its efforts to discover those of which it has not been notified.

Table 6.3. Commission state-aid activity, 1992–1994

	1992	1993	1994
Total number of decisions	502	455	498
no objection after preliminary examination	435	399	440
proceedings initiated	30	30	40
positive final decisions	31	19	15
negative final decisions	6	7	3
New cases	558	561	594
measures notified	452	475	526
measures not notified	102	85	68
existing measures	4	1	0

Note: The figures do not necessarily tally; some proceedings are not decided in the same year.
Source: European Commission, Community Competition Policy in 1993, and European Commission, General Report on the Activities of the European Union in 1994.

The Commission's regulation of state aids in a rapidly changing economic environment is likely to play an even more important role in the work of DG IV in the future, particularly if controls over mergers and restrictive practices are transferred to an independent agency. Member governments continue to spend around 100 billion ecus a year on assistance to companies, but this money is spent within an increasingly regulated environment. Effective controls are not easily compatible with a division of responsibilities between the Commission and the member governments. If the various liberalization programmes come to a successful fruition in the 1990s without any major distortions to the European market, this will count as a major achievement for the Commission and DG IV.

The proposal for a European Cartel Office

We have repeatedly stressed the extent of the Commission's powers and scope for exercising its own discretion in the area of competition. This has led to recurrent unease about the criteria applied and the procedures followed by the Commission, an unelected body, acting as policeman, prosecutor, judge, jury, and prison warden. The idea of transferring some of these discretionary powers to an independent agency is not a new one, but it has come increasingly to the fore recently, partly as a result of German interest in establishing such an agency, and partly because it may be put on to the agenda of the 1996 IGC. The German government first mooted the proposal for an independent agency shortly after the Treaty of Rome was signed. When the Merger Treaty was concluded in 1965 the member states even went so far as to agree that, if a ECO was set up, it would be based in Luxemburg.

The revival of the idea in the 1990s owes much to the German government's enthusiasm for the purest possible application of competition-policy principles, untainted by political or other economic policy considerations. Like the Bundesbank, the Federal Cartel Office in Germany is regarded as a shining example of the advantages to be gained from an independent authority. In both cases the contention may be challenged, hence some critics have questioned the effectiveness and independence of the Federal Cartel Office (Wilks and McGowan 1995a). Nevertheless, the issue of reforming the competition practices of the Commission is clearly on the agenda (House of Lords 1993) and the German proposal for an ECO will have to be taken seriously.

Alternatives to an independent agency were set out by Claus Dieter Ehlermann (1995), until spring 1995 Director General of DG IV. He drew attention to a number of potential disadvantages of an independent ECO and suggested instead that DG IV's work could be more effectively reformed by setting up an independent consultative body at Community level along the lines of the Monopolies and Mergers Commission in the UK. Such a body would be consulted by the Commission before it took decisions and its opinions would be published, thus stimulating public debate and transparency. In addition, Ehlermann suggested that the position of the Commissioner responsible for competition could be strengthened in order to bolster his position within the college as a whole. One way to achieve this would be to extend the scope for delegating powers to the competition Commissioner, so that competition-policy decisions would not have to run the gauntlet of the whole college. In this way the political manœuvres that surrounded Sir Leon Brittan's narrow victory within the Commission on the de Havilland case would be avoided. Ehlermann's proposals aimed to meet the major German objection to the current process, which stresses the risk that competition policy may be sullied by a growing politicization and by conflicts between competition objectives and other policy aims being pursued by the Commission.

The Commission's White Paper on competitiveness (Commission 1993*d*) is often cited as an example of potentially changing priorities within the Commission. As Wilks and McGowan (1995*b*) point out, this barely mentions competition policy, even though the Commission's 1993 Report on Competition (Commission 1993*g*) attached great significance to the White Paper's objectives. Karel van Miert, the responsible Commissioner from 1993, gave the competition purists cause for concern because of his 'pragmatic' style and interventionist tendencies. Van Miert's approach was different from those of both Peter Sutherland and Sir Leon Brittan in seeing competition policy less as an end in itself and more as an instrument for maximizing the economic potential of the EU.

Supporters of an independent ECO would like to protect competition policy by removing it from the direct control of a Commission. They see it as increasingly open to political compromise and vulnerable to opposition from within the member states across the range of competition-policy issues. Following the example set by the Federal Cartel Office, the ECO would be independent from the Commission and would be specifically charged with applying a strict competition test to dominant positions, restrictive practices, and mergers, with scope for appeal to either the Court of Justice or possibly the Commission. Although the German proposal did not include state aids, others (for instance the *Financial Times*) have argued that it should. There have been discussions within DG IV about the possibility of turning the present Merger Task Force into an independent agency, in order to satisfy the

German demands, whilst retaining as many powers as possible within the Commission. The debate around the ECO proposal is in effect a debate about the purposes and future of competition policy in the EU.

Conclusion

As long as there has been an active competition policy at the European level, there has been controversy and discussion about whether rules derived from the model of perfect competition can be applied in isolation or whether competition objectives must be considered alongside other economic, political, and social objectives. This argument has tended to ebb and flow with the prevailing economic circumstances. During periods of relative optimism, integrative dynamism, and prosperity in the EU, competition policy has flourished and the Commission has been able to exercise a uniquely independent role. At other times, marked by relative pessimism, integrative sclerosis, and economic hardship, a strict interpretation of the competition rules has been deemed unwise and the Commission has found both its powers and procedures under attack. However, it is beyond dispute that some sort of competition policy will continue at the European level, since it has to be seen, along with the internal market which it is designed to protect and advance, as part of the irreducible core of Union activity. It does not therefore seem to be a logical candidate for 'variable geometry' solutions or 'renationalization'. Although many member states developed their own national legislation only after the establishment of European rules, the existence of the single market severely limits its applicability. The continuing debate is thus mainly about how best to develop and to implement an effective policy, both in its own right and *vis-à-vis* other policy objectives. The absence of an active industrial policy at the EU level leaves competition policy to bear the weight of industrial as well as competition considerations.

Notes

1. Articles 85 and 86 roughly correspond to Articles 65 and 66 of the Treaty of Paris (ECSC), although with the significant difference that the High Authority of the ECSC has much greater powers over the control of mergers, exercised still since 1967 by the Commission for coal and steel.
2. Council Regulation (EEC) 17/62, *Official Journal*, L13, 21 Feb. 1962, p. 204.
3. In 1993 DG IV opened 404 new cases in thirty-two different sectors.

4. Council Regulation (EEC) 4064/89, *Official Journal* L395, 30 Dec. 1989.
5. The Commission distinguishes between general aid, sectoral aids, and regional aid.
6. See the *22nd General Report on the Activities of the European Communities*, 1988, para. 431.

Further Reading

For a survey of competition policy before the single-market programme see Swann (1983). For detailed accounts of policy developments consult *Common Market Law Review* and *European Law Review*. For an excellent overview of the merger regime see Bulmer (1994*b*). For the debate over the reform of competition policy see Wilks and McGowan (1995*b*).

Bulmer, S. (1994*b*), 'Institutions and Policy Change in the European Communities: The Case of Merger Control', *Policy Administration*, 72/3: 423–44.
Swann, D. (1983), *Competition and Competition Policy* (London: Methuen).
Wilks, S., and McGowan, L. (1995*b*), 'Disarming the Commission: The Debate over a European Cartel Office', *Journal of Common Market Studies*, 33/2: 259–73.

Stephan Leibfried and Paul Pierson

Despite the widespread assumption that EU involvement in social policy has been minimal, the dynamics of market integration have led to a substantial spillover on to the Community level. Resistance by national governments to loss of autonomy and conflicts of interests between rich and poor regions or between employer and employee interests present formidable obstacles to activist EU policies. Member states have lost more control of national welfare policies, in the face of the pressures of integrated markets, than the EU has gained in transferred authority. The multi-tiered pattern which has emerged is largely court-driven, marked by policy immobilism at the centre and by market integration, which imposes tight constraints on national policies.[1]

Introduction

Accounts of European social policy generally present a minimalist interpretation of European Union (EU) involvement. The sovereign nation state, so goes the argument, leaves little social policy role for the European Community (EC). The Community is regarded as a matter of 'market-building' only, leaving an exclusive, citizen-focused, national welfare-state, its sovereignty formally untouched, though perhaps endangered indirectly, by growing economic interdependence. 'Welfare states are national states' (de Swaan 1992: 33; cf. also Lange 1992). On the face of it, the European welfare-state indeed looks national. There is no EC welfare law granting individual entitlements against Brussels; there are no direct taxes or contributions funding a social budget which would back such entitlements; there is no Brussels welfare bureaucracy to speak of. 'Territorial sovereignty' in social policy, so conventional wisdom holds, is alive and well (for an overview cf. Collins 1975; Kenis 1991). We disagree. The process of European integration has eroded both the *sovereignty* (by which we mean legal authority) and the *autonomy* (by which we mean *de facto* regulatory capacity) of member states in the realm of social policy. National welfare states remain the primary institutions of European social policy, but they do so in the context of an increasingly constraining multi-tiered polity (Pierson and Leibfried 1995*a*,*b*).

While the quite extensive barriers to EC action have prevented any true federalization of European social policy, the dynamics of creating a single market have made it increasingly difficult to exclude social issues from the EC's agenda. The emergence of a multi-tiered structure, however, is less the result of welfare-state-building ambitions of Eurocrats than a result of spillovers from the single-market initiative. By 'spillovers' we mean the process through which the completion of the internal market leads the EC to invade the domain of social policy (see initially E. Haas 1958; Lindberg and Scheingold 1970). Recall that the single-market initiative was based on a deregulatory agenda and assumed that initiatives to ensure 'free movement of goods, services, capital, and labour' could be insulated from social policy issues, which would remain the province of member states. This is a dubious assumption, and it is worth noting that it runs directly contrary to the central tenets of political economy, which stresses precisely that economic action is embedded within dense networks of social and political institutions (cf. Hall 1986; North 1990). Already there is significant evidence that the tidy separation between 'market issues' and 'social issues' is unsustainable. Irrespective of the results of 'high politics' struggles over social charters and treaty revisions, the movement towards market integration carries with it a gradual erosion of national welfare-state autonomy and sovereignty, increasingly situating national regimes in a complex, multi-tiered web of social policy.

This transformation of sovereign welfare states into parts of a multi-tiered system of social policy occurs through three processes (see Table 7.1). 'Positive', activist reform results from social policy initiatives taken at the 'centre' by the Commission and Council, along with the European Court of Justice's (ECJ) often expansive interpretations of what those initiatives mean. 'Negative' reform occurs through the ECJ's imposition of *market compatibility requirements* that restrict and redefine the social policies of member states. Finally, the process of European integration creates a range of indirect pressures that do not legally require but none the less strongly encourage adaptations of national welfare states. We consider each of these processes in turn. In the first section of this chapter we briefly review the EC's modest direct efforts to develop an activist social policy. Like others, we stress the formidable obstacles: institutions that make reform difficult, limited fiscal resources, jealous member-state protection of 'state-building' resources, and an unfavourable distribution of power among interest groups. In the remainder of the chapter, however, we turn to the processes through which European integration is gradually transforming national welfare states into components of a unique 'multi-tiered' social policy framework. The second section reviews the development of what we call *market compatibility requirements—*

Table 7.1. Social policy transformation: processes, key actors, and examples

Processes	Key actors	Illustrative examples
'positive' social policy initiatives to construct areas of competence for uniform social standards at EC level	Commission, expert committees, EJC (background actors: European Parliament, ETUC, UNICE, ESC)	gender equality (Art. 119, EEC); health & safety (Art. 118a, EEC); 1989 Social Charter; 1992 Social Protocol ('Maastricht')
'negative' social policy reform via imposition of market compatibility requirements	European Court of Justice, Commission; Council (national governments)	labour mobility: 'coordination'; regulation (Reg. 1408/71, 574/72); freedom of services (Arts. 7a, 59–66, EEC); regional as well as sectoral subsidies
indirect (*de facto*) pressures of integration that force adaptation of national welfare states	market actors (employers, unions); Council (national governments)	'social dumping'; harmonization of tax systems; stages of EMU

legal challenges to those aspects of national welfare states that conflict with the single market's call for unhindered labour mobility and open competition for services. In the third section we consider the *de facto* rather than *de jure* pressures on national regimes. These pressures result from factors such as competitive demands for adaptations of national economies to a single market and, potentially, a single currency area. In the final section we pull together these arguments to highlight some of the distinctive features of this emerging multi-tiered system of social policy.

'Positive' initiatives from the centre go hand in hand with major and visible social conflicts and have accompanied the EC from its beginnings. The Rome Treaty, for instance, met stiff resistance in the French National Assembly in part because of concern that weak social clauses endangered the well-developed French welfare state (Marjolin 1989: 284–97). 'Negative' integration efforts were historically less visible but are just as old. The coordination of rules governing labour mobility was enshrined in one of the earliest EC legislative acts, although similar action on services dates only from the 1980s onwards. Naturally, 'indirect pressures', the third level distinguished in Table 7.1, are of more recent vintage, since such pressues could only build up as integration intensified.

The limited success of activist social policy

Discussions of social policy generally focus on such prominent actors as the Council, the Commission, and Parliament, and the representatives of business and labour interests. The European Commission, in particular, has been a central actor in direct attempts to construct a significant 'social dimension'—areas of social policy competence where uniform or at least minimum standards are set at the EC level. These attempts have occurred in fits and starts during the past few decades. It has been a saga of high aspirations and modest results, marked by a plentitude of 'cheap talk', produced in the confident knowledge that the requirements of unanimous Council votes meant that ambitious blueprints would remain unexecuted. This story has been well told elsewhere (Mosley 1990; Streeck and Schmitter 1991; Vogel-Polsky and Vogel 1991; Lange 1992). Here we review only the broadest outlines, for our main argument is that the analytical focus on the efforts of Euro-federalists to foist an activist 'social dimension' on a reluctant Council has been misleading. European integration has indeed altered the making of European social policy, but largely through quite different mechanisms.

The obstacles to an activist role for Brussels in social policy development have always been formidable (Leibfried and Pierson 1992). These obstacles include both institutional constraints and the balance of power among relevant social interests. EC institutions make it much easier to block reforms than to enact them. Generally, only narrow, market-related openings for social legislation have been available. Even on this limited terrain reform requires a super-majority. The member states themselves, which serve as gatekeepers for initiatives that require Council approval, jealously protect social policy prerogatives. Economic and geopolitical changes since the Second World War have gradually diminished the scope of national sovereignty in a variety of domains. The welfare state remains one of the few key realms of policy competence where national governments still appear to reign supreme. Given the popularity of most social programmes, national executives will usually resist losses of social policy authority.

A further barrier is the relative weakness of the 'social democratic' forces most interested in a strong social dimension. Unions and social democratic parties have become far weaker in the past fifteen years in most of Europe, especially among the founding members of the EU. At the European level, organizational difficulties and profound conflicts of interest between high-wage and low-wage areas of the EU limit labour's influence, sometimes leading to new regional inter-class alliances. At the same time, business power has grown considerably, in part because of the increasing capital mobility that European integration has fostered. The balance of power among social interests has further hindered efforts that institutional blockages, limited fiscal resources, and the tremendous difficulties of harmonizing widely divergent and deeply institutionalized national social policies would have rendered highly problematic in any event.

In this context, substantive policy enactments have been rather rare. Expansive visions of Community social policy have had far lower priority than initiatives for an integrated market. Knowing that British opposition rendered serious initiatives impossible (on the background cf. Kleinman and Piachaud 1992a), member states, EC officials, and interest groups have made rhetorical commitments to the construction of a social dimension (Lange 1992; Ross 1995; Streeck 1995). Over the past two decades, there have been loud public fights over a series of initiatives to increase the social policy competence of the EC. In most cases, the vigour of the rhetorical battle far exceeded the true implications of the proposals at hand. In any case, these initiatives invariably were either defeated outright or radically scaled back. The struggle over the Social Charter in the 1980s was typical (Sylvia 1991). Efforts to establish an ambitious programme for European-level legislation were rebuffed by the British, who eventually refused to agree to even a much watered-down version that was signed by the other member states. The exercise allowed various actors to adopt politically useful public postures, but gave only modest momentum to actual policy initiatives.

Legislative reform has been limited to a few areas where the Treaty of Rome or the single-market project allowed more significant latitude. A notable example concerns the gender-equality provisions of Article 119 of the Treaty of Rome (EEC). This provision, offered as a face-saving concession to France, lay dormant for almost two decades. In the 1970s, however, the Council agreed to a number of directives which gave the 'equal treatment' provision some content. Over the past twenty years, the ECJ has played a crucial activist role, turning Article 119 and the directives into an extensive set of require-ments and prohibitions related to the treatment of female (and occasionally male) workers (cf. Falkner 1994a: 77–122; Ostner and Lewis 1995). These rulings have required extensive national reforms—although not always to the benefit of women. To take just one example, ECJ decisions have had a dra-matic impact on private and, indirectly, also on public pension schemes. The Court's insistence on equal retirement ages in occupational pension schemes forced Britain to level ages up or down. By choosing to raise the retirement age for women the government will save billions of pounds, while avoiding much of the blame for the cuts. When in the *Barber* Case the ECJ made a sim-ilar ruling for occupational pensions, fear that the ruling might be applied retroactively to private pensions (at a cost estimated at up to £40 billion in Britain and DM 35 billion in Germany) fueled 'what is probably the most intense lobbying campaign yet seen in Brussels' (Mazey and Richardson 1993a: 15). While this pressure led the negotiators of the Treaty on European Union (TEU) explicitly to limit retroactivity, the prospective impact of the Court's rulings remains dramatic.[2] Thus the EC has come to play a consider-able role on gender issues, although the market-oriented nature of the Community and its restricted focus on the paid-labour market has circum-scribed the EC's interventions.

A second example concerns the extension of regulations governing health and safety in the workplace according to Article 118a (cf. in detail Eichener 1993). The single-market initiative allowed qualified majority voting in this area out of fear that national regulations could be used as non-tariff barriers to trade. Surprisingly, policy-making has produced neither stalemate nor lowest-common-denominator regulations. Instead, extensive regulations have generally produced quite a high level of standards. Furthermore, EC reg-ulators moved beyond the regulation of *products* to the regulation of produc-tion *processes*, where the concerns about barriers to trade would seem inapplicable. As Eichener has documented, the Commission's role as 'process manager' appears to have been critical in this low-profile environment. Much of the crucial decision-making took place in committees composed of policy experts. Some of these experts were linked to business and labour groups, but business interests did not have the option of simply refusing to participate, since regulatory action was likely to proceed without them. Representatives within these committees were often interested in innovation, having gravitated towards Brussels because in regulatory issues it seemed to be 'where the action is'. In this technocratic context, 'best practices' from many member states (and from then other countries such as Sweden) were pieced together to form a quite interventionist structure of social regulation. The Commission played a central part in joining together the work of different committees and incorporating concerns of other actors such as the European Parliament—all the while actively promoting particularly innovative proposals.

This development provides one indication that the institutional restrictions on social policy initiatives have loosened somewhat in the past decade. The Single European Act's (SEA) introduction of qualified majority voting in some domains has made social policy the focus of sharp conflict. There have been significant struggles to determine the range of issues that can be treated on a majority-vote basis, either under Article 100a (covering harmonization of legislation to avoid distortions of competition) or under the SEA's exception for proposals governing the health and safety of workers. Members of the Commission, the European Parliament, and the European labour organization ETUC have pushed with some success for expansive readings of these clauses. UNICE, the main employers' organization, has strongly opposed such a move (Lange 1992: 235–56). While many proposals connected to the Social Charter have been watered down or stalled, the combined impact of what has been passed is far from trivial (Addison and Siebert 1993). One sign of the growing room for social policy initiatives was the enactment of the Maternity Directive of October 1992, passed under the 'health and safety' provisions allowing qualified majority voting. This legislation requires more generous policies in several EU countries. At the same time, it introduces a policy 'ratchet', prohibiting other countries from cutting back their existing regimes.

The TEU may also have lowered the institutional barriers to social policy enactments, although the creation of a separate institutional track for social policy has also generated tremendous uncertainty. In the run-up to Maastricht, a number of countries expressed a desire to expand EC involvement in social policy. Recognizing the likelihood of continued British resistance, however, the member states seemed to be moving towards another symbolic exercise. Yet with John Major needing a 'victory' to placate anti-EC Tories at home, even a symbolic agreement proved elusive. Faced with an impasse between British unwillingness to sign a social agreement and French

refusal to sign a treaty that did not contain one, the Dutch and German negotiators, together with Jacques Delors, engineered a complex compromise
(Lange 1993; Ross 1994). All twelve member states agreed to allow the eleven
member states other than Britain to go forward on social policy issues under
a new 'Social Protocol' which expands the scope for qualified majority voting.

Although the possibility for majority voting has been expanded, some
social policy changes (as in social security) still require unanimity, while
some areas (like wage-bargaining) are completely exempted from EU competence (see Social Protocol Article 2 (3, 6)). A qualified majority vote will
now be sufficient on issues pertaining to the 'improvement in particular of
the working environment to protect workers' health and safety; working conditions; the information and consultation of workers; equality of treatment
between men and women with regard to labour market opportunities and
treatment at work; the integration of persons excluded from the labour market' (Social Protocol, Article 2 (1, 2)). Britain will not participate in such
decision-making, nor will it be governed by any policies taken in this framework.

How this unprecedented solution will affect social policy-making remains
unclear. The ECJ will need to rule on the legality of an arrangement that
employs the Union's governing apparatus for a subset of the EU membership.
Britain may eventually rejoin the eleven, either as a result of a change in government in London or because of growing perceptions that the costs associated with being the 'odd one out' are too high.[3] Once Britain opted in—say,
if a Labour government were elected—it would not be free to 'opt out' again.
It is, however, far from clear that being outside the new social policy framework will impose significant costs; it may in fact confer certain competitive
advantages. British Prime Minister John Major, for one, claimed a victory:
'Europe can have the social chapter. We shall have employment . . . Let
Jacques Delors accuse us of creating a paradise for foreign investors; I am
happy to plead guilty' (Agence Europe, 3 Mar. 1993: 13–14). Yet so far UK-
based Euro- or multinational firms seem to be implementing the one measure so far adopted under the Protocol, in order to preserve intra-firm
uniformity and also to ward off possible consumer conflicts.

In principle, the Social Protocol should facilitate efforts to expand EU
social policy. Britain's capacity to obstruct legislation has been diminished,
and the four 'poor' states (Greece, Ireland, Portugal, and Spain) do not command enough votes to block reform under the Social Protocol's qualified
majority vote rules. In 1995 a long-delayed European Works Council directive was approved under these procedures (cf. Rhodes 1995). Yet while some
aspects of the Social Charter may move forward, further initiatives seem

likely to be modest in the near future. The Commission is itself involved in intensive soul-searching concerning its proper social policy role (Ross 1994: 221–6). Efforts to combat stubbornly high European unemployment have taken centre stage, and the Commission seems to have accepted at least some of the British case about the need to promote 'flexibility'. The White Paper on 'Growth, Competitiveness, Employment' revealed a change in emphasis towards reducing labour costs, calling for tax reforms that would generate 'a substantial reduction of non-wage labour costs (between 1 and 2 percentage points of GDP), particularly for the least-skilled workers' (Commission 1993d: 116 ff.). The member states seem unlikely to allow the Commission to take the lead on such issues, suggesting that the immediate prospect is for consolidation, with the completion of some current agenda items but few new initiatives.

The dismissal of claims that a significant EU role now exists in social policy are largely based on examination of 'high politics'—the widely publicized struggles over positive, centre-imposed social policies through devices like the Social Charter and Social Protocol. Developments such as the Maternity Directive suggest that there may now be room for some European initiatives. EU legislative activity is probably as extensive as, for example, federal social policy activity in the United States on the eve of the New Deal (Robertson 1989). Yet if member states have lost considerable control over social policy in the EU this is primarily because of processes other than the efforts of Union officials to develop social policy legislation.

European integration and market compatibility requirements

Lost amidst the noisy fights over Social Charters and Social Protocols has been the quiet accumulation of EC constraints on social policy connected with market integration. The past three decades, and especially the most recent one, have witnessed a gradual if incremental expansion of Community-produced regulations and, especially, court decisions that have seriously eroded national welfare-state sovereignties. Political scientists have paid scant attention to this area of 'low politics', entranced by the world of 'high politics' and 'high conflicts'. The topic has been left to a small set of European welfare lawyers who have monitored another centre of policy-making: the courts (cf. Weiler 1991; Burley and Mattli 1993; on the general myopia about courts: Shapiro and Stone 1994).

The ECJ has delivered (see Table 7.2) more than three hundred decisions

Table 7.2. ECJ social policy cases

Years	Social security (coordination)	Other social policy (Arts. 117–125, EEC)
1959–70	28	1
1971–80	108	8
1981–90	113	69
1991–4	74	63
Total	323	141

Source: Information from Jochen Streil, Legal Data-Processing Service, ECJ, 14 Mar. 1995.

on social security coordination and more than one hundred decisions on other social policy matters within the scope of Articles 117–25 (EEC)— enough to incite pleas for a specialized EC welfare court for which the ECJ would function as a court of appeals. The ECJ's overall case-load has been growing rapidly, from thirty-four cases filed in 1968 to 280 in 1980 and 553 in 1992. Social policy cases account for a growing share within this rising total. A comparison with core common-market topics like customs union and free movement of goods, competition (including taxation), and agriculture is instructive. While social cases accounted for only 6.3 per cent of the total in these four categories in 1968, that share had increased to 22.8 per cent in 1992. By 1992 only competition cases arose with greater frequency, and the rate of growth for social policy cases was far greater (figures compiled by Caporaso and Keeler 1993: Table 1, after 24).

The four hundred plus social policy cases have a distinct pattern of origin. Even after the series of enlargements, cases have emerged mostly from the original EC6 countries and from the UK. Remarkably, Belgium and the Netherlands have produced the most cases. Since plaintiffs usually may not appeal to the ECJ directly, the national legal profession plays a critical 'intermediary' role. Indeed, the activist stance of the legal profession as a whole seems crucial. In Germany, for instance, lower welfare and labour courts often try to outmanoeuvre their national courts, employing the ECJ to overturn firm national precedents 'from above'.

The EC's social dimension is usually discussed as a corrective or counter to market-building, but it has instead proceeded largely as part of the market-building process. It is this market-building process that primarily spurred the demand for court decisions. The nexus between the market and social policy was at least partially acknowledged at the outset, where social policy in the Community was addressed largely in relation to the problem of reducing restrictions on labour mobility. Articles 48–51 (EEC) deal with the freedom of movement (for workers), with Article 51 providing: 'The Council shall, acting unanimously on a proposal from the Commission, adopt such measures in the field of social security as are necessary to provide freedom of movement for workers . . .'.

Freedom of movement for workers

That a labour mobility regime of 'coordination' would restrict welfare-state sovereignty was on the minds of Treaty-makers in 1957 (Romero 1993).[4] At the time, however, such impacts were neither very visible nor contentious. An already entrenched *intergovernmental* consensus existed on which the Treaty could build: bi- and multilateral social security treaties, drafts of a European Coal and Steel Commmunity Social Security Treaty for miners and steelworkers, and standards of the International Labour Organization (Schulte 1994*b*: 408, 422). These embedded *inter*national legal norms facilitated fast and silent *supra*nationalization. The new regulations, along with the obligations they created for member states, gradually became more deeply institutionalized—mostly in the quiet of the Court's chambers. It was not until the end of the 1980s that member states began to wake up to the full import of 'coordination' and to struggle with it.

In line with the Community's agenda of market integration, the EC also had from the outset competence to regulate and ensure the freedom of movement of services under Articles 59–66 (EEC). At first sight this competence does not seem to allow much room for social policy. In contrast to coordination, the Treaty's signatories saw no real connection between the freedom of services and their sovereign welfare-state-building. But developments in recent times have shown that this constitutional principle and its implementation have potentially far-reaching consequences for national social policy regimes, guaranteeing both the freedom of movement to consumers 'of social policy' to shop where they want, and the right of service providers to deliver their services 'across the border' into another welfare state. In the future this spillover may become a major terrain for European conflicts over social policy reform. In the rest of this section, we briefly indicate how creating a free market for labour and services has directly intruded on the sovereignty of national welfare states.

One of the crucial points of tension between national welfare states and the developing common market has concerned regulations governing the mobility of labour across the jurisdictional boundaries of member states. Intra-European migration—compared with the US—is small; there are only five million workers, including their dependents, who actually exercise this freedom.[5] But these numbers have surpassed the 'critical mass' necessary to generate continuously increasing litigation at the ECJ level. In legal terms, the adaptation of social policy to a developed context of 'interstate commerce' does not require a quantum leap in European migration itself. Individuals as litigants and national courts who refer cases to the ECJ are, together with the ECJ itself, the central actors in shaping this multi-tiered EC policy domain. They have instigated a large corpus of national and, especially, supranational adjudication since 1958.

A detailed review of this case-law cannot be attempted here (Leibfried and Pierson 1995a). Over a period of thirty years a complex patchwork of regulations and court decisions has partially suspended the principle of member-state sovereignty over social policy in the interest of European labour market (and now consumer) mobility, limiting national capacities to contain transfers 'by territory' (Maydell 1991: 231; Eichenhofer 1992). To take just one example, attempts to create a minimum pension benefit in Germany during the 1980s foundered in part because of concerns that the benefit would be 'exportable' to non-German EC citizens who had worked for some time in Germany (Zuleeg 1993).

To summarize the key implications of European law:

- A member state may no longer limit most social benefits to its citizens. Regarding 'foreigners' from within the EU, the state no longer has any power to determine whether foreign people have a right to benefits or not. Benefits must be granted to all or withheld from all. This development is remarkable since 'citizen-making' through social benefits—demarcating the 'outsider'—was a watershed in the history of state-building on the Continent, especially in France and Germany.

- A member state may no longer insist that its benefits only apply to its territory and thus are only consumed there. As a result, today's state can exercise its power to determine the territory of consumption only to a limited extent—basically when providing in-kind or universal means-tested benefits.

- A member state is no longer entirely (though still largely) free to prevent other social policy regimes from directly competing with the regime it has built on its own territory. In Germany, for instance, there are many 'posted' construction workers from other EC countries, who work for extended periods while covered by many of their home country's social regulations. Thus, the state has lost its exclusive power to determine how the people living within its borders are protected.

- Member states do not have an exclusive right to administer claims to welfare benefits. Rather, the authorities of other nations may also have a decisive say in adjudicating benefit status in individual cases.

If complete *de jure* authority in these respects is what sovereignty in social policy is all about, it has already ceased to exist in the EC. This has been a complex process, in which supranational efforts to broaden access and national efforts to maintain control go hand in hand, and are calibrated from conflict to conflict, and court-case by court-case.

This transformation has not occurred without member-state resistance. Individually, member states have baulked at implementing particular facets of coordination, although they have been effectively taken to task for this by the ECJ. Collectively, the member states have recently sought to roll back some aspects of coordination, unanimously agreeing to revisions that will allow member states to restrict portability in a somewhat broader range of cases following proper 'notification'.[6] The impact of this shift remains unclear, though it may partially offset some loss of sovereignty. The notification provision must still pass muster with the ECJ.

Coordination, however, has become the entering wedge for an incremental, rights-based 'homogenization' of social policy. Neither 'supranationalization' nor 'harmonization' seems an appropriate label for this dynamic, since each implies more policy control at the centre than currently exists. The process is more like a market-place of 'coordination', with the ECJ acting as market police, enforcing the boundaries of national autonomy. It structures the interfaces of twelve national social policy systems, with potentially far-reaching consequences for the range of policy options available to national welfare states.

Free movement of services

The 'free movement of services' doctrine also directly affects national welfare states, though the scope of this influence remains relatively opaque. While labour mobility issues have been worked out in hundreds of ECJ decisions spanning almost four decades, the influence of the 'free movement of services' really surfaced only with the passage of the SEA of 1986. So far, it has generated only a few leading cases and comparatively little secondary Community law. Nevertheless, judging from the pervasive influence the single market now has and potentially will have on the now common private insurance market, there appear to be significant prospects for a remoulding of national policies in the social services, especially in the area of health care (cf. Schulte 1991: 250; for a contrasting view: Altenstetter 1992).

In principle, at least so it used to be thought, each state may choose its own policies for social services. However, the 'freedom of services' doctrine may have considerable effects on national service delivery systems. Here, the divergent characteristics of member-state policy structures become crucial

(Bieback 1991: 929). For example, some member states (e.g. Britain, Italy) have national health-care systems (with marginalized markets or non-market systems). Others (e.g. Germany, France, and the Netherlands) have insurance systems, where the state only supplies funds for goods and services which are bought from private providers. As Bieback (1991: 932) observes, 'According to the ECJ the free movement of services clause in Article 59 of the Treaty of Rome applies only to services dealt with in markets and supplied for money, but not for services which are usually delivered as part of a "national service". Thus countries which organize their health-service systems on the basis of private markets where public and private suppliers may compete and access is free, like the systems of social insurance in France and Germany, are open for competition from suppliers from other countries, whereas the "National Health" systems are closed, except if they incorporate competitive structures like parts of the British system.'

Whereas access of foreign providers in 'non-market systems' is a non-issue, in 'quasi-market systems' access of foreign private deliverers is buttressed by EC law.[7] This may create a deregulatory dynamic, especially when regimes rely on closed national producer markets for social services.[8]

The emerging free market of services can be seen in most dramatic form in the rapidly evolving private insurance market. As of 1 July 1994, national private insurance has been drawn into the common market of the EC. The furious pace of cross-border mergers and acquisitions is creating a heavily concentrated insurance sector operating at the European level. Integrated European insurance markets allow for a greater differentiation of policy-holders by risk groups (cf. Stone 1989), and thus for cheaper, more profitable policies with lower operational costs. Such an integrated private sector would confront now fifteen national, internally segmented, public insurance domains. Insurance providers with the option of relocating to more lenient member states will have increased influence over national social regulations. At the same time, the clash between particular national regulatory styles and the quite different traditions of competing insurers from other member states is likely to be intense.

The results of 'public/private' interplay in the context of a radically altered private sector are difficult to anticipate. There is, however, considerable evidence from studies of national welfare states that the reform of private-sector markets can have dramatic effects on public-service provision (see e.g. Rein and Rainwater 1986). Public and private insurance compete mainly in areas like occupational pensions, life insurance, and supplemental health insurance. Permanent turf quarrels between public and private seem likely concerning where 'basic' (public) coverage should end and 'additional' (private) insurance may begin. This is part of a broader process in which movement towards the single market challenges existing demarcations between the public and private spheres. The welfare state, which has traditionally been a key area for establishing these demarcations, is bound to be affected by the gradual and often indirect redrawing of boundaries.

The balance between a free market (of services) and institutionally autonomous national welfare states, two principles embedded in the EU constitution, thus is not a static but a dynamic one. Here is open terrain for Brussels, with a large potential for restructuring national services delivery regimes. This problem is likely to be particularly severe for Sweden and Finland, the two Nordic countries which entered the Union in 1995, since they have systematically pursued a policy of marginalizing competitive pressures in social-service provision. As Kärel Hagen (1992: 289) argues,

Political ambitions of providing high- and equal-quality health care to all segments of the population, have required the extensive use of public monopolies that may militate against enterprise freedoms guaranteed by Community legislation. The same applies to state restrictions on private pension insurance and on how their funds are to be managed. In general, any kind of state welfare policy which is deliberately designed to prevent private purchasing power from being reproduced in the consumption of welfare goods supplied by the market, will run counter to the freedoms of the common market.

Since the two principles of a free market in services and national autonomy over social policy contradict each other, it is up to the EU to fix, again and again, an ever-shifting demarcation line. While labour market coordination has been the main social policy item for the ECJ in the past three decades, issues concerning freedom to supply social services (Article 59, EEC), freedom to consume social services, and freedom of settlement are all coming to the fore. As Bieback (1991: 932) notes,

As long as the Community opens the free market for social services there are only three options. Either the competence of the EC is extended to control and regulate the market for social services, or the Member States coordinate and harmonise their systems voluntarily, or finally all national systems of social services opt out of market systems into 'national' systems. Evidently, all options tend to increase the pressure towards harmonisation.

We can add, then, two further general points to our list of restrictions on member-state sovereignty and autonomy:

- Member states may no longer mix market and state components in providing welfare at will, and impose a preferred 'welfare mix' of monetary transfers, in-kind payments, and services. The power to determine the make-up of the welfare state is being reduced.

- Member states can no longer exclusively decide who may provide social services or benefits. They no longer exclusively organize social-service occupations, since mutual recognition of degrees and licences from other member states intervene. And they have more limited capacity to protect their national service organizations from the competitive efforts of service organizations in other member states.

To summarize, even by looking only at issues of labour mobility and freedom of services, one can see a wide range of market compatibility requirements, through which either EC regulations or ECJ decisions impinge on the design and reform of national social policy. Examples related to the single market could easily be multiplied—for example, restrictions related to firm subsidies in regional policy. In Italy, for instance, the central government has been using abatements of social-insurance taxes as a strategy to attract investment to the Mezzogiorno. While the Commission agreed to permit this until the end of 1993, it then initiated ECJ proceedings against the continuation of the practice on grounds of 'unfair competition' (*Euroreport* 5, 1994). Similarly, changes in Germany's social-insurance system for farmers require Brussels's approval since such insurance is considered a sectoral subsidy. The broader point is clear: a whole range of social policy designs that would be available to sovereign welfare states are prohibited to member states within the EU's multi-tiered polity.

European integration and *de facto* pressures on national welfare states

We have argued that the EU now intervenes directly in the social policies of member states in two ways: by enacting significant social policy initiatives of its own, and by striking down features of national systems that are deemed incompatible with the development of the single market. In addition, the process of European integration also has less direct, but none the less significant effects on member-state social policies, as both the economic policies of the EU and the responses of social actors to those policies put pressures on national welfare states. Because these effects are indirect they are difficult to measure, but they none the less add to the general picture of increasing supranational influence over the design of national social policy.

The most frequently cited source of pressure on welfare states within the EU is the possibility that heightened integration may lead to 'social dumping'. The term refers to the prospect that firms operating where 'social wages' are low may be able to undercut the prices of competitors, forcing higher-cost firms either to go out of business or relocate to low social wage areas, or pressure their governments to reduce social wage costs. In extreme scenarios, these actions could fuel a downward spiral in social provision, eventually producing very rudimentary, 'lowest common denominator' national welfare states. Supporters of social policy in the EU countries with well-developed welfare states—for example, labour confederations like the German DGB—have particularly stressed this concern.

There is some evidence that these kinds of pressures have restricted social expenditures in the United States, where labour (and capital) mobility is far greater than is currently the case in the EU (P. E. Peterson and Rom 1990; for a comparative background see Boltho 1989). Despite widespread attention to this issue, however, the evidence that European integration will fuel a process of social dumping remains limited. As a number of observers have noted, the 'social wage' is only one factor in investment decisions, and firms will not invest in low social wage countries unless worker productivity (relative to wages) justifies such investments. Neo-classical trade theory suggests that high social wage countries should be able to continue their policies as long as overall conditions allow profitable investment. One sign of the ambiguous consequences of integration is the fact that northern Europe's concerns about 'sunbelt effects' are mirrored by southern Europe's concerns about 'agglomeration effects' in which investment would flow towards the superior infrastructures and high-skilled work-forces of Europe's most developed regions.

Social dumping may generate greater fears than current evidence warrants; the opposite could be the case for some of the other ways that economic integration creates pressures on national social policy systems. The single market is encouraging a gradual movement towards a narrowed band of value-added tax rates. In theory, governments finding that their VAT revenues have been lowered will be free to increase other taxes, but this may be no simple task. Because it is politically easier to sustain indirect taxes, the new rules may create growing constraints on member-state budgets, with clear implications for national social policies (Wilensky 1976; Hibbs and Madsen 1981). This is likely to be a particular problem for Denmark, which relies heavily on indirect taxes rather than payroll taxes to finance its generous welfare state (J. H. Petersen 1993; 1991: 514–22; Schulte 1994*b*: 409).[9]

The move towards monetary union, with its tough requirements for budgetary discipline, may also encourage downward adjustments in welfare provision. For example, to participate in the final stage of monetary union, Italy would have to reduce its budget deficit from 10 per cent to 3 per cent of GDP by the end of the decade. This served to legitimate the Berlusconi and the Dini governments' intense pursuit of major cuts in old-age pensions and other social benefits in 1994 and 1995. While other countries face less radical adjustments, the convergence criteria present formidable problems for almost all of them.[10] Here again, the significance of the EU's indirect effects is hard to ascertain. Monetary union remains an uncertain proposition at best, and governments would have faced pressure for austerity in any event. The convergence criteria do not, of course *require* budget reductions—tax increases would also be possible—but they considerably strengthen the hand of those seeking such cuts.

Monetary union would not only put pressures on national social programmes; it could prod the Community into a more active role in efforts to combat unemployment. Analysis of the prospects for monetary integration in Europe has frequently been coupled with discussion of the need for accompanying social policies to address the likely emergence of regional imbalances.[11] Monetary union would strip national governments of significant macro-economic policy levers, and a Community-wide macro-economic stance will create significant regional unemployment problems. Somewhat flexible exchange rates allow local adaptations to local economic conditions. Once these instruments are dismantled, combating pockets of regional unemployment at the national level will be more difficult (Eichengreen 1992: 32–7).

It is difficult to evaluate the consequences for national welfare states of these various indirect pressures. Many of the possible problem areas lie in the future, and some of the others, such as social dumping, are difficult to measure even if they might be occurring now. One has to weigh the pressures for reform against the welfare state's considerable sources of resilience (Pierson 1996). Yet the picture that emerges is one where national governments possess diminished control over many of the policies that have traditionally supported national welfare states—macro-economic policies, tax policies, perhaps industrial-relations systems. Again, these developments challenge the dominant view that European integration is a 'market-building' process that advances relentlessly while leaving the development of social policy a purely national affair.

Social policy in Europe's emerging multi-tiered system

Scholarly attention has focused largely on Commission efforts to establish a 'social dimension' of Community-wide policies or at least minimum standards. While far from trivial, these efforts have modified member-state social policies in relatively few areas. More important, though much less visible, have been the social policy effects of the single market's development. These have occurred either directly, as the Commission, national courts, and the ECJ have sought to reconcile member-state policy autonomy with the effort to create a unified economic space, or indirectly through pressures on the support structures of national welfare states.

We are living through a period of rapid change in the relations between nation states and an increasingly global market system. '[T]he central question to pose is: has sovereignty remained intact while the autonomy of the state has diminished or has the modern state faced a loss of sovereignty?' (Held 1991a: 213). In the EU, both member-state sovereignty and autonomy have diminished in tandem (Leibfried 1994). The process has been subtle and incremental, but developments within the Community as a whole increasingly constrain national welfare states. Member states now find their revenue bases under assault, their welfare-reform options circumscribed, many of their service regimes under threat of new competition, and their administrators obliged to share control over policy enforcement.

What is emerging in Europe is a unique multi-tiered system of social policy, with three distinctive characteristics: a propensity towards 'joint-decision traps' and policy immobilism; a prominent role for courts in policy development; and an unusually tight coupling to market-making processes. First, policy-makers at the European level are tightly hemmed in by the scepticism of the Council, the density of existing national-level social policy commitments, and the limited fiscal and administrative capacities of the EU. Compared with any other 'multi-tiered' system, the EU social policy-making apparatus is extremely bottom-heavy (Kleinman and Piachaud 1992b; Pierson and Leibfried 1995a). The quite weak 'centre' has limited capacity to formulate positive social policy. As a result, social policy evolution is likely to be more the result of mutual adjustment and incremental accommodation than of central guidance. From the centre come a variety of pressures and constraints on social policy development, but much less by way of clear mandates for positive action.

Yet there has been a considerable weakening of the member states' position as well (Pierson 1995). With the gravitation of authority, even of a largely negative kind, to the European level, the capacity of member states to design their welfare states as they choose is also diminishing. The role of the member states at this juncture is multi-faceted if not contradictory. Significant losses of autonomy and sovereignty occurred without member states paying a great deal of attention. In some cases—such as Italy's role in pushing for enhanced labour mobility—member states actively pursued sovereignty-eroding initiatives. While member states currently resist some of the single market's implications for their own power, their capacity to do so is limited by their fear of jeopardizing the hard-won benefits of European integration. Member-state resistance is further limited by an *institutional ratchet* effect. Once you become an EU member you are bound by ECJ decisions, and can pursue reforms only through the slow and difficult procedures available under Community rules. The combination of diminished member-state authority and continued weakness at the EU level is likely to restrict the room for innovative policy. As Fritz Scharpf observes, 'the policy making capacities of the Union have not been strengthened nearly as much as capabilities at the level of member states have declined' (1994a: 219; and also 1994c).

Member states still 'choose', but they do so from an increasingly restricted menu. At a time when control over social policy often means responsibility for announcing unpopular cut-backs, member-state governments sometimes are happy to accept arrangements that constrain their own options. Given the unpopularity of retrenchment, governments may find that the growing ability to blame the EU allows changes which they would otherwise be afraid to contemplate. The movement towards a multi-tiered political system opens up major new avenues for the politics of 'blame avoidance' (Weaver 1986). It has been suggested that this dynamic strengthens national executives at the expense of domestic opponents, and this may be the case (Milward 1992; Moravcsik 1994). Yet in the process of escaping from domestic constraints, national executives have created new ones that profoundly limit their options. Decision-making bodies at *both* the national and supranational level face serious restrictions on their capacity for social policy intervention, since they have partly 'locked themselves in' through previous steps towards integration.

The second distinctive characteristic of social policy-making in the EU is that the constraints and requirements which do develop from the centre are unusually court-driven. It is as much a series of rulings from the ECJ as the process of Commission and Council initiative that has been the source of new social policy. While the Council and Commission are prone to stasis, the ECJ's institutional design fosters activism. Once confronted with litigation, the ECJ cannot escape making what are essentially policy decisions as a matter of routine. The Court also relies on simple majority votes, taken in secret, sheltering it from the political immobility typical of the EU. In most cases, only a unanimous vote of the Council can undo ECJ decisions. The structure of EU institutions puts the ECJ on centre stage. Attempts at corporatist policy-making have generated much of the drama surrounding Europe's social dimension, but businesses and unions have had little direct involvement in the decisions that have actually created legally binding requirements for the social policies of member states.[12]

Legal strategies have had the advantage of leaving taxing, spending, and administrative powers at the national level. It should be emphasized, however, that such a court-led process of social policy development is likely to have its own logic. Decisions are likely to reflect demands for doctrinal coherence as much or more than substantive debates about the desirability of various social policy outcomes. The capacity of reforms built around a judicial logic to achieve substantive goals may be limited. Furthermore, courts may have less need to consider political constraints in prescribing solutions. A possible danger is that court initiatives may exceed the tolerance of important political actors within the system. After all, centralized policy-making was made difficult in the EU for a reason, and ECJ activism may generate resentment. This is, of course, one aspect of the current disquiet over the 'democratic deficit'.

Finally, Europe's emerging multi-tiered system of social policy is uniquely connected to a process of market-building. Of course, social policies in mixed economies always intersect in a variety of complex ways with market systems. In the past, however, social policy has generally been seen as part of what Karl Polanyi described as a 'protective reaction' against the expansion of market relations (Polanyi 1994). Social policies have grown up in response to the shortcomings of market arrangements.

At the European level, however, interventions in the traditional spheres of social policy have generally not taken this Polanyian form. Indeed, initiatives like the Social Charter have usually been dismal failures. Even in areas such as gender issues where the EU has been activist, policies have been directly connected to labour-market participation, while broad issues of family policy have been ignored. Instead, as the centrality of decisions regarding labour mobility and free service markets reveals, EU social policy interventions have grown up as part of the process of market-building itself. Never before has the construction of markets so visibly and intensively shaped the development of social policy initiatives. The overall scope of EU interventions has been, we emphasize, extensive. These interventions reveal that national welfare-state regimes are now part of a larger, multi-tiered system of social policy. Member states profoundly influence this structure, but they no longer fully control it. While the governance of social policy occurs at multiple levels, however, the EU's peculiar arrangement is also different in many respects from traditional federal welfare states, distinguished by a weak policy-making centre, court-driven regulation, and strong links with market-making processes. The EU's unique political arrangement is producing a pattern of policy-making quite different from that of any national welfare state (cf. Streeck 1995).

Notes

1. This essay is a modified and condensed version of Leibfried and Pierson 1995*b*. We are indebted to Michelle Everson for her intensive advice on the role of the EC in regulating private insurance.

2. The retroactivity of *Barber* was thought sufficiently important (and costly) to warrant the attention of the Maastricht Treaty writers. In Maastricht a somewhat less costly, but not the least costly, version of retroactivity was chosen. It took a lot, a unanimous Treaty change, to recoup the initiative from the ECJ, and the ECJ upheld the solution found in Maastricht in *Ten Oever* (Case 109/91 of 6 Oct. 1993). Estimated costs of full retroactivity for Germany are from Berenz (1994: 437); for Britain, from Mazey and Richardson (1993*a*: 15).

3. Both Labour and the Liberal Democrats made an issue of Britain's status in the last European Parliament elections, asking why Portuguese and Greek workers should have access to higher standards of social protection than British workers. Public opinion in Britain has been consistently much more favourable towards the Social Protocol than that of the Conservative government.

4. Romero (1993) traces the history of the 'mobility of labour' clause in the Rome Treaty and the specific contributions of Italian politics to it, with Italy being the permanent labour exporter of that epoch. The example shows that under particular circumstances member states may push hard for expanded European competence. Without Italy the European Union might be missing the freedom of labour mobility altogether. To take another example, without France the gender aspects of the social dimension, based on Article 119, would be absent.

5. While immigration is quantitatively more significant than intra-European mobility, the latter is far more relevant to the expansion of EU competence. Immigration is still primarily a member-state prerogative, coming under Maastricht's intergovernmental pillar and not under its supranational headings (cf. Ireland 1995).

6. 'Notification' must be approved by unanimous vote of the member states, although so far there seems to be a 'gentlemen's agreement' to allow such self-exemptions. To date, these have been exercised largely by the 'Latin Rim' countries and by Britain (cf. Schulte 1994*a*).

7. Some producers are more likely to take this route than others, especially private international service organizations involved in hospitals, drug markets, and the provision of medical equipment (Bieback 1993: 171). Such producers are likely to become strong actors at the EU level, be it *vis-à-vis* the Commission or in the Courts.

8. The process seems to have developed furthest with national drug markets. Directive 89/105 has already set minimum standards for all national systems of price control and price-fixing. Since equivalent pharmaceuticals are much cheaper in some other European countries, German sickness funds are quite interested in 'importing' such drugs. In addition, the Commission has developed

proposals for a single drug market, which would strongly, though indirectly, harmonize national-health insurance systems, for example by undoing price-controls and introducing significant co-payments. In 1994 the European Agency for the Evaluation of Medicinal Products started surveillance and licensing work under European law (*Euroreport* 2 and 5, 1994).

9. On the other hand, some see Denmark's low payroll taxes as a model for future European competitiveness (cf. Scharpf 1994*b*: 29 ff.).

10. This argument is explored by Hans-Jürgen Krupp (1995: 9–10), who points towards EMU functioning as a cap on further welfare-state expansion at the national level.

11. As Ross notes (1994: 152–3), the Spanish stressed this implication of monetary union during the run-up to Maastricht, and made it clear that they would not sign a new Treaty unless a major expansion of regional redistribution was included.

12. This is not to say that the activities of economic actors are irrelevant to the development of Community social policy. The influence of business has, for example, been considerable in restricting Commission efforts to pursue a more activist social dimension, and in advancing the deregulatory agenda that has set the framework for ECJ decisions.

Further Reading

For the main features of the subject see Leibfried and Pierson (1995*b*); the first and last chapters provide a guide to theoretical explanations. For a 'social policy' view see Kleinman and Piachaud (1992*a*), and for the overall legal framework see Bieback (1991). For the history of labour mobility consult Romero (1993). For contrasting interpretations of the topic see Kenis (1991) and Swaan (1992). Regular updates may be found in the *Journal of European Social Policy*.

Bieback, K.-J. (1991), 'Harmonization of Social Policy in the European Community', *Les Cahiers de Droit*, 32/4: 913–35.

Kenis, P. (1991), 'Social Europe in the 1990s: Beyond an Adjunct to Achieving a Common Market?', *Futures*, 23/7: 724–38.

Kleinman, M., and Piachaud, D. (1992*a*), 'Britain and European Social Policy', *Policy Studies*, 13/3: 13–25.

Leibfried, S., and Pierson, P. (1995*b*) (eds.), *European Social Policy: Between Fragmentation and Integration* (Washington, DC: Brookings Institution).

Romero, F. (1993), 'Migration as an Issue in European Interdependence and Integration: The Case of Italy', in A. S. Milward, F. M. B. Lynch, F. Romero, and V. Sørensen (eds.), *The Frontier of National Sovereignty: History and Theory, 1945–1992* (London: Routledge, 1993), 33–58 and 205–8.

Swaan, A. de (1992), 'Perspectives for Transnational Social Policy', *Government and Opposition*, 27/1: 33–52.

CHAPTER EIGHT

COHESION AND STRUCTURAL ADJUSTMENT

David Allen

Structural fund spending, less than 5 per cent of the 1975 EC budget, is projected at 35 per cent of the 1999 budget, under the SEA commitment to economic and social cohesion. The debate is over whether this increase has promoted a new pattern of 'multi-level governance', or reflects side-payments to facilitate package deals. High-level bargaining among governments has characterized decision-making on the size and objectives of the structural funds, with the Commission playing a subordinate role. Since the 1988 reforms the Commission has attempted to exploit implementation of these objectives to build links with sub-national bodies. National governments' concerns for fair shares clash with the Commission's promotion of Community-wide programmes. The prospect of eastern enlargement will shape structural-fund development after 1999, pitting the four main recipient states against the demands of applicants and the payer countries. 'High politics' will probably continue to predominate.

Introduction

In this chapter I examine the relationship between the structural funds and the European Union (EU) objective of 'economic and social cohesion'. In so doing I trace the evolution of the structural funds from their separate and relatively insignificant origins to the anticipated situation in 1999, when they will constitute around 35 per cent of the overall Union budget, accounting for the expenditure of some 30 billion ecus (at 1992 prices). For some time now there has been a direct correlation between the decline in agricultural expenditure, through the European Agricultural Guidance and Guarantee Fund (EAGGF) Guarantee Section, and the rise of structural-fund expenditure. This shift in emphasis reflects the growing priority that is given, not just to the reform of the common agricultural policy (CAP) for its own sake, but also to the objective of 'economic and social cohesion'. In this context the CAP is not only wasteful, it also reinforces regional disparities by channelling funds to the wealthier parts of the Union.

Regional disparities,[1] both within the Union as a whole and within the member states, have always been regarded as barriers to what was initially referred to in the Treaty of Rome as 'harmonious development' but which, since the Single European Act (SEA), has become known as 'economic and social cohesion'. Whilst those who agreed the treaties have always seen the co-ordination of national economic policies and the development of Community policies as two ways of achieving this objective, successive enlargements of the Community have also focused attention on the relationship between the structural funds and cohesion.

It is, however, one of the central contentions of this chapter that the search for economic and social cohesion serves mainly as a rationalization, albeit a credible one, rather than as an explanation for the rapidly expanding system of structural support. It is argued that the decision to establish the various funds, and after 1988 to increase significantly the resources at their disposal, can best be understood in terms of a series of side-payments designed to facilitate general packages related to the advancement of the Community, now the Union, as a whole. The basic assumption is, then, that the essential framework for structural fund activity has been determined by a process of high-level interstate bargaining, assisted by the European Commission, but primarily resolved by the central governments of the member states.

Despite this rather firm restatement of a 'state-centric' explanation for policy development, I shall also examine the contention that the way in which the structural funds are managed has led, either deliberately or inadvertently, to a situation in which the autonomy of the central governments of the member states is challenged, both by the development of the Commission's supra-national authority and by the growing role of sub-national actors at the regional and local level. Many enthusiasts for further integration, particularly those who advocate a 'Europe of the regions', have taken heart from what they see as exciting new roles for supranational actors—most obviously the European Commission and, to a lesser extent, the Committee of the Regions—involved in the development of structural-adjustment policies. Taking the concept of subsidiarity to its logical conclusion, they are also inter-ested in the apparently important role that management of the structural funds seems to give to both regional and local government, as well as to pri-vate-sector participants.

It is obviously important to examine the stated aspirations of the policy-implementation processes that have been established as a result of the 1988 and 1992 reforms of the structural funds. In this chapter I shall also attempt a critical evaluation of their real significance to see if the 'state-centric' expla-nation of policy-making in this area is seriously challenged. In evaluating cohesion policy and the role of the structural funds, I shall attempt therefore to draw some conclusions about the impact of this area of policy on the Union's institutional arrangements and about the philosophy of develop-ment on which the current policy would appear to be based.

Finally, two questions concerning the future will be briefly touched upon. Within the Union as it is presently constituted arrangements for structural intervention in pursuit of cohesion have been agreed up to 1999. However, the philosophy of economic integration embodied in the treaties would sug-gest that these structural interventions are designed only for the short-to-medium term and that in the long run the operations of the market itself can be relied upon to provide increased welfare for all, whilst at the same time eradicating significant regional disparities. There is already clearly a conflict between the competition objectives of the treaties, designed to enhance the working of the market, and the idea of using the structural funds to achieve cohesion by overcoming regional disparities, but in way that has to be seen as inhibiting the working of the market (Frazer 1994).

The second question about the future is more fundamental and relates to the inevitability of further enlargement to include the countries of central and eastern Europe. To date the growth of the structural funds has been directly related to previous Community enlargements, but few people believe that the current structural-funds arrangements for a Union of fifteen could be extended to encompass new member states such as Poland or Hungary, let alone Bulgaria and Romania. If the structural funds and the concept of cohesion are best understood as essential facilitators for broad Union agree-ments, then they may not be able to play such a role indefinitely. Given the fundamental importance to the Union of being able to conclude such broad agreements, and given the similar importance of developing a substantial relationship with the states of central and eastern Europe, the next stage in the evolution of the structural funds would appear to be critical for the well-being of the Union.

Cohesion and the structural funds: the basic framework

As noted, the preamble to the Treaty of Rome established that the member states of the EEC were 'anxious to strengthen the unity of their economies and to ensure their harmonious development by reducing the differences existing between the various regions and the backwardness of the less favoured regions'. The original idea was that this, and indeed most of the other objectives stated in the preamble, would be achieved by the 'establishment of a common market and by the progressive approximation of the economic policies of the member states' (Article 2, Treaty of Rome). Whilst most of the relevant activity involved the 'negative' process of removing barriers to the free operation of the market, provision was also made for more positive intervention. There was to be a common agricultural policy (Article 3(d), Treaty of Rome) which would require 'one or more agricultural guidance and guarantee funds' (Article 40(4), Treaty of Rome) to be set up, a European Social Fund (Article 3(i) and Articles 123–8, Title III, Chapter 2, Treaty of Rome), and a European Investment Bank (Article 3(j) and Articles 129–30, Title IV, Treaty of Rome). There was, however, no specific provision for a regional policy or fund, and so when, in the early 1970s, their establishment was deemed desirable, recourse had to be made to Article 235[2] and to high-level interstate bargaining at two European summits, both held in Paris in 1972 and 1974 (H. Wallace 1977). As Helen Wallace makes clear, there were a number of reasons for establishing a Regional Development Fund at this time, including the Commission's desire to combine an extension of competence with an attack on regional disparities. However, in moving towards her conclusion that 'the first allocations from the RDF proved little more than an exercise in pork barrel politics' and that it remained to be seen 'whether this will provide a viable basis for a coherent common policy', H. Wallace (1977) clearly establishes that the European Regional Development Fund (ERDF) mainly owes its existence to the fact that it played a major part in facilitating a broad deal between the EC member states. This deal involved both the enlargement of the EC and the beginning of moves towards economic and monetary union (EMU). As we continue to trace the evolution of the structural funds we shall note that, as in 1969–75, broad Community deals involving both enlargement and EMU are always significant explanatory factors.

The basic ERDF deal in 1975 established the Fund, agreed a figure (1.3 billion ecus) for 1975–8, and also established a set of national quotas to inform the division of the Fund. The Fund was too small to have any significant impact on regional disparities and thus cannot be seen either as the basis for a serious common regional policy or as a significant contribution towards the sort of convergence that would have aided the EMU process. Nevertheless, within the parameters discussed above, rules still had to be established for the day-to-day management of the funds. These were developed by the Commission and rationalized in terms of the logic of regional policy and the enhancement of cohesion.

Regardless of the way that the basic deal was put together, the Commission, and in particular Directorate General XVI, responsible for regional policy, hoped to exert its influence over the selection of projects submitted by national governments by ensuring that national governments did not use ERDF money simply to replace national expenditure. At first the Commission was constrained by the ability of member states to limit their project applications to the size of their quotas and by the reluctance of the member states to make their finances transparent so that 'additionality' could be ensured. Nevertheless the enabling regulations that established the ERDF and various institutions, including the Regional Policy Committee, provided the guidelines for future reforms based on the notions of partnership, additionality, and a concern to focus assistance on the most needy regions. Further stimulus for reform was provided by the system that evolved for the dispersal of the European Social Fund (ESF), which from the start sought to involve economic and social 'partners'.

However, despite their aspirations and despite minor reforms in 1979[3] and 1984,[4] both the Commission and the various regional authorities found themselves to be essentially marginalized in a policy process that was quickly to become an instrument of national policy-making (J. Scott 1995). The fact that the ERDF found its treaty base in Article 235 meant that the Commission's desire to make functional progress towards a more ambitious regional development policy was inhibited by the need for unanimity. The European Parliament too saw opportunities in the fact that ERDF expenditure came into the 'non-compulsory' area of the budget and therefore was a potential source of expanded influence, particularly if the fund as a whole was increased and if the Commission gained more flexibility over the manner of its dispersal.

Despite these minor advances and glimpses of the integrative potential of an expanded regional development policy, this was not to come about by neofunctional incremental progression. Instead it required the stimulus of major change and the need to construct another fundamental Community deal to bring about a significant reform of the structural funds. The stimulus was provided by the twin impact of further enlargement (to include Greece and later Spain and Portugal) and deepening in the guise of the SEA and its centre-piece, the single-market programme. Once Greece had been admitted to the Community in 1981, it was able to join France and Italy in demanding concessions from the rest of the Community in return for accepting the applications of Spain and Portugal—whose Mediterranean orientation, especially in agriculture, might be seen as a potentially costly challenge to the advantages that the other three 'southern' member states already enjoyed. The problem was solved by the introduction of the impressively named Integrated Mediterranean Programme (IMP), rationalized as a contribution to cohesion but best understood as a side-payment designed to ease further enlargement.[5]

Those who negotiated the SEA successfully implemented a strategy that was designed to facilitate a complex Community bargain by separating out the fundamental deal, as enshrined in the SEA, from the means of financing it. These were later incorporated in the 1988 Brussels agreement (Delors-1) on the 1988–93 'financial perspective'. This procedure was exactly and successfully reproduced at the time of Maastricht with the fundamentals of the Treaty on European Union (TEU) being agreed in December 1991, and the second, 1993–99, financial perspective (Delors-2)—the Maastricht 'tab', as it became known—being agreed one year later at the Edinburgh European Council (Shackleton 1993b; see also Ch. 4). This two-stage process, whereby distributive issues are separated out from, but clearly linked to, decision-making over fundamental Community deals, has been explored generally by Majone (1994a) and with specific reference to the development of the structural funds by McAleavey (1994).

The SEA amended the Treaty of Rome in a number of ways relevant to our enquiry. Although there was no reference to cohesion in the preamble nor any changes to the eleven objectives listed in Article 3 of the Treaty of Rome, the SEA did insert into the EEC Treaty a new Title V (Articles 130a–e). This (in Article 130a) for the first time linked a concern to 'strengthen economic and social cohesion' with the aim (lifted from the original Rome Treaty preamble) of 'reducing the disparities between the levels of development of the various regions and the backwardness of the least favoured regions, including the rural areas'.

Article 130a thus establishes that there is a connection between reducing regional disparities and strengthening cohesion; Article 130b then goes on to consider how this might be achieved. Under this Article action was to be taken through the three structural funds, while the European Investment Bank (EIB) was seen as only 'supportive'. Its loans could contribute to the workings of the internal market (what might be seen as the 'competitive' approach to removing regional differences), the formulation and implementation of the Community's policies (competition policy, EAGGF—guarantee, transport, etc.), and the conduct and coordination of member states' economic policies. In other words the structural funds alone were not designed to bring about cohesion, but to supplement the workings of the free market and the general economic policies of the member states—a factor that might be worth bearing in mind when considering the future of these funds, and clearly a factor that can be used by the member states to rationalize the fact that the overall size of the structural funds, even when they reach their 1999 target, is insufficient to make any significant impact on either regional or national disparities within the Union.

From the Commission's perspective there is better news in Article 130c, which provides a treaty base of its own for the ERDF, and Article 130d, which called on the Community to rationalize and reform the objectives and implementation of the structural funds. It was clear, therefore, that broadly agreed changes to the structural funds were fundamental to the package that facilitated the signing of the SEA, but that the exact details of how the funds might be reformed and how much money would be devoted to them were left to separate and subsequent negotiations. The problem is that there is very little empirical evidence to support the claim that a deal about the size of the future structural funds was done at the time of the signing of the SEA or again at Maastricht, although it seems obvious to observers of the subsequent behaviour of the member states that such a deal was struck in both cases. George Ross, in his study of Jacques Delors, argues that in the case of the SEA it was 'a matter of proposing a programme to make good on the social and economic cohesion clauses in the SEA which would combine some serious commitment to preventing the single market from penalising the EC's less developed members and "paying off" new members Spain and Portugal' (Ross 1995: 40).

The Delors-1 package provided for a doubling of the structural funds, as the three individual funds were now to be known, so that by 1993 they would account for 25 per cent of the Community budget. The details of the proposed reforms that were provided for in the SEA will be dealt with in the next section, but the point needs to be made that these reforms concerned the detail of implementation within the basic parameters that had been established by high-level intergovernmental bargaining.[6] Between 1988 and 1993 the enlarged structural funds and the reformed procedures began to function. This stimulated a great deal of optimism amongst those who advocated a joint and concerted attack, from above and below, by the European Commission and by the various regional and local authorities involved in the new structural-fund procedures, on the powers of the central governments of the member states. The term 'multi-level governance' (Marks 1993), when used in the context of the policy-making process associated with structural funds, and in particular when related to the notion of 'partnership' introduced by the 1988 reforms, was as much an aspiration as an accurate analytical description. In any case, it referred more to policy-implementation than to policy-making.

The major revolution in European affairs that occurred throughout 1989, which culminated in the admission of the five eastern Länder into the EC, as a result of the unification of Germany, inevitably stimulated pressures for both a deepening of the Community and further enlargement. This pressure led to the two Intergovernmental Conferences (IGCs) from which emerged agreement on the new TEU at Maastricht in December 1991. In the mean time it was agreed that the special needs of the five new Länder would be met by the additional allocation of 3 billion ecus for the 1991–3 period, after which a new package for the EC as a whole would have to be negotiated. The money for the eastern Länder had to be 'additional', so as not to upset 'the delicate balance in the Funds' allocation of resources and their breakdown by objectives and regions' (Commission 1990f).

The Maastricht agreements were built around a complex intergovernmental bargain, much of which was negotiated at the last moment; of the details we have only scant record (Ross 1995: 188–93). However, it was clear that further changes to the structural funds were central to the bargain that was struck. Article B of Title I (Common Provisions)[7] of the TEU lists as an objective of the Union the promotion of '. . . economic and social progress which is balanced and sustainable, in particular through the creation of an area without internal frontiers, *through the strengthening of economic and social cohesion* and through the establishment of economic and monetary union . . .' (my emphasis). Further changes relevant to our investigation concern amendments and additions to the Treaty of Rome (Title II of the TEU). Article 3 records a considerably expanded list of objectives, including 'the strengthening of economic and social cohesion (Article 3j) and the encouragement and development of Trans-European Networks (TENs) (Article 3n). Title XIV contains an amended version of Articles 130(a) to 130(e). Special note should be taken of Article 130(d); this provides the basis for a further reform of the tasks, objectives, organization, and implementation of the Funds and also for the establishment of a new Cohesion Fund to provide financial contributions in the field of environment and trans-European transport networks. The TEU also provided for the establishment of a new

Committee of the Regions and required that, like the Economic and Social Committee (and presumably with a similar impact!), it should be consulted on all aspects of ERDF allocation and implementation.

Thus, as with the SEA, the member states successfully employed a strategy of detaching agreement on the objectives of the Treaty from the exact details of how those objectives would be financed or implemented. It was clear, however, that these questions, in particular those involving finance, would have to be resolved by a similar process of intergovernmental bargaining before Maastricht could be ratified by all the member states. For some member states even the assurance that these details would be resolved in their favour in the future was not enough, and many of their worries were dealt with in a Protocol on Economic and Social Cohesion annexed to the TEU. This Protocol is important as it represents an interim agreement between the High Contracting Parties (member states in intergovernmental mode) that provides the basis for most of the changes that were agreed either at Edinburgh or in the enabling legislation for the new structural funds that came into force in September 1993.[8]

The Protocol restates all the cohesion objectives and implicitly links them to economic and monetary union. It goes on, in an interestingly directive fashion, to set the basic parameters for the subsequent structural-fund arrangements. Thus the Protocol records that the doubling of the structural funds between 1987 and 1993 *implies* 'large transfers, especially as a proportion of GDP of the less prosperous member states'.[9] It also notes that the EIB is 'lending large and increasing amounts for the benefit of the poorer regions' and that there is 'a desire for greater flexibility in the arrangements for allocations from the structural funds' (I take this to be a reference to the desire of the member states to free themselves of some of the bureaucratic controls introduced by the 1988 structural-fund reforms). Similarly, there is a reference to a 'desire for modulation of the levels of Community participation in programmes and projects in certain countries' and a proposal 'to take greater account of the relative prosperity of member states in the system of own resources'. Both these latter suggestions seem to refer to some sort of a deal whereby future arrangements for financing the EU in general, and for financing and organizing the structural funds in particular, will suit the interests of the poorer EU member states—for instance by directing compensatory funds towards them and by increasing their national (as opposed to European) impact by playing down the principle of additionality.

Thus, having given the forthcoming negotiations a considerable 'steer', the Protocol goes on to reinforce this by reaffirming the importance of cohesion and the role of the structural funds in its achievement (thus negating the argument of any member state that might try to suggest that cohesion could be achieved by other, less expensive, means). It further reaffirms a conviction that the 'EIB should continue to devote the majority of its resources' to the promotion of cohesion and declares a willingness to 'review the capital needs of the EIB as soon as this is necessary for that purpose' (more promises!). The need for a 'thorough evaluation of the operation and effectiveness of the structural funds in 1992' is established, along with the need to review 'the appropriate size of these funds in the light of the tasks of the Community in the area of economic and social cohesion'. The Protocol reveals more of the basic deal done at Maastricht in its reference to an agreement that the Cohesion Fund will go to member states with a per capita GNP of less than 90 per cent of the Community average and 'which have a programme leading to the fulfilment of the conditions of economic convergence as set out in Article 104c'. In other words, the Cohesion Fund is an additional side-payment to Greece, Spain, Portugal, and Ireland, but its rationale is to be the need for (national) convergence for EMU purposes rather than regional development. The basic principle of the Cohesion Fund therefore would seem to relate to the strengthening of central governments, rather than regions and regional actors within the member states. Conditions are clearly attached to the Cohesion Fund, but they are not conditions designed to enhance the power of either the European Commission or sub-national regional governments in the way that the conditions imposed by the 1988 reforms did.

This tendency to move away from Commission-based criteria for the allocation of the structural funds is not confined to the suggested arrangements for the Cohesion Fund. The Protocol is referring to the funds in their totality when it records the intention of the member states to allow 'a greater margin of flexibility in allocating financing from the structural funds to specific needs not covered under the present structural funds regulations'. Concerns about transparency and efficiency are reflected in a recognition of the need to 'monitor regularly the progress made towards achieving economic and social cohesion'. More significantly, member states' budgetary, as opposed to cohesion, concerns are reflected in their declaration that they intend to take greater account 'of the contributive capacity of the individual member states in the system of own resources' and that they propose to examine 'means of correcting, for the less prosperous member states, regressive elements existing in the present own resources system'.

Thus it is true that the TEU itself leaves much for further negotiation, and there is indeed scope for what Marks (1993) refers to as 'post treaty interpretation and institution-building'. However, it is argued here that, in the case of the structural funds, the treaty provisions, the Cohesion Protocol, and the Edinburgh budget agreements (themselves clearly the result of high-level intergovernmental bargaining) significantly set the parameters for the subsequent activities of both the Community institutions and sub-national actors. It may be strictly true to say that the Treaty alone is just a starting-point for negotiation amongst the interested parties and that it does indeed contain no overall spending commitments. None the less I have tried to show, following Majone (1994a), that the important subsequent negotiations in 1992 (in Edinburgh) and again in 1993, when the new structural-fund regulations were drawn up, represented the final working-out of a basic deal, agreed at Maastricht in 1991, but sequentially implemented for reasons of decision-making efficiency.

At the Edinburgh European Council in December 1992 the member states agreed to increase the size of the structural funds from 18.6 billion ecus in 1992 (the 1988 financial perspective having provided for a doubling of the funds between 1988 and 1993) to 30 billion ecus in 1999 (at 1992 prices) (Shackleton 1993b). This means that the share of the EC budget devoted to the structural funds has increased from 4.8 per cent in 1975 to 9.1 per cent in 1987 to 28 per cent in 1992 to a projected 35 per cent in 1999. At Edinburgh there was also agreement on how the structural funds should be allocated between the various objectives, but there was only vague agreement about the shares to be guaranteed to each member state. Nevertheless it is important to note that it was clearly accepted that such calculations were relevant and would condition the nature of the implementing policies adopted. The final agreement about how the spoils would be divided up between the member states was a condition of and thus came before the implementing regulations could be passed (Allen 1993).

At Edinburgh it was also agreed that the total amount available via the Cohesion Fund would be 15.1 billion ecus and that the annual totals would rise from 1.5 billion ecus in 1993 to 2.6 billion ecus in 1999. Because at the time the TEU had not been ratified, provision was also made for interim arrangements for 1993, based on Article 235 (Commission 1994b). The member states were primarily interested in the total amount of structural funding that they would each receive and it is clear that some promises were made.[10] Annexe 3 of the Edinburgh communiqué indicated a rough division of the Cohesion Fund alone (up to 1996), whereby Spain would receive between 52 and 58 per cent, Greece and Portugal between 16 and 20 per cent and the island of Ireland 7–10 per cent.[11]

The rest of the structural funds were also divided up at Edinburgh between the various objectives determined by the Commission in its 1988 reforms (see next section). It was determined that 93.8 billion ecus would go to Objective 1 areas and that a significant percentage of that would go to the 'Cohesion 4'. The exact determination of which regions would be eligible for funds from the various objectives was left until later, although it was clear that, on this too, promises were made in order to facilitate agreement at Edinburgh. It has been suggested, for instance, that Germany, because of significant internal pressure from the Länder, was prepared to accept the extension of the British budget rebate to 1999 only once it had obtained assurances that the five eastern Länder would be eligible for at least 13 billion ecus from Objective 1 funds. Similar divisions were agreed for the other objectives (see Table 8.1). As noted above, there had to be further negotiations in the Council to establish the final parameters of each member state's 'take' before the necessary enabling legislation could be passed in late 1993.

In this section I have tried to establish that, whilst the increase in the size of the structural funds played an important role in facilitating the basic or 'history-making' decisions that led to the SEA and the TEU, much of the haggling over distributive implications was effectively left until later, because 'the resolution of both types of decision simultaneously would be practically impossible' (McAleavey 1994).[12] However, the critical point is that both types of decision are linked together by the initial 'historic' deal; the secondary 'distributive' arrangements both reflect previous high-level agreements and are themselves concluded by a similar process of high-level bargaining between the representatives of the central governments of the member states. It is only once these two stages have been completed that other actors come to play a role in the way that these 'historic' agreements are rationalized and implemented in terms of 'Community' objectives such as cohesion, concentration, programming, partnership, and additionality— although competition criteria are often conveniently overlooked. The process of implementation of these fundamental agreements does indeed provide potential scope for a challenge to the dominant role of the central governments of the member states. But it is necessary to be clear about exactly what central governmental powers are being challenged here. Some analysts clearly feel that this is an important area and that to understand the integration process it is necessary to 'go beyond the areas that are transparently dominated by member states; financial decisions, major pieces of legislation and the Treaties', and to understand that 'beyond and beneath the highly visible politics of member state bargaining lies a dimly lit process of institutional formation (integration?) and here the Commission plays a vital role' (Marks 1993). Others of course have argued that the implementation of the structural funds involves a mobilization of sub-national actors (both public and private) which is significant, even if it does not represent a major challenge to central governments, because it has led to an 'enhancement of the governing capacity of the system as a whole' (Hooghe and Keating 1994).

Table 8.1. Structural funds: breakdown by member state and objective, 1994–1999[a] (in million ecus at 1992 prices)

Country	Objective 1	Objective 2	Objectives 3 & 4	Objective 5a	Objective 5b	Community initiatives	Total	(%)
Belgium	730	160	465	191.6	77	177.9	1,821.5	(1.3)
Denmark	—	56	301	262.5	54	86.6	760.1	(0.5)
Germany	13,640	733	1,942	1,133.8	1,227	1,264.6	19,940.4	(14.6)
Greece	13,980	—	—	—	—	990	14,970	(10.9)
Spain	26,300	1,130	1,843	431.6	664	2,241.5	32,610.4	(23.9)
France	2,190	1,765	3,203	1,912.7	2,238	1,231.8	12,540.5	(9.1)
Ireland	5,620	—	—	—	—	374.4	5,994	(4.3)
Italy	14,860	684	1,715	798.6	901	1,504.9	20,463.5	(15)
Luxemburg (0.05)	—	7	23	40	6	5.9	81.9	
Netherlands	150	300	1,079	159.2	150	211.9	2,050.1	(1.5)
Portugal	13,980	—	—	—	—	1,232.6	15,212.6	(11.1)
UK	2,360	2,142	3,377	439.3	817	814.3	9,949	(7.2)
Total	93,810	6,977	13,948	5,369.3	6,134	10,136.7	136,395[c]	
(%)	(68.7)	(5.1)	(10.2)	(3.9)	(4.5)	(7.4)[b]		

[a] Objective 2 total and breakdown for 1994–6 only.
[b] These figures are for the 9 Community initiatives agreed by July 1994.
[c] The actual total for 1994–9 will be 141,471 million ecus. The difference is made up by the 1997–9 tranche of Objective 2 funding, the other Community initiatives and a small amount to be spent on innovative measures.

Whilst we must now turn to examine the significance of the implementa-tion stages of the structural-fund arrangements, it should be made clear that it is the contention of this chapter that—despite a great deal of observable behaviour (and of recorded observations)—this does not constitute a significant stage in the policy process. This is because of the restrictive para-meters set by the basic or 'historic' agreements that I have described above. Thus whilst there is much to describe in the detail of how the structural funds are implemented, it is not clear that this is very significant either in terms of integration in general or regionalism and cohesion in particular.

The implementation of the structural funds

Between 1975 and 1988 the implementation of the structural funds gave the Commission little discretion other than over the minuscule Social Fund. The ERDF was distributed via a system of national quotas. This meant that the member states merely applied for funding up to their quota limit for projects that were already planned and in some cases completed. Some progress was made in 1979 when the Commission got the Council to agree to a small non-quota section, and in 1984 the quotas were relaxed into 'indicative ranges'. These, by setting a maximum and minimum allocation for each state, meant that the Commission effectively gained some discretion over around 11 per cent of the ERDF budget. The Commission used this new freedom to develop its own regional priorities and to introduce its own preferred 'programme' approach to regional assistance. It was agreed in 1984 that 20 per cent of the Fund would be devoted to 'programmes' rather than individual projects, and that, whilst most of these programmes would be national, some would be ini-tiated and developed by the Commission. Furthermore, steps were taken to coordinate better the activities of the three separate funds so that the IMPs and the Integrated Development Programmes (IDPs) were the first that might be said to involve the structural funds as opposed to each individual fund.

The Commission's major opportunity to expand its own role flowed from the decision, discussed above, to place the notion of 'cohesion' on to the Community agenda and to link this with the doubling of the structural funds. The SEA (Article 130d) had laid down that the implementation of the struc-tural funds should be reformed on the initiative of the Commission. Once the first financial perspective (1988) had been agreed, the Commission duly pro-posed, and the Council passed, a package of regulations that amounted to a major reform.[13] Similarly, the TEU amended Article 130d so as to provide the opportunity for further refinement of the 1988 reforms. In 1993 these further reforms were built into the package of enabling regulations that were agreed, once the member states had resolved their differences over their individual shares and the criteria by which they would be allocated.[14]

In the 1988 reforms the Commission established four principles for relating the structural funds to the objective of economic and social cohesion. These principles were clearly designed to provide for the development of a coherent Community policy, covering both specific regions and the Community as a whole, involving a more autonomous role for the Commission and challenging the stranglehold of national central governments by also seeking to involve sub-national partners. Although all four principles were maintained in the 1993 reforms, the member states made much of the TEU principle of subsidiarity to claw back some of their autonomy in this area. The four principles were:

- *concentration* of measures around five priority objectives;
- *partnership*, involving the closest possible cooperation between the Commission and the 'appropriate authorities' at 'national, regional and local level' in each member state at every stage in the policy process from preparation to implementation;
- *additionality*, such that EC funds complement rather than replace national funding; and
- *programming*, whereby multi-annual, multi-task, and occasionally multi-regional programmes are funded, rather than uncoordinated individual national projects.

The implementing regulations in both 1988 and 1993 provided for the structural funds to be allocated to IDPs, rather than individual projects, each of which has its own budget. These programmes can be initiated at the national level or the Community level and can be financed by one or more of the Community's structural funds. In the period from 1994 to 1999 90 per cent of the structural funds will go towards nationally initiated programmes and 9 per cent towards Community initiatives. The final 1 per cent is to be spent on a series of innovative measures designed by the Commission; these mainly involve the creation of networks such as Dionysus, in which ten French, Italian, Spanish, and Portuguese wine-growing regions pool resources and share information.

The programmes initiated at the national level used to be adopted by the Commission on the basis of Community Support Frameworks (CSFs). These were separately negotiated between the Commission and each member state on the basis of either national or Regional Development Plans (RDPs), drawn up by that member state supposedly acting in conjunction with its regional authorities. The 1993 reforms streamlined this process; a member state can now choose to submit just one Single Programming Document (SPD). These are not negotiated with the Commission like the CSFs, and they contain proposals for programmes within them from the outset, thus shortening what was a rather bureaucratic three-stage process of negotiation. Because of time constraints this often meant the effective exclusion of 'partners' and the domination of the process by central governments. Instead there is now just one proposal (the SPD) from the member states, followed by Commission decisions on the individual programmes.

Once adopted, the programmes are monitored and assessed by monitoring committees at national, regional, and operating level. These monitoring committees are meant to bring together representatives of the Commission, the member states, the EIB, representatives of local and regional authorities, and, more recently, the voluntary and private sectors as well. The regulations leave it to the member states themselves to designate the 'competent authorities and bodies'. In Britain the monitoring committees are usually chaired by one of the government's ten Regional Directors (*Independent on Sunday*, 5 Feb. 1995). At the start of the process, however, the time constraints remain tight. The package of implementing regulations for the 1994–9 period were not agreed until towards the end of 1993, but the Commission had still adopted all the relevant CSFs and SPDs by the end of 1994. However, as in the 1988–93 period, the member states were not as speedy in their appointments for all the monitoring committees.

Community initiatives are drawn up on the basis of guidelines established by the Commission alone, although the Regional Policy Committee must be consulted. For the 1994–9 period the Commission is pursuing twelve initiatives concentrating on seven themes, each with a named programme.[15] It had agreed on all these by June 1994. By the end of 1994 the member states had proposed programmes in line with these initiatives. These will be implemented and monitored in a fashion similar to the national initiatives.

Theoretically, then, the 1988 and 1993 reforms set up a policy-implementing system within which the Commission could use the structural funds to develop coherent and coordinated programmes designed to enhance cohesion in line with the four principles of concentration, additionality, programming, and partnership. The development of these four principles will now be briefly examined.

Concentration

The original five objectives (a sixth was added on enlargement to cover the sparsely populated regions of the Nordic countries) represented the Commission's attempt to impose consistent geographical and functional criteria on the management of the Funds and thereby to concentrate spending on the most needy regions and states.

- *Objective One* was originally meant to cover regions where the GDP per capita was less than 75 per cent of the EU average. But to meet the demands of some member states this strict definition has been loosened recently, thus ensuring the eligibility of areas such as Merseyside in the UK, all five eastern Länder in Germany, and several French *arrondisse-ments*. In the 1994–9 period the Objective 1 areas will receive almost 70 per cent of the total 1994–9 structural fund package (excluding the Cohesion Fund) drawn from the ERDF, ESF, EAGGF, the ECSC, and the EIB.
- *Objective Two* covers regions affected by industrial decline, where the unemployment level is above the EU average. Objective 2 regions are to be funded by the ERDF, the ESF, the EIB, and the ECSC and will receive around 11 per cent of the total package.
- *Objective Three* is aimed at combating long-term unemployment, and *Objective Four* is specifically aimed at facilitating the occupational integration of young people. These two Objectives were effectively merged after 1993 and are serviced by the ESF, the EIB, and the ECSC. Between 1994 and 1999 they too will receive about 11 per cent of the total package.
- *Objective Five* is subdivided between 5*a*, funded by the EAGGF-Guidance Section and earmarked specifically for agricultural and forestry assistance, and 5*b*, which involves not only the EAGGF-Guidance Section but also the ERDF, ESF, and EIB. This is aimed at promoting the development of rural areas mainly via diversification away from traditional agricultural activity (A. Smith 1995). In the 1994–9 period Objectives 5*a* and 5*b* will each receive just over 4 per cent of the total package.
- *Objective Six* will deal with the special problems facing the very thinly populated regions of the Nordic countries.

The focusing of the structural funds on objectives whose criteria are defined by the Commission has led to both a geographical and a functional concentration of resources (mainly via Objective 1) in the poorer member states and regions. Thus, when the Cohesion Fund is taken into account as well, Greece, Spain, Ireland, and Portugal will see their take from the structural funds double in the 1993–9 period. Nevertheless, there is a constant tension between the Commission's desire for more concentration and the concern of all the member states to get their fair share of the structural funds; so national allocations persist. In the 1993 reforms new regions from within the richer states (such as Hainault in Belgium, the Highlands and Islands in the UK, and Flevoland in the Netherlands) were also given Objective 1 status (Hooghe and Keating 1994). The Commission's ability to launch its own initiatives has led to further increases in functional concentration across regions, but here the resources involved are very small.

Additionality

The Commission was successful in using the 1988 reforms to insist that structural-fund payments were additional to those already planned by the member states and that they did not lead to a reduction in national expenditure, as they had tended to do in the past. The Commission has pushed the member states to 'account openly and transparently for the structural funding they receive and for the continuing implementation of national expenditure' (J. Scott 1995). Member states are now required to demonstrate additionality when they submit either their RDPs or their SPDs to the Commission, and the Commission has successfully threatened recalcitrant states with the withholding of funds if they do not comply.

Despite this relative success for the Commission, it has been noted (J. Scott 1995) that the additionality rules for the new Cohesion Fund are more lax. This is partly because the cohesion money is seen as a 'convergence' aid and as such is specifically designed to assist the poorer national governments increase levels of structural assistance (covering the environment and TENs) without increasing the burden on government expenditure. In a conflict between the objectives of cohesion and convergence, it may well be that the hard fought-for principle of additionality will lose some of its force.

Programming

The Commission has been successful in moving away from the unco-ordinated funding of nationally selected projects towards the more coordinated funding of programmes designed by both member states and the Commission using Commission-determined criteria. Whilst the overall level of funding is determined intergovernmentally, the Commission has gained a high degree of control over the designation of both structural-fund objectives and the 'designated areas' (note not regions) that they will apply to. As we have seen in relation to the designation of Objective 1 areas, the member states naturally seek to influence the Commission's programming criteria and they continue to argue about the latitude that they have within and between programmes. Throughout 1994 the UK Government refused to submit plans for spending under the Objective 4 programme, but argued (unsuccessfully) instead that its Objective 4 allocation should be transferred to its Objective 3 programme (*Financial Times*, 17 Nov. 1994).

The Commission has been able to push its programming principle furthest in the areas where it has most autonomy: Community initiatives and innovative measures, which add up to 10 per cent of the overall structural funds. Here the Commission is in the best position to develop Community-wide programmes such as Rechar or Regis.[16] However, Hooghe and Keating (1994) have pointed out that there has been some friction caused between the Commission and national authorities over the Commission's exercise of its autonomy—friction that has often been exacerbated by coordination failures within the Commission itself, resulting in conflicts and contradictions between national and Commission-initiated programmes. In the 1993 reforms a new Council Committee on Community Initiatives has been established to enhance the influence of the member states over this area of relative Commission autonomy (an example of 'spillback', perhaps).

Partnership

As noted above, a great deal of excitement has been generated by the Commission's advocacy of the principle of partnership in the management of the structural funds. The call in the 1988 reforms for the close involvement of regional and local bodies with the Commission and the national authorities in the planning, decision-making, and implementation of the structural funds has led to suggestions that a form of multi-level governance is emerging within the European Union (Marks 1993).

In the 1988 reforms the Commission certainly sought to involve regional and local authorities in all stages of the programme-implementation process from planning through monitoring to assessment. An advisory Consultative Council of Regional and Local Authorities was established (to be replaced in the TEU by the Committee of the Regions). Regional and local authorities began individually, and occasionally collectively, to lobby in Brussels and in national capitals (McAleavey 1994; Mazey and Richardson 1993*b*). Whilst some have argued that the impact of the principle of partnership has led to significantly enhanced regional and local governmental involvement in the policy process (Landaburu 1990; Laffan 1996), others contend that, despite a great deal of mobilization, much of the activity is symbolic and that very little power or autonomy has been wrested from the still dominant national governments (Hooghe and Keating 1994). McAleavey's (1994) study of the extensive activity of the Objective 2 lobby, which sought to preserve Objective 2 funding in the 1994–9 structural-funds programme, reinforced this view when it concluded that the impact of the lobbying had been minimal and that it was not possible to detect sub-national access to the intergovernmental bargaining that culminated in the Edinburgh agreements. Similarly, the actual influence of the many regional offices now based in Brussels has been questioned, although it is certainly the case that regional and local authorities are not slow to claim the credit for any structural funds that come their way.

In developing the principle of partnership the Commission has been criticized for inconsistency in its choice of partners. It is certainly true that for some programmes the Commission prefers regional authorities, for some local authorities, and for others partners from the private sector. There are often tensions between these groups, particularly between regional and local authorities. There is also an unresolved tension between the Commission's desire for partnership and its desire for consistent, coherent, and coordinated Community-wide programmes. If the principle of partnership really did lead to significant sub-national involvement in actual structural-fund implementation decisions, then the enormous regional variety within the Union, both in terms of organization and influence, would lead to many different solutions to similar problems being pursued by different regions in different ways (A. Smith 1995). That this is not the case and that we were able to point to an emerging Community policy on cohesion suggests that the involvement of sub-national actors in the implementation and monitoring process remains essentially symbolic.

This is not to deny that the structural funds policy has led to a great deal of regional mobilization, although this varies considerably throughout the Union, depending on the governmental structure of individual member states. Thus in Belgium regions like Flanders, or in Germany the Länder, are fully involved in the preparation, financing, monitoring, and assessment of programmes. It was, however, the Belgian government (admittedly a coalition of regional parties) that took part in the intergovernmental bargaining that determined the overall size of the funds and the initial allocations between objectives. On the other hand, states like Greece, Ireland, and Portugal have no elected regional tiers and so the powers of the central governments are, if anything, increased by their involvement in the implementation of policy. In Britain and France, where there are less significant regional tiers of government, the central governments have fought hard to preserve their autonomy of decision by exploiting their power to determine the nature of regional, local, and private participation by designating the partners and their control over which programmes are submitted to the Commission for agreement.

Thus, although the pursuit of partnership has mobilized sub-national actors and drawn them into the policy process, it is hard to make a case that this represents a challenge to the autonomy of the central governments of the member states. Regionalism may well be on the agenda as a result of the structural-fund arrangements, but developments to date do not suggest that any significant governance changes have yet taken place, or at least not consistently across the EU.

Conclusions

The most striking feature of the structural funds is the enormous increase in overall expenditure on them since 1988. Structural-fund expenditure has risen in direct correlation with the decline in agricultural expenditure as a percentage of the EU budget. It is now clearly one of the Union's core activities. We have established that policy-making can be understood in two stages, with the 'historic' or constitutional decisions best understood in the classic intergovernmental mode described by Moravcsik (1993). Whilst it is clear that deals about the size and allocation of the structural funds have been used to facilitate wider Community developments, it is not easy to present or explain these side-payments in quantifiable terms. No real effort has ever been made to base the cohesion policy on an estimate of the costs for the poorer states of accepting either the single-market programme or EMU. We have also suggested that, despite the presence of numerous regional and local authorities both lobbying and participating more formally in the policy process, there is little evidence that they have made any impression at all on the intergovernmental stages of negotiation. However, it is certainly the case that in some states (most obviously Germany and Belgium) regional interests must be given serious consideration by the central government when considering its negotiating stance in the EU. However, there is even more evidence to suggest that other member states have used their role in the intergovern-mental bargaining stage to strengthen their position *vis-à-vis* their sub-national authorities.

Nevertheless, I have argued that this 'top down' approach to structural-fund policy-making could be challenged at the implementation stage by the determination of the Commission to involve sub-national actors in the pol-icy process. Here we have found a considerable mobilization of such actors. Such activity clearly varies between the member states, and is primarily deter-mined not by the structural-fund policy itself but by the constitutional arrangements within any particular member state and by the different poli-cies within countries on issues of regional development. There is some evi-dence that some states have been forced to create new tiers of government in order to comply with the implementing regulations, but there is no evidence to suggest that these developments or the enhanced role of the Commission represent a significant threat to the autonomy of the central governments of the member states. Indeed there is some evidence that in the 1993 reforms the member states, armed with the principle of subsidiarity, have regained what-ever ground they might have lost in 1988. Thus Hooghe and Keating (1994) point out that the member states have reasserted their influence over the selection of eligible regions, and that the introduction of the SPD has sim-plified the process in their favour, by reducing the points of influence for both the Commission and the sub-national partners. Furthermore, the rules on additionality have been relaxed, especially as far as the Cohesion Fund is con-cerned. In these ways national control over the policy process has been strengthened.

However, it should be noted that, whilst some advocates of further inte-gration would seem to be interested in any policy process that can be shown to diminish the power of the member states, it should perhaps be remem-bered that the EU was built and sustained by member states that had rela-tively strong national governments. If the structural funds and the concept of cohesion are increasingly used to reinforce the power and legitimacy of the EU's governments, this is not necessarily to the detriment of integration. A genuine 'Europe of the regions', encouraged and nurtured by a growing supranational/sub-national alliance over access to increased structural funds, might not in the long run lead to harmony.

The implementation of the structural funds has led to rivalries and conflicts, both between regional and local authorities and between public and private 'partners'. Not all of this conflict has enhanced the power of the Commission, despite the fact that the increase in the size of the structural funds has led to a considerably increased role for DG XVI.

Finally, we should consider the likely future of the structural funds. It is clear that the Union will be under pressure to enlarge in the near future to include the states of central and eastern Europe. It is also clear that the present member states are not in the mood to increase significantly the overall size of the Community budget so as to extend the present system of structural funding to the new applicants on the same basis that the present member states receive support. This is a major change; in past enlargements the structural funds have not only proved stretchable but have been considerably expanded, partly because they have been seen as facilitators of enlargement. Although this chapter has concentrated on policy-making rather than examining the substance of the cohesion policy, it is clear that even at present levels the funds allocated are not large enough to make a significant difference either to regional disparities or to convergence, even though the income from the funds is significant for the poorer states (Tsoukalis 1993).

The EU economy could probably therefore survive without the redistributive effect of the structural funds. But it is not clear that the EU system of governance could survive. If it is accepted that the structural funds are essential facilitators and legitimizers of basic Union deals, then it is not clear how these deals might be brokered in the future if the structural funds are not available. Whilst the applicant states are probably not expecting that much in the way of structural-fund hand-outs, they will not tolerate second-class membership. Yet this is what they will have if the funds are kept for the present member states but not extended to others. Whilst the major net contributors to the EU budget are in no mood to consider increased levels of funding, the poorer states may not be prepared to support further enlargement if the result is a loss to their structural-funds support.

If the structural funds were to be run down, then, as we have argued above, the objectives of cohesion and convergence would not necessarily be endangered. It might be possible to envisage economic and social cohesion being sought via the operation of the market, assisted by competition policy, with the Commission acting only to prevent the member states engaging in a competitive policy of national or regional subsidies.

Notes

1. In the Treaty of Rome these were 'differences existing between the various regions and the backwardness of the less favoured regions' (Preamble, EEC Treaty) and in the Treaty on European Union they are 'disparities between the levels of development of the various regions and the backwardness of the least favoured regions, including the rural areas' (TEU, Article 130a).
2. Article 235 of the Treaty of Rome reads: 'If action by the Community should

prove necessary to attain, in the course of the operation of the common market, one of the objectives of the Community and this Treaty has not provided the necessary powers, the Council shall, acting unanimously on a proposal from the Commission and after consulting the Assembly, take the appropriate measures.'

3. A small non-quota section under the exclusive control of the European Commission was built into the process for allocating the ERDF.

4. Instead of rigid national quotas a system based on 'indicative ranges' was introduced, giving the Commission a bit more flexibility (around 12 per cent of the ERDF was now effectively within the Commission's discretion) in the allocation of the ERDF to the various member states. In 1984 the Commission also made some initial progress towards replacing the funding of individual projects by the establishment of 'programmes', which again were expected to give more scope for a Commission perspective on regional policy to be injected into expenditure plans.

5. The Integrated Mediterranean Programmes (IMPs) were first established in 1985 and involved the transfer of some 6.6 billion ecus in three- to seven-year programmes to parts of France and Italy and all of Greece. The Commission sought to develop its own agenda with the IMPs by insisting on funding integrated programmes rather than individual projects.

6. For instance, both Britain and France showed some last-minute reluctance to sanction the doubling of the structural funds, but this was overcome by the persistence of the German presidency of the Council, Delors, and the willingness of Chancellor Kohl, representing the Community's major net budget contributor, to foot the bill. There is no evidence to suggest that any of these leaders spent much time discussing the advantages and disadvantages of regional policy as the deal was struck.

7. It will be recalled that there was no reference to cohesion in the Preamble to the SEA. It can be argued that the fact that there is no reference to the regulation of competition in the TEU Common Provisions suggests that cohesion has primacy over competition. For the development of this argument see Frazer (1994).

8. Council Regulations (EEC) Nos. 2080/93, 2081/93, 2082/93, 2083/93, 2084/93, and 2085/93, all of 20 July 1993 and published in the *OJ* L193/1.

9. This is the method of allocating the funds' resources as preferred by the poorer member states.

10. The Irish Prime Minister claimed that he had 'firm pledges of I£8billion' from the Edinburgh European Council (*Financial Times*, 2 July 1993). The Spanish government obtained written assurances that it would receive Pta 6,400 billion (*Financial Times*, 5 July 1993), and the Portuguese government claimed that it was promised Esc 3,500 billion (*Financial Times*, 9 July 1993).

11. In 1993 under the interim measures Spain received 54.9%, Portugal 18.1%, Greece 17.9%, and Ireland 9.1%.

12. For an elaboration of the concept of 'history-making decisions' see J. Peterson (1995a).

13. Regulation 2052/88 *OJ* 1988 L185/9 (the parent regulation), Regulation 4253/88

OJ 1988 L374/1 (the coordination regulation), Regulation 4254/88 *OJ* 1988 L374/15 (ERDF), Regulation 4255/88 *OJ* 1988 L374/22 (ESF), and Regulation 4256/88 *OJ* 1988 L374/25 (EAGGF).

14. See above, n. 8.
15. Interreg (cross-border cooperation; 2.9 becus), Leader (local development in rural areas; 1.4 becus), Regis (support for the most remote regions; 600 mecus), Employment (the integration into working life of women, young people, and the disadvantaged; 1.4 becus), Adapt, SME, Rechar, Konver, Resider and Retex (adaptation to industrial change; 3.8 becus), Urban (urban policy; 600 mecus), and Pesca (restructuring the fisheries sector; 250 mecus).
16. See note 15.

Further Reading

For excellent overviews of the subject see Hooghe and Keating (1994) and Marks (1992). For a review of the theoretical debate about regional development in the EU see J. Scott (1995). For surveys of regionalism see Jones and Keating (1994) and Hooghe (1995).

Hooghe, L. (1995) (ed.), *The European Union and Subnational Mobilization* (Oxford: Clarendon Press).

—— and Keating, M. (1994), 'The Politics of European Union Regional Policy', *Journal of European Public Policy*, 1/3: 367–93.

Jones, B., and Keating, M. (1994) (eds.), *The European Union and the Regions* (Oxford: Clarendon Press).

Marks, G. (1992), 'Structural Policy in the European Community', in A. Sbragia (ed.), *Euro-Politics: Institutions and Policymaking in the 'New' European Community* (Washington, DC: Brookings Institution, 1992), 191–224.

Scott, J. (1995), *Development Dilemmas in the European Community* (Buckingham: Open University Press).

CHAPTER NINE

ENVIRONMENTAL POLICY: THE 'PUSH–PULL' OF POLICY-MAKING

Alberta Sbragia

European environmental policy is driven by a tension between the proponents of stringent standards and the reluctant, within and between countries. A few 'leader' states—usually Germany, Denmark, and the Netherlands—push their progressive environmental policies on to the agenda in Brussels. Although their legislation is not wholly adopted by the EU, enough is accepted to push the 'laggard' countries to adopt more stringent legislation than their domestic audiences demand. The impact of domestic politics on European environmental legislation results in part from fears among the 'leaders' that their competitiveness will decline within the single market, unless other EU members are forced to meet the costs of similarly high standards.

Introduction

Environmental policy achieved much greater prominence during the 1980s and early 1990s than it had had in the earlier period of Community policy-making. The European Union (EU) is now a central actor in the making of environmental policy in Western Europe. In fact, 'it cannot be repeated too often that it is impossible to understand the environmental policy of any of the EC member states without understanding EC environmental policy' (Haigh 1992b: Preface).[1]

The Community's adoption of a significant legislative programme for environmental protection, as well as its transformation into a key policy-maker *vis-à-vis* the member states, is striking. The odds against that accomplishment would seem at first glance to be rather formidable. Since the core mission of the European Community (EC) had been to enhance economic growth, the inclusion and expansion of a policy area viewed by many of the member states as inhibiting such growth would seem to be unlikely.

Furthermore, given the economic costs imposed on both firms and public authorities by environmental legislation, one might presume that national governments would not willingly transfer their authority over this policy area to the EC. Yet national governments did agree to allow the EC to act in the environmental area long before they formally included environmental protection in a treaty, and they agreed to hundreds of pieces of legislation adopted in Brussels. What kind of policy process led to environmental protection becoming so important at the European level?

In this chapter I argue that the policy process in the environmental arena is typically driven by a small number of member states which are significantly more environmentally progressive than the rest. Their domestic politics are such that these states project 'green' policies on to the EU. The policy process in Brussels is therefore characterized by the need to balance the politics, interests, and norms of the 'green' countries with the pressures for economic development and financial restraint felt by the sceptics. The EU's institutional framework provides the means by which that 'balancing act' is carried out.

The politics of environmental policy within the Union is multi-layered. The Commission and the Council of Ministers play a critical role *vis-à-vis* both the activist and the reluctant member states. The Union's policy process operates independently of the domestic politics of any member state. The Commission and the Council of Ministers and at times the European Parliament (EP) and the European Court of Justice (ECJ) participate in an independent inter-institutional political dynamic. 'Green' member states must operate within the Union's complicated policy-making machinery. And yet at a fundamental level the political game in Brussels is entangled with the domestic politics of the member states: the importance attached to environmental protection in Brussels is determined in significant measure by the domestic politics of the 'green' member states.

Environmental policy-making features a 'push–pull' dynamic. The internal politics of the environmentally progressive states 'pushes' the process in Brussels along. The actual process is itself pivotal in 'pulling' most of the member states towards levels of environmental protection which, left to their domestic devices, they most probably would not adopt.

In this chapter I consider the politics of the Union's environmental policy by briefly analysing domestic politics and then turning to the political process which takes place in Brussels. In the first section, I shall sketch the differences of approach towards environmental protection across the member states as well as the domestic politics of the most important of the 'leader' member states—Germany. The second main section of the chapter focuses on the political action in Brussels.

The leader–laggard dynamic

Environmental policy-making is most often driven by the fact that one or more member states approve significant new environmental protection laws. The formulation of environmental policy in the EU often resembles a process which Peter Haas has termed the 'leader–laggard' dynamic (see also Héritier 1994). In his words,

Leader governments are pressed by both industry and public opinion to draw other governments up to their levels of environmental protection. 'Leaders' are countries with stringent sectoral environmental measures, which promote their sectoral standards for universal adoption. . . . 'Laggards' are countries with relatively weak measures, which are reluctant to accept more stringent measures (P. Haas 1993: 138).

Typically, an environmentally progressive state passes national legislation which is more stringent than that found in the EU generally, and the pressures for 'Europeanization' begins. The Union's policy-making process 'pulls' the reluctant states to agree to higher levels of environmental protection.

The politics of environmental policy must therefore be analysed at the national level as well as at the European level. Given the 'leader–laggard' dynamic, it is the policy dynamics of the leader states which typically serve as the 'motor' of the EU's efforts to protect the environment. It is their domestic politics which 'push' the Community's policy process along.

'Leaders': at home and in Brussels

Only three of the Union's member states—the Netherlands, Denmark, and Germany—approve environmental legislation across a range of issue areas in a sustained fashion. (The three new states will each fit into this 'green' category as well). They are the environmental 'leaders'. The 'laggard' category includes Greece, Italy, Spain, Portugal, Ireland, and Belgium which typically pass national environmental legislation only within the context of applying directives adopted in Brussels. The United Kingdom, France, and Luxemburg are in a middle category, but would typically be considered 'laggards' if compared with the 'leaders' (L. Kramer 1992: 52–3).[2]

As Kramer perceptively points out, 'In the absence of Community rules no relevant environmental rules would come into being at all in large parts of the Community' (L. Kramer 1992: 53).[3] In fact, EC directives are 'the single main factor' in the adoption of national environmental legislation within the Mediterranean member states (La Spina and Sciortino 1993: 222). Whereas 'leaders' along with environmentalists often find Community policy unsatisfactory, 'laggards' argue that because of that same policy they are taking significant steps towards environmental protection, steps for which there may be little domestic political support.

Although member states certainly pressure the Commission to force environmental issues on to the Union agenda, it is their adoption of national legislation that triggers the 'Europeanization' of regulation. Laws to protect the environment also tend to restrict trade, and that restriction brings in the Community (Brinkhorst, 1991: 92). Secondly, firms within a country which has approved progressive environmental legislation will be anxious to avoid being put at a disadvantage *vis-à-vis* their competitors elsewhere.

The 'green troika'—or one of its members—are often the 'leaders' in an issue area. For example, the draft directive on large combustion plants was very closely modelled on German legislation. In fact, Anthony Zito suggests that 'a German Ministry official met the Commission official responsible, gave him a copy of the German regulation, and asked him to develop it into a Community directive' (Zito 1995*b*: 27). Zito argues that the influence of Germany was critical in pushing the Community to respond to the issue of acid rain: the initial directive was proposed under the German presidency of 1983 and the final directive was adopted under the German presidency of 1988. In the area of water policy, the Commission's initial focus on that issue area is likely to have reflected the attention given to water pollution by policy élites in the Netherlands and (West) Germany in the late 1960s and early 1970s (Bresserts *et al.* 1994; OECD 1993*b*: 67; Rudig and Krämer 1994: 64–6). Directive 78/611 on lead in petrol was inspired by German laws (Golub 1994:

113). More recently, German legislation on packaging had a major impact on the Commission's initial work on packaging regulations (Mazey and Richardson 1992: 113).

It is important to remember, however, that environmental policy is segmented into sectoral policy and that political dynamics vary considerably across sectors. And no member state is the 'leader' in all environmental sectors at all times. Therefore, states outside the 'green troika' act as 'leaders' in certain areas.

Of the troika, Germany is the key member state driving the progression of environmental policy. None the less, both the Dutch and the Danes have had an impact far beyond what would be expected from small member states. Typically, the Dutch and Danes need German support to move an issue along the policy process, and the Germans need the support of the other two states to have a sustained impact on the policy process. These three states share similar 'norms of environmental behavior' (Skjærseth 1994: 38), which lead them to support 'green' policies even when their national interests are not served (Skjærseth 1994: 37).

The broad contours of environmental politics within the Community cannot easily be divorced from environmental politics within Germany. We shall now take a brief look at German domestic politics so as to understand why Germany takes such a progressive role in a European context.

The key 'leader': Germany

The first movement towards environmental protection in Germany came in 1969.[4] Upon coming to power, the Social Democrat/Free Democrat coalition gave environmental protection high priority. Environmental ministries began to be established at the level of the German Land (Rudig and Krämer 1994: 65). However, with the first oil crisis of 1973–4 and the subsequent recession, the environmentalist impulse began to falter. With the resignation of Willy Brandt as Chancellor in 1974, internal party politics changed in ways which disadvantaged environmental policy.

By 1980 the legislation originally proposed by the coalition in 1971 had been passed, but the restrictions imposed were generally much weaker than those originally proposed (Hucke 1985: 160–2; Boehmer-Christiansen and Skea 1992: 163). Although Britain and Germany were to approach environmental protection very differently in the 1980s, the two countries were very similar in 1979. They were both 'unwilling to become pace-setters in a comparative context' (Weale 1992b: 68).

In the 1980s, however, Germany became a pace-setter, galvanized by the realization in 1982 that German forests were dying due to the effects of acid rain (Boehmer-Christiansen and Skea 1992: 6). Before the Social Democrat/ Free Democrat coalition fell in October 1982, it had agreed to 'Europeanize' German domestic policy. The coalition's programme agreed

to begin an initiative at the intergovernmental level (i.e. through the OECD (Organization for Economic Cooperation and Development), UNECE (United Nations Economic Commission for Europe), and the EC to make German emission and product standards the guidelines for international policy . . . [and] there was to be an effort to get vehicle emission standards tightened at the European level . . . (Boehmer-Christiansen and Skea 1992: 193)

These goals were adopted more or less completely by Chancellor Kohl's government, which came to power in December 1982. Germany was to take a new leadership role in Brussels, and EC environmental policy was to gain a higher profile.

In 1983 the large combustion plant legislation was adopted, strictly limiting sulphur dioxide emissions from large furnaces (Weale 1992a: 68). In 1986 Chancellor Kohl created the Federal Ministry for Environment, Nature Protection, and Reactor Safety (Bundesministerium für Umwelt, Naturschutz und Reaktorsicherheit) (Weale 1992b: 166). The success of the Green Party (founded in 1980) in being elected to the Bundestag in 1983 perhaps most dramatically represented the return of a 'green' consciousness to German politics. That consciousness has remained more or less stable (even though the Green Party's fortunes have fluctuated), making Germany the most 'green' by far of the Union's large member states (Kolinsky 1994: 164).

Once 'the political agenda of Germany had been decisively changed' by environmental issues (Weale 1992b: 159), tough environmental laws threatened to undermine the competitiveness of German firms as their competitors in other EC states did not face the same regulatory burdens. 'Europeanizing' the burden of regulation became a way of keeping the playing-field level for German firms (especially chemical firms and automobile manufacturers) (W. Grant 1993: 64).[5] Von Weizsäcker has pointed out that 'emission control abroad . . . implies a cost burden on the industrial competitors' (von Weizsäcker 1990: 51).

Simultaneously, German environmental regulations threatened to keep goods out of the German market, thereby jeopardizing the goal of a single market. Environmental regulation decided in Brussels was thus a way of keeping the new demand for environmental protection from threatening the achievements and aspirations of economic integration. As Frank Boons writes, 'environmental programmes that are adopted in one country can have substantial consequences for economic actors in other countries' (Boons 1992: 85).

The debate in Germany evolved to the point where a number of ideas loosely categorized as belonging to the school of 'ecological modernization' became important in the German policy-making process. The principle of precaution—*Vorsorgeprinzip*—became central to environmental policy. Furthermore, the relationship between the economy and the environment began to be reinterpreted. Pollution control came to be viewed by many as contributing to economic growth rather than as inhibiting it (Weale 1992*a*: 75–90).

The 'leaders' and EU environmental policy

Although Germany, the Netherlands, and Denmark often approve environmental regulations at home and then try to have such regulations adopted throughout the EU, European legislation does not neatly reflect the levels of protection found in the 'troika'. In some issue areas, regulations have not imposed the degree of stringency they desire. At other times and in other issue areas, however, the European regulations have been *more* stringent than those of the 'leader' countries (Maasacher and Arentsen 1990: 76; von Weizsäcker 1990: 51). Final legislative outcomes are by no means a copy of the 'leader's' national legislation.

Once a 'leader' country becomes involved in 'Europeanizing' its legislation, it is drawn into a complex institutional framework and policy-making process. The dynamics of that process cannot be controlled by any single member state or even by a coalition made up of the 'green troika'. The outcome is a contingent outcome. It is to that process that we now turn.

The Union's rules for environmental policy-making

Environmental policy is now a major EU policy area and contributes significantly to the view of the Union as a 'regulatory state' (Majone 1994*a*). Yet environmental protection was not one of the original aims of the Treaty of Rome, and the adoption of environmental legislation has not fitted easily into established decision-making procedures. All directives passed before the Single European Act (SEA) came into force in 1987 had to be justified in relation either to the creation of the common market or under Article 235, which allowed the Community to move into new areas to accomplish its goals. The creation of the common market was frequently used as a legal base, even though trade was only tenuously involved. Decisions had to be taken unanimously, and the Parliament was minimally involved.

Both the Single European Act and the Treaty on European Union (TEU) have created complex decision-making procedures for environmental policies. The SEA introduced qualified majority voting and the use of the cooperation procedure in the European Parliament for environmental directives linked to trade harmonization (Article 100a). Such regulations pre-empted national regulation and established a Community-wide standard.

Regulations designed to protect the environment, but unconnected to the single market (Article 130s) required unanimity, but did allow national standards to be more stringent than those agreed to in Brussels. Given that the delineation between legislation linked and not linked to trade harmonization is often ambiguous, the Commission and the Council often differed as to the appropriate legal basis for legislation.

In the period preceding the Maastricht negotiations, a broad consensus developed that the competence of the EU in the area of environmental protection should be both extended and clarified. The extension of the use of qualified majority voting to issues not linked to trade harmonization was broadly accepted. However, as a result of Spanish pressure during the negotiations, the use of qualified majority voting was limited to some extent (Kim, unpublished). Four different decision-making procedures now exist for environmental legislation, with the result that the decision-making process is extremely complex and differs depending on the environmental issue area concerned (see Box 9.1).

Box 9.1. Voting procedures under the TEU

1. Qualified majority voting by the Council and the cooperation procedure with the Parliament (under Article 130s(1)).

2. Qualified majority voting by the Council and co-decision with the Parliament for internal market harmonization measures, proposals in the areas of public health, consumer protection, trans-European networks, and 'general action programmes' relating to the environment (under Articles 100a, 129, 129a, 129d and 130s(3)).

3. Unanimous voting by the Council and consultation with the Parliament (in certain cases under Article 130s(2)).

4. A unanimous decision by the Council under Articles 100a or 130s(2) to adopt a measure by qualified majority. This implies consultation with the European Parliament.

Although the environment was not an issue area assigned to the EC by the Treaty of Rome, the ECJ initially argued that environmental protection was an implied power and in 1985 ruled 'that environmental protection was one of the Community's essential objectives' (Koppen 1993: 133). The

Commission was thus given strong legal backing for its proposals by the Court in the period before the SEA gave the EC explicit legal authorization to protect the environment.

The EC's First Action Programme, approved in November 1973, seems to have been triggered partially by Dutch and German anti-pollution legislation, which could distort trade, as well as pressure from national and international environmentalists (Hildebrand 1992: 25; Vogel 1993b: 117). But the Commission was also interested in directly promoting environmental protection as such.

Directives were proposed and approved which had no links to trade and lacked a transfrontier dimension (directives on bathing waters and drinking water fit in this category). In Nigel Haigh's words,

The EC never confined itself to the two classic justifications for international environmental policy, first, that some issues . . . are not confined by national frontiers and may indeed be global in character and, second, that international trade is impeded by differing standards. (Haigh 1992a: 235–6)

The EC passed a considerable amount of legislation even before the SEA (Majone 1992: 304), and after the SEA came into effect in 1987 the EC became even more active. As David Vogel has pointed out, 'between 1989 and 1991, the EC enacted more environmental legislation than in the previous twenty years combined' (Vogel 1993b: 125). In fact, environmental policy has been the area in which the Community has increased its activity the most (Liefferink et al. 1993b: 4).

By the early 1990s Community legislation addressed all areas of environmental protection, although those policy issues concerned with industrial pollution had received more attention than those dealing with nature protection. By the 1990s 'the main lines of national environmental policy are, to a large extent, prescribed by the EC' (Bennett 1992: 94).

The ECJ, for its part, ruled in 1988 that environmental protection was such an important area that measures to ensure it could legally create obstacles to trade. The case in question, known as the 'Danish bottle case', pitted Danish regulations requiring recyclable beverage containers against the Commission, which argued that the regulation was unfair to exporters to the Danish market. The Court's ruling indicated that environmental protection would, under certain conditions, be allowed to inhibit intra-EC trade.

Politics in Brussels

The EU provides the forum in which the green member states can try to export their environmental standards to others within a European institutional framework (Boehmer-Christiansen 1992; Weale and Williams, 1992). The Commission may indeed propose regulations which are inspired by the 'green' laws of one of the 'leaders', but the Commission, supported by the Parliament, often also acts as a 'policy entrepreneur' when it comes to the substance of the proposed legislation. The Council of Ministers, the ultimate decision-maker, is the European institution in which the member states attempt to make changes to the draft directive proposed by the Commission so that it will be more attuned to their national interests.

The EU's institutions interact with each other in such a way that the final provisions of a directive cannot be predicted either from the position of a 'leader' country or from the 'entrepreneurial' position of the Commission. However, during the inter-institutional dynamic which characterizes policy-making in this area, the Parliament usually takes the most 'green' position, the Commission typically is allied with the Parliament, although its environmental stances are more moderate, and the bargaining within the Council of Ministers focuses on issues such as how stringent specific restrictions should be or how pollution should be measured. The Council often relaxes at least some of the restrictions proposed by the Commission.

The Commission

Although the Commission is often referred to as a unitary actor, it is in fact made up of different groups, often with very different views on environmental protection. The key groups are the various Directorates General (DGs) within the Commission, the commissioner responsible for the environment, and the college of commissioners (which must approve proposals to be sent to the Council of Ministers).

Environmental proposals which do not intrude on other policy sectors are the province of DG XI, the Directorate General for Environment, Nuclear Safety, and Civil Protection. DG XI, for example, is largely responsible for water policy (Richardson 1994: 147). One of the striking features of DG XI is its small size. For example, only fifteen officials are charged with the supervision of chemicals, whereas 500 officials are so charged in the American Environmental Protection Agency. In general, officials within DG XI are dependent on external sources of technical expertise, whether that be provided by national officials or environmental groups (Mazey and Richardson

1992: 115). DG XI has been influenced by the ideas linked to 'ecological modernization' primarily developed in Germany, but it has been unsuccessful in convincing other DGs to view pollution control as an economic good rather than as a burden to the economy.

Environmental interest groups have found DG XI very receptive. In fact, officials from other DGs often complain that it has been 'captured' by 'green interests'. In Richardson's words, DG XI 'has been assisted by having a politicized, mobilized, and effective constituency of environmental groups. These groups have, in turn, been backed by a no less effective and political constituency of scientists and other experts' (Richardson 1994: 147). The European Environmental Bureau and the World Wildlife Fund for Nature have particularly close ties with DG XI.

Environmental proposals which come out of the Commission have at times been through a bruising internal battle, particularly those which are linked to the single market. In general, the DGs concerned with industrial and economic affairs will have tried to weaken or block proposals introduced by DG XI. The DGs concerned with economic affairs typically have far greater resources and technical expertise than does DG XI, and the interest groups linked to such DGs have much greater influence and status than do the environmental groups which have access to DG XI.

In the area of greenhouse gases, the conflict between DG XI and DG XVII (energy) led to an impasse in 1989. The former favoured carbon/energy tax, while the latter opposed it on the grounds that it would hurt economic growth. They thus could not agree on recommendations for a Communication on environment and energy (Jachtenfuchs and Huber 1993: 43). DG III (industry) is beginning to threaten DG XI's traditional dominance in water policy (Richardson 1994: 147). Even the well-known Danish bottle case was the result of an internal struggle within the Commission: DG XI did not want to take Denmark to the ECJ whereas DG III did. DG III won, and Denmark was taken to the Court on the grounds that its packaging law violated the requirements of the single market (Weale and Williams, 1992: 59).

Typically, the environment portfolio of the responsible minister has not been viewed as one of the most desirable; certainly commissioners involved with industrial and financial affairs have been perceived as more important than commissioners for the environment (Weale and Williams 1992: 58). However, Carlo Ripa de Meana, commissioner for the environment from 1989 to 1992, was unusually successful in raising the political profile of environmental issues—not always to his colleagues' liking (Ross 1995: 161, 197; C. Grant 1994: 107–9; Zito 1995a).

Ripa di Meana was widely viewed as being particularly effective in using the European Parliament to support his position within the Commission, and benefited from a climate of public opinion which favoured increased environmental protection (Judge 1992: 199–204). Other Commissioners have played a less flamboyant but none the less important role. Stanley Clinton Davis, for example, is credited with continuing the negotiations which led to the approval of the large combustion plant directive, the Community's major attempt to control acid emissions (Bennett 1992: 130). Commissioner Clinton Davis's intervention echoed that of a predecessor, Commissioner Karl-Heinz Narjes, who played an important role in the for- mulation of the draft directive on large combustion plants (Zito, 1995*b*: Ch. 3).

European Parliament

The EP plays a complex role in the policy process. While the Parliament has arguably been more of an actor in the environmental arena than in any other, its impact is typically not as great as that of the other actors in the process (Judge, 1992). The Parliament's new role in the co-decision procedure, how- ever, gives its Environment Committee more weight.

The Parliament's substantive impact depends on the decision-making pro- cedures in force for any single proposal, the interest of the commissioner for the environment in cooperating with it, and the degree of consensus which can be obtained within the Council of Ministers. The Parliament is most effective when it works with a Commissioner such as Ripa de Meana who is sympathetic to parliamentary power, knows how to use the Parliament's powers to extract concessions from the Commission, and who is very recep- tive to amendments proposed by the Parliament. The Parliament is much more effective operating under the cooperation than under the consultation procedure.

The Parliament has strongly supported the Commission's use of Article 100a as the legal base for major environmental proposals because it triggers the cooperation procedure. In the case of car emissions, the Parliament used the cooperation procedure so as to force the Council to adopt a much more stringent regulation than it would have if it had been using the consultation procedure, but the conditions under which such a result can be obtained are not often present (Tsebelis 1994). The Parliament is important in agenda-set- ting broadly defined (Judge 1992: 191; Tsebelis 1994). Although the impact of the new co-decision procedure established by the TEU is still unclear, it is likely that the new procedure's 'real potential rests in the enhancement of Parliament's informal, and positive, influence' (Judge and Earnshaw 1994: 269).

The Council of Ministers

Although the Commission is an important actor in the policy process, and its draft directives set the parameters of debate, it is the national environment ministers, acting within the Council of environment ministers, who are the ultimate decision-makers within the EU. It is they who typically negotiate the final agreements. Given that much environmental legislation has been passed unanimously, the dynamic is one in which countries with relatively little interest in environmental protection are drawn into a policy-making process in which ministers of the environment play a more prominent role than they may in the national government.

The 'ratcheting' upwards of environmental protection therefore is a product of an institutional policy-making process which empowers actors such as ministers of the environment who are typically weaker in national capitals than are, for example, ministers of economics or ministers of transport. The process allows ministers from even the 'leader' countries to propose restrictions at the European level which they might well be unable to have accepted at home. In other words, a minister of the environment is more important in Brussels than at home. That fact helps to explain why environmental protection progressed as far as it did under conditions of unanimity and widely held views in national governments which treated environmental protection as a brake on much-desired economic growth.

The environment minister holding the Council Presidency can shape the agenda during her term. For example, the Danish environment minister, under pressure from the Danish World Wildlife Fund, used the Presidency to support the integration of environmental and regional policy (Mazey and Richardson 1994: 37). However, a Presidency has to demonstrate skill and subtlety to achieve success during that term. The German Presidency of 1994 was not skilful when dealing with the draft directive on integrated pollution prevention and control which had been shaped by British influence and experience (Schnutenhaus 1994: 324–8). Once Germany took over the Presidency, it changed the directive to reflect German practice. In fact German influence was so obvious that one analyst concluded that Germany was using 'the Presidency as a Trojan horse for introducing German national air pollution control legislation' at the European Union level (Schnutenhaus 1994: 328). The revised draft was unacceptable to many member states, and, at the time of writing, negotiations were continuing.

Although typically the Council of Ministers weakens the restrictions proposed by the Commission (Rehbinder and Stewart 1985: 261), a great deal of intergovernmental bargaining is needed to reach a compromise. The member states are far from united in their response to a Commission proposal. In fact, the member states are deeply divided over environmental policy (Weale and Williams 1994: 12–13). These differences underlie the bargaining which takes place within the Council and slow down decision-making.

Given the differences among the member states, the early environmental directives—primarily in the area of water pollution—were so far-reaching and have been so costly and difficult to implement that the question arises as to why the Council of Ministers unanimously approved them in the first place. ('Ministers of the environment' as such did not even exist in the member states until the mid-1980s.) Certainly, the fact that implementation rests in the hands of the states helped the Council reach agreement. Furthermore, Joseph Weiler argues that the Commission was able to act so successfully as a 'policy entrepreneur' because environmental policy was not taken seriously in national capitals and was not given the kind of careful attention other EC policy initiatives received (Weiler 1991: 2449).

For example, after British accession, Britain was typically not represented by the senior minister for the environment at Council meetings; junior ministers were dispatched instead (Golub 1994: 100–1). Furthermore, British negotiators ignored warnings from their lawyers that the compromise language they accepted was open to alternative interpretations. Finally, the British, as a new member state, simply did not take directives on the environment very seriously. In the words of one official, 'it was a shattering blow when it became apparent that we were going to have to spend very considerable sums of money' (Golub 1994: 126).[6]

As the implications of environmental policies became clearer, the politics changed. The Council of Ministers took much longer to find a compromise. The 'leader' countries, driven by domestic concerns, typically influenced the Commission's draft directive in the direction of high levels of environmental protection. That pressure was reinforced by environmental interest groups with access to DG XI, but often opposed by groups linked to industry. The 'leader' states then had to negotiate within the Council to safeguard such a level against the efforts of the 'laggards' who view the Commission's draft as too stringent. (See Fig. 9.1.)

The case of acid rain is illustrative of this dynamic. In that case, the German government (along with the UK) had questioned the link between air pollution and acidification. However, when the political climate changed, the German government changed its position. Germany subsequently became the most aggressive supporter of emission reductions within the EC, while the UK remained a sceptic about the link between air pollution and acid rain (Bennett 1992: 96; Boehmer-Christiansen and Skea 1992). The UK, joined by other 'laggards', fought the tough standards for vehicle emission controls and power plants championed by Germany. Finally, in order to

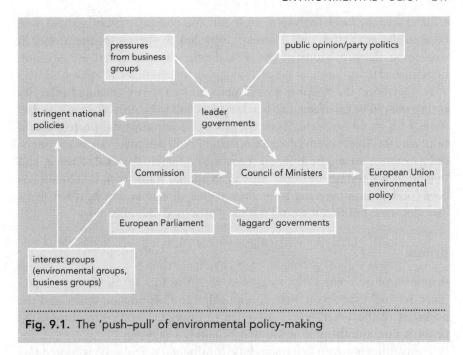

Fig. 9.1. The 'push–pull' of environmental policy-making

obtain the 1988 agreement on power plants, the German environment minister, Topfer, had to agree to less stringent car emission limits than he had wanted (Bennett 1992: 129–30). However, while the Community's limits on power plants were less strict than the German ones, they were none the less much stricter than those in the 'laggard' states. In the case of vehicle emission standards, the intervention of the European Parliament and the use of qualified majority voting led to the adoption of much tougher emission standards than the 'laggards' had wanted (Arp 1993; Kim 1992; Tsebelis 1994).

The 'laggards'

Although the position of any single state outside the 'green troika' varies across issue areas, the British and Spanish positions are fairly consistently sceptical about the Commission's draft directives. Ironically, both countries have excellent records in transposing EC directives into national law, although Spain leads the member states in citations for infringements (Pridham 1994: 92–3).

Spain's position is driven by an overriding desire to increase its standard of living and to generate employment. It views EU environmental legislation as imposing the standards of the 'rich north' on the 'poor south' (Aguilar 1993: 232–3). The long-serving Prime Minister Felipe Gonzáles has demonstrated a 'dismissive approach to the environment', and political parties in Spain have been indifferent to the issue (Pridham 1994: 86). Spain was the main actor in weakening the provisions of the TEU in the environmental arena (Kim, unpublished).

Furthermore, the Spanish government insisted that if qualified majority voting were to be extended, financial support had to be provided for member states who could not easily bear the costs of environmental protection. In response, the Treaty established a Cohesion Fund, designed to provide financial assistance for environmental measures to Spain, Portugal, Greece, and Ireland. It is a testament to the EU's institutional framework that Spain has been 'pulled' into agreeing to environmental directives which, if left to its own domestic politics, it would not have proposed at home.

Britain

Britain is more commonly associated with the role of an environmental 'Euro-sceptic' than Spain. British opposition to Commission drafts and the positions of the 'green troika' have led to dramatic stand-offs, convoluted negotiations, and the perception that Britain is a 'laggard'.

The conflict between Britain and its Continental neighbours is rooted in a 'definition of pollution which [has] differed fundamentally from Continental conceptions of pollution' (Golub 1994: 37). That difference in definition has influenced British assessments of numerous directives proposed by the Commission, which have been influenced by Continental traditions and regulatory systems rather than by those found in Britain.

German and British approaches towards environmental protection differ particularly sharply, and often negotiations within the Council focus on reconciling their two positions. As mentioned, German thinking on environmental protection has been influenced by the ideas associated with 'ecological modernization' (Weale 1992b: 79).[7] Such ideas were given little prominence in Britain, and the divergence in views between the two countries within the EU is at least partially rooted in this difference.

The 'precautionary principle' is central to German environmental policy and justifies going ' "beyond science" in the sense of being required to make decisions where the consequences of alternative policy options are not determinable within a reasonable margin of error and where potentially high costs are involved in taking action' (Weale 1992b: 80).[8] By contrast, British policymaking has been driven by a strong reliance on scientific evidence of cause-and-effect relationships. Furthermore, while the cost of environmental protection has been of major concern to Britain, the financial burden of environmental protection has been given much lower priority by Germany. The

German system of environmental control tends to be technology-forcing, and the Germans favour 'best available technology' (BAT) at the European level. The British, by contrast, have favoured requiring the use of 'best available technology not entailing excessive economic costs' (BATNEEC). Whereas the Germans tend to think high environmental standards are compatible with, and even produce, competitiveness, the British see a trade-off between protection and economic growth.

Finally, Britain's geographic characteristics have led the British to favour 'monitoring the air or water itself, instead of monitoring the source of the pollutant' (Golub 1994: 37). They therefore support the use of environmental quality standards, whereas Germany favours the use of strict emission standards. Britain therefore has often opposed German efforts to impose European-wide emission standards. Much of the conflict within the Council of Ministers is rooted in that difference of approach. Such a conflict coloured the politics of water-pollution policy and hindered the passage of relevant directives for years. In the eyes of the British, the fact that they enjoyed short, fast rivers and the sea gave them a comparative advantage which they wished to exploit to increase competitiveness. (See Box 9.2.)

In addition to disagreement rooted in basic differences of approach, the British (along with other countries) have strenuously opposed environmental directives which ran counter to the interest of the affected industry. The

---

Box 9.2. Differences in environmental perspectives

GERMAN VIEWS	BRITISH VIEWS
• High standards help economic competitiveness—ideas linked to 'ecological modernization' are important	• High standards may hurt competitiveness —ideas linked to ecological modernization are not widely accepted
• Precautionary principle—policy-makers are willing to go 'beyond science'	• Distrust of 'precautionary principle'— policy-making is influenced by scientific data
• Emphasis on 'Best Available Technology' (BAT)	• Emphasis on 'Best Available Technology Not Entailing Excessive Cost' (BATNEEC)
• Use of uniform emission standards	• Use of environmental quality standards
• Legalistic regulatory structure with little administrative discretion	• Flexible regulatory structure with much discretion in enforcement
• Belief in articulating general principles	• Policy relies on pragmatic case-by-case approach
• Generally, cost is *not* an overriding consideration	• Cost *is* a very important consideration

long battle over car emissions was rooted in the desire to protect the investment made by British (as well as Italian and French) industry in 'lean burn'
technology, whereas the German position supported the use of catalytic converters, in which German auto manufacturers had invested in order to export
to the American market. Although the Green Party in Germany and environmental groups called for speed limits as a way of minimizing environmental
damage, German auto manufacturers responded very harshly. They viewed
such limits as damaging their hold on the global market for high performance
cars. The German car industry successfully kept the political debate centred
on their preferred technology (catalytic convertors) rather than on changes in
personal behaviour which might have resulted in the same degree of environmental protection (Boehmer-Christiansen and Weidner 1992: 48).
Interestingly, they successfully ruled out of consideration alternative technologies, such as lean-burn, being developed outside Germany, and insisted
on their own as the base for the European directive.

One of the most persistent and chronic undercurrents in the environmental policy process has to do with British complaints about a directive *after*
Britain has voted for it in the Council. Such complaints have been particularly
important in the area of water-pollution policy (Haigh 1992*b*: 4.2.1). The
complaints have centred on the cost of implementing the directives, a cost
which the British government often argues is much higher than had been
forecast. Such complaints have now become an almost regular feature of the
policy process. Since the ratification of the TEU, the British have used the
concept of 'subsidiarity' to argue for repeal or simplification of water directives (*ENDS Report* 1994: 17–21).

The directive on urban waste-water treatment (91/271) is a recent example
of that dynamic. The directive's scope is much broader than that of the
bathing-waters directive which itself caused considerable controversy in
Britain. None the less, the directive was agreed to unanimously in 1991.
Shortly after the directive became operative, however, Britain began to claim
that compliance would be excessively costly. In November 1993 the British
Chancellor of the Exchequer, at a meeting of the EU finance ministers, asked
the Commission to delay implementing the deadlines in the legislation. He
argued that the cost for Britain was five times greater than the original estimates. The Commission refused. The general impression within the
Commission was that Britain had known the cost of compliance would be
roughly eight billion pounds (Haigh 1992*a*: 4.6–8; *Environment Watch:
Western Europe*, 4 Feb. 1994: 18–19).

The cost of compliance with environmental directives, and water directives
in particular, fuelled the British (and French) demand that subsidiarity
required a review of such directives. In June 1993 the 'Franco-British' docu-

ment was released which asked for a review of twenty-two existing and proposed pieces of legislation. The water directives were included in the list. In response, the Commission agreed to 'simplify' directives adopted in the 1970s—and to allow the member states more flexibility—but not to lower the standards embedded in those directives (Commission1993*a*: 17).

The outcome of the Franco-British demand for a 'simplification' of water directives may be an unwelcome one for the countries concerned. The Commission's first proposed revision—that of the bathing-waters directive adopted in 1976—would actually add a bill of one billion pounds to the two billion pounds already being invested to bring British bathing waters in England and Wales up to existing EC standards (*ENDS Report* 1994: 17). It would not be the first time that British environmental policy-making in Brussels had had unanticipated consequences!

Conclusion

The policy process in the environmental arena illustrates how a minority of states can use the EU to force an upgrading of standards in the rest of the Union. Rather than 'environmental dumping', the Union's policy-making process has led to 'up-market environmentalism'. It also illustrates how important is, to use Helen Wallace's term, the 'magnetic field' of the EU for environmentally progressive states. Just as it is said that no country can adopt Keynesian economic policies unilaterally for long, so it may be said that no trade-dependent country can adopt environmentally progressive policies unilaterally for long either. Given the extent of economic integration within the Union, an environmentally progressive state cannot easily protect its firms without resorting to the international arena. Once environmental policy became integral to the political economy of 'leader' states, it moved from being what Joseph Weiler has termed a 'side game' into a game with high stakes for all parties concerned (Weiler 1991: 2449).

The comprehensive scope of the Union—comprehensive in comparison with traditional international organizations—and the binding nature of decisions adopted there allow states to 'externalize' many more environmental policies across more issue areas in that forum than in other international fora. Once the idea of environmental protection as separate from the harmonization of trade barriers was accepted as a legitimate goal, and once the notion that EC policy should focus exclusively on transboundary issues was implicitly rejected, it was but a relatively short step to the regulation of urban waste water. It is nearly inconceivable, however, that any other international organization would be allowed to concern itself with urban waste-water treatment or the quality of drinking water. If 'domestic' issues exist at all, one would expect those two to be considered 'domestic' rather than 'international'. Once twelve member states have unanimously agreed to deal with the problems of urban sewage and drinking water at the Union level, it is fair to ask whether we are dealing with transnational politics or with something akin to Wolfgang Wessel's notion of 'fusion'—in this case a fusion of national and transnational policy-making.

The EU's institutions—including the Council—play a critical role precisely because EU-wide 'epistemic communities' are less developed than might be expected, and policy networks rarely cross the 'leader–laggard' divide. Europe is divided in rather fundamental ways on environmental questions. Yet consensus has been obtainable hundreds of times across a large number of specific and controversial issue areas because the division of Europe on environmental issues is embedded in an institutional structure premissed on the assumption that consensus is in principle possible.

The dynamic outlined in this chapter is one in which the domestic politics of environmentally progressive states is critical for the momentum of the Union's institutional process. The 'leader' states are under constant pressure at home to 'Europeanize' domestic legislation. Once Brussels becomes involved, the other member states become engaged in a process which upgrades their domestic standards. Their environment ministers are able to agree to policies in Brussels which would be very difficult to adopt in national capitals. The Union's policy-making process therefore is influenced by domestic politics in 'green' countries and in turn affects domestic policy-making in countries reluctant to initiate environmental policies on their own.

Notes

1. See also (Vogel 1993a).
2. Roughly the same analysis is given by Laurens Jan Brinkhorst, who served as the Director-General of DG XI until his election to the European Parliament. Brinkhorst included Belgium in the 'middle' category with the UK, France, and Luxemburg, but otherwise his analysis echoes that of Ludwig Krämer (who headed the enforcement division of DG XI). See Brinkhorst (1991: 91).
3. It is important to note that on any specific environmental issue, the Netherlands, Germany, and Denmark may be joined by one or two other countries which, while probably less ambitious in their policy goals, none the less can be viewed as strong allies of the core three. In the case of climate policy, for example, Italy and Belgium joined the pro-protection group. See (Skjæreseth 1994: 34–5).

4. For an excellent description of German environmental policy in a variety of areas, see OECD (1993*b*).
5. Car manufacturers strongly opposed the setting of speed limits within Germany and supported the mandatory imposition of catalytic converters throughout Europe (von Weizsäcker, 1990: 51).
6. The information on the British role in the two years after accession is taken from Golub (1994: ch. 4).
7. See e.g. the article written by the long-serving German environment minister Klaus Topfer (1992: 278).
8. The other two main principles of German environmental policy are the 'polluter-pays' principle and the 'cooperation' principle. See Rudig and Krämer (1994: 52).

Further Reading

For an overview see Hildebrand (1992) and Liefferink *et al.* (1993*a*). For a political evaluation see Weale (1992*a*). For more detailed illustration see Baker *et al.* (1994), Haigh (1992*a*), and Zito (1995*a*).

Baker, S., Milton, K., and Yearly, S. (1994) (eds.), *Protecting the Periphery: Environmental Policy in Peripheral Regions of the European Union* (Essex: Frank Cass).

Haigh, N. (1992*a*), 'The European Community and International Environmental Policy', in A. Hurrell and B. Kingsbury (ed.), *The International Politics of the Environment: Actors, Interests, and Institutions* (Oxford: Clarendon Press, 1992), 228–49.

Hildebrand, P. M. (1992), 'The European Community's Environmental policy, 1957 to "1992": From Incidental Measures to an International Regime?', *Environmental Politics*, 1/4: 13–44.

Liefferink, J. D., Lowe, P. D., and Mol, A. J. P. (1993*a*), 'The Environment and the European Community', in eid. (eds.), *European Integration and Environmental Policy* (London: Belhaven, 1993).

Weale, A. (1992*a*), *The New Politics of Pollution* (Manchester: Manchester University Press).

Zito, A. (1995*a*), 'Integrating the Environment into the European Union: The History of the Controversial Carbon Tax', in C. Rhodes and S. Mazey (eds.), *The State of the European Community*, iii (Boulder, Col.: Lynne Riener, 1995).

CHAPTER TEN

ENERGY POLICY: FROM A NATIONAL TO A EUROPEAN FRAMEWORK?

Janne Haaland Matlary

Two of the three founding treaties dealt with energy, reflecting its importance to economic reconstruction and concerns about security of supply. But national monopolies and divergent national interests blocked Commission attempts to move towards common policies until the mid-1980s. Shifting assumptions about the policy framework, and the privatization of some national utilities, pushed for inclusion of energy in the single-market programme. The Commission used its competition powers to promote an internal energy market and to float proposals for a common energy policy. Changes in national policy and links to environmental policy shifted the attention of interest groups to the EU level. The Commission has been an active policy entrepreneur as well as regulator, using changes both in national assumptions and in the character of the market to press towards common policies.

New thinking about energy policy

Energy policy seldom made the headlines in the European Community (EC); it attracted attention only when there was a crisis of supply at hand, or, worse, a nuclear accident. During the 'oil shock' in the winter of 1973 everyone was concerned about the supply of oil and its vulnerability to political conditions. In the early 1980s a major diplomatic conflict erupted between the US and western Europe over gas imports from Russia. Energy became, for a time, 'high politics'. And the political effect of Chernobyl still lasts. But crises pass, markets normalize, and energy consumers do not think about the origin of their supply in normal conditions. As long as gas and oil flow and there are no nuclear accidents, energy issues remain largely a matter of 'low politics'.

This may seem an odd starting-point, given that two of the three original European communities were specifically focused on energy policy. The European Coal and Steel Community (ECSC) and Euratom were both designed to provide building-blocks for a collective energy policy. Yet even the 1973 crisis did not provide a catalyst for a really common policy, even though most EU members were heavily dependent on imported energy supplies, as Table 10.1 shows (Black 1977). However, in 1980 one major energy-producing country started to reform its energy sector, namely the UK. The energy sector was to be privatized, along with other important industrial sectors which had traditionally been publicly owned and managed, such as telecommunications. The thorough-going reform in the UK was symptomatic of new thinking about the organization of energy markets in other countries as well. A debate began on whether energy really was such a national concern after all, and on whether the introduction of free-market rules might make the energy sector and energy trade more efficient. In the 1980s this discussion surfaced in the Scandinavian countries, in France, and in the Netherlands. There was an emerging 'paradigm shift' in thinking about the state's role in the traditionally public sectors of the economy. Why should the state continue as the owner in these sectors? If not, why should energy policy be the responsibility of state?

This development occurred in a period of stable political conditions: there were no major supply crises, and little attention was focused on the question

Table 10.1. Imports of coal, oil, gas, and electricity in EU countries (projected million tonnes of oil equivalent for 1995)

Country	Coal	Oil	Gas	Electricity
Austria	3,5	10,6	4,9	0,6
Belgium	10,3	24,4	11,3	
Denmark	7	0,6		
Finland	5,5	12,7	3,3	0,6
France	15,4	99,3	26	−5,8
Germany	15,8	134,2	47	0,5
Greece	1,4	22,4	0,6	
Ireland	2	6,3		
Italy	14	84,7	33,4	3
Luxemburg	0,6	1,8	0,5	0,45
Netherlands	7,2	22,5	−27,9	0,8
Portugal	3,4	13,2		0,2
Spain	9,5	68,3	7,2	0,1
Sweden	2,7	25,3	0,8	0,6
UK	9,6	−12,7	7,4	1,1

Adapted from the IEA's 1993 review, *Energy Policies in IEA Countries* (Paris: OECD, 1994)..

of import dependence in most western countries. The role of the EC in this policy field was thus 'prepared' by a policy discussion that increasingly took place primarily between national governments and their relevant interest groups in the energy field.

The 'givens': the importance of structural interests

Energy policy is, however, a policy area in which there exist very clear structural interests. A country which is primarily an energy importer has different interests from a country which is an exporter, and these givens are difficult to manipulate. Table 10.2 shows the considerable differences in production and imports among EU member states. Thus, the freedom of action in energy policy may be very small. Since energy, necessary for a country's industrial base, is in many ways a 'strategic good', its procurement has been regarded as vital to a nation's well-being and security. For this reason energy supply has generally been a highly national concern. There has been little of a 'free market' in products other than in oil, and even for oil the political aspects of supply have been very important. Governments have sought to diminish their dependence on imported energy, and thus developed distinct national approaches to energy policy. As Black underlined in an earlier edition of this

Table 10.2. Key energy indicators for EU countries (projected million tonnes of oil equivalent for 1995)

Country	Production	Imports	Total
Austria	9	18.7	27.7
Belgium	11.4	42.7	54
Denmark	14.7	5.3	19.8
Finland	12.3	19.7	32
France	118.9	124.4	243.3
Germany	156.4	194.3	353.1
Greece	9.3	15.8	25
Ireland	3.3	7.4	10.8
Italy	29.7	132.8	162.5
Luxemburg	0.03	3.5	3.5
Netherlands	66.5	2.6	68.8
Portugal	2.1	17.2	19.4
Spain	30.5	68.4	98.9
Sweden	30.4	196	50
UK	212.5	5.5	220

Adapted from the IEA's 1993 review, *Energy Policies of IEA Countries* (Paris: OECD, 1994).

volume, policy cooperation within the EC has therefore been limited (Black 1977).

Thus, whenever a crisis upsets the normal functioning of the market, political issues tend to intrude, a factor that should be kept in mind when analysing European energy policy. Recently the emphasis has been on introducing market rules across Europe, but the traditionally strong role of the state in energy policy could be reactivated. Even if energy policy gradually becomes less of a national concern and more of a shared European concern, the political nature of issues such as supply security or the environment is never far from the agenda.

The history of EC energy policy

The EC was conspicuously absent as an international actor in the energy area during the international political crises about energy policy in the 1970s. The oil shock of 1973–74 induced a strong reaction among western countries that resulted in new multilateral cooperation and in the creation of the International Energy Agency (IEA), as a buffer against price hikes. The IEA was linked to part of the OECD.[1] In the EC there was a concern about import dependence, but little agreement on how to tackle the problem beyond bilateral arrangements between member states and the Organization of Petroleum Exporting Countries (OPEC). They managed only to decide to develop so-called guidelines for energy policy in response, the weakest form of policy co-ordination, and not legally binding in any way. In the period 1973–88 European energy policy did not develop beyond guideline recommendations which represented 'lowest common denominator' compromises.

The EC lacked the overall competence for energy policy, despite the existence of the ECSC and Euratom. The former sought to create a common coal and steel policy, but no coal policy emerged. Euratom aimed at developing the basis of a common approach to nuclear energy, but this did not succeed. The Treaty of Rome (EEC) did not mention energy policy specifically, although the general provisions and rules on competition apply to the energy sector, as to other industrial sectors.

It is not surprising that the conventional judgement has been that energy policy is among the 'weakest' policy areas in the EC, because all major efforts to create a common energy policy have failed. Stephen George thus concludes that energy policy may develop beyond national policies in the 1990s, but that it remains an area where national policies have been very strong indeed (George 1991). Stephen Padgett finds that 'the strategic economic importance of the energy sector meant that policy autonomy was guarded jealously by national governments' (Padgett 1992).

However, as we shall see, the role of the Commission in creating the elements of an energy policy has been greater than these studies assume. After 1985 the energy sector was increasingly drawn into the drive to create the single market. The new thinking about a natural-energy policy, which we regard as the beginning of a 'paradigm shift', also had consequences for energy policy at the European level. Even without a formal competence for energy policy the Commission could take advantage of both developments to open up the policy debate.

Furthermore, the Single European Act (SEA) brought a procedural advantage. While member states can veto most proposals in energy, proposals about the internal energy market (IEM) are decided by majority voting. This has been a dynamic element in the decision-making process which the Commission has been skilled at exploiting. Also, recourse to the European Court of Justice (ECJ) has been an important strength for the Commission. In the period since the inception of the IEM in 1988, DG IV, the Commission service responsible for competition, has been active in applying the competition rules to the energy sector, hitherto relatively free from such interference. However, in spite of these assets for the Commission, member governments and business groups have had strong and entrenched interests. There is still a long way to go to the achievement of the IEM as the Commission would like to see it, and even further to the creation of a truly common energy policy.

The 1980s brought a general emphasis on privatization in previously public sectors, especially in the UK. Interest grew in looking at the possible benefits of deregulation as such. In the USA, for example, the gas sector was already privately owned and had begun to be deregulated. In Europe views about energy policy changed, and it was no longer a foregone conclusion that energy would remain a concern of national policy or that supply security would remain at the top of the energy-policy agenda. Instead market actors, as well as politicians, started to explore ways of privatizing and deregulating the energy sector. This trend coincided with the development of the ideas behind the single market: the latter was a consequence of the need for freer markets in general in Europe, and thus not unconnected with the debate on energy-sector reform as such.

From 1988 the EC had a mandate to develop an internal energy market as part of the general single-market mandate. However, there was no mandate to develop a common energy policy in terms of, for example, supply security. The internal market was about creating common-market rules for the trans-portation, sale, and other elements of trade in energy products. But the emerging 'paradigm shift' as regards energy policies enabled EC develop-ments to acquire much more importance. EC proposals set the tone for car-rying forward much of the new thinking, and in this sense ideas emanating from the IEM debate carried a significance in themselves, even though not all the proposals made were adopted. I shall return to the importance of ideas in this policy field in the final section.

It is unclear what member governments expected from the IEM. On the one hand, they endorsed the single-market mandate; on the other hand, they resisted its impact whenever there was some disadvantage to their national energy sectors. Logically, the freer the energy markets, the less political power resides with member governments and the greater the scope for the Commission as regulator. The member governments were thus wary of the Commission's agenda, concerned lest the IEM entail the gradual unfolding of a common energy policy (CEP). As we shall see, the Commission tried to develop both an IEM and a CEP at the same time, but it was severely con-strained by entrenched national and business-group interests.

The Commission used the threat of applying the competition legislation in order to put pressure on the member states in the negotiations over proposed directives, and partly succeeded. DG IV had applied the competition legisla-tion to other sectors, such as transport and telecommunications, and drew on this precedent in dealing with cases in energy. The ECJ has often proved an important influence on policy. DG XVII of the Commission, responsible for energy, for a long time tried the negotiated approach, based on submitting draft directives to the Council, since this carried more political legitimacy than the 'Court route', but it turned to the Court when the negotiated approach failed. But an energy policy could not be premissed on the internal market alone. The Commission therefore used external 'windows of opportunity' to propose policy that would respond to a certain 'demand' in the member states. It also tried very hard to establish a formal competence for a CEP in the Treaty on European Union (TEU).

From the internal energy market towards a common energy policy

In 1988 a working paper from the Commission listed all the obstacles to the creation of an internal energy market.[2] The main White Paper on the Internal Market (Commission 1985) did not mention energy at all. It was thought to be a policy field where it would be next to impossible to develop a market according to competition rules. The energy sector in Europe was (and is still in many ways) typified by national monopolies of imports, exports, sales, and transmission (Matlary, forthcoming; McGowan, 1993; Stern, 1990). None the less it was agreed that there could be no real internal market in the sense of a 'level playing-field' unless the energy sector was included. Industry in some countries clearly enjoyed the benefit of cheap energy, and the transportation of energy across Europe remained the prerogative of national companies, each operating on different terms. There was thus a clear *rationale* for attempting to establish common rules for transit and transit tariffs, as well as trying to break up energy monopolies. DG XVII was charged with the difficult task of creating an IEM. This work started in 1989 with the first directive proposals, to be discussed below.

The internal energy market proposals

DG XVII of the Commission introduced a package of proposals for the IEM during 1989 including transparency of electricity and gas prices; less restrictive rules on transit for gas and electricity; and plans to monitor large investments in the energy sector.[3] These proposals met with varying degrees of resistance: the proposal for electricity transit was adopted without too much resistance, but the transit question for gas was more difficult. It was the subject of negotiation and reformulation for almost two years, before being adopted in October 1990 with Germany and the Netherlands voting against. The final directive had been modified considerably, and safeguarded the predominance of the established transmission companies.

The reason for this controversy was that gas transmission in Europe traditionally has been dominated by a handful of companies that own the national grids (pipeline systems). They set the conditions for the transportation of gas through their pipelines, including the tariffs for this service. These companies did not want to lose control of the transmission system, and rejected the idea that the transportation system should be seen as another 'highway' to transport a good. Gas trade is typified by long-term contracts of up to thirty years' duration, the normal lifetime of a power plant. It was feared that open access to transit might upset this long-term character: customers might start to 'shop around' for the cheapest gas, instead of being bound to a contract with a single supplier/transmitter. This in turn might mean that the large and distant gas fields in Siberia and in the North Sea would not be developed, as they were more dependent on long-term contracts to back investment decisions. A 'freer' gas market would thus undermine the then market structure in Europe, something which entailed advantages for some, disadvantages for others. In particular Ruhrgas and Gasunie opposed the plan. In this matter national governments largely defended the positions of their national companies. Only the UK, which had already deregulated its gas transportation system, fully supported the Commission's proposal.

Some other IEM proposals fared better, although it proved difficult to take big steps to make energy prices transparent. This is also connected with the diverse taxing structures, which make it difficult to compare prices. The Commission hopes that, with time, prices will converge as a result of other measures to create an internal energy market.

However, the open access issue was not abandoned, despite the opposition. The Commission reopened the issue in summer 1991 with a draft directive on the further liberalization of gas and electricity transit. This proposal had been discussed in working groups for over two years, and in early 1995 there was still no agreement. The French and German governments do not want an open transportation system for electricity and gas, as they stand to lose political control over the energy sector, while only the British can easily support this proposal, since their domestic legislation already anticipates it. Here we see that the member governments are wary of losing their control over energy policy, although they might find it hard to argue against a freer energy market as part of the internal market effort. Their economies would benefit from lower energy prices, but their embedded energy interests dictate that they still think in terms of national policy. For instance, in France the large share of electricity production from nuclear energy makes it advantageous to have access to other countries' transportation systems in order to facilitate exports, but the French none the less want to retain the monopoly role of Electricité de France (EdF) within France, traditionally safeguarded by public ownership. This same tendency can be observed also in other countries: governments would like to reap the benefits of an internal energy market, but not to bear the uncertainties of losing political control over the energy sector. Governments' responses to the IEM have thus been ambivalent, most clearly reflected in attitudes to the open-access proposals.

A second area related to the IEM is a new legal provision for developing infrastructure in the Treaty on European Union (TEU). This provides for a common programme for the coordination and development of infrastructure, thus granting new competence to the EU in this field, with a 'Cohesion Fund' to be used to finance new trans-European networks (TENs).[4] The inclusion of an EU competence in this area does not represent a new area of policy; it echoes the objectives of the IEM in aiming both to open up the existing energy infrastructure and to aid the construction of infrastructure in the 'cohesion' countries. A fully developed IEM might logically require an integrated infrastructure across the EU and beyond. By early 1995 the Commission had defined several energy infrastructure projects and assigned them the status of TENs,[5] eligible for special funding from the Cohesion Fund and from the European Investment Bank. There are some hopes that planning for new infrastructure will increasingly take place at the European level, thus making it possible to develop energy networks in southern and in eastern Europe.

A third area of the IEM is tax policy. Here little progress has been made, although it is the aim of the internal market policy gradually to arrive at a common tax structure. This is not an area easily amenable to harmonization, as the work on value added tax has shown.[6] One very controversial proposal has been the energy/CO_2 tax, which so far has been outvoted (see below).

Finally, it should be added that the development of the European Energy Charter ('the Charter') has been linked to the IEM. The Charter is an international legal treaty that aims at stabilizing investment rules for energy exploration, production, and transport in the signatory countries. It was conceived by the Dutch government in 1990 as a way of making investment in the energy sector in Russia, what is now the Commonwealth of Independent States (CIS), and eastern Europe more attractive. In one sense the Charter can be read as an extension of the IEM to all of Europe and beyond, in so far as its provisions insist on the application of free-market principles in the CIS and in eastern Europe. The legally binding Charter treaty[7] was signed in Lisbon on 14 December 1994 by most western and industrialized countries. The

Charter itself is not legally binding, but the accompanying treaty and specific protocols will be binding on its signatories.[8] The latter will cover the areas of environment and energy, hydrocarbons, and nuclear safety. The aim of the Charter is to promote an efficient energy market through the price mechanism, with due attention paid to the environment, as the section on objectives reads.[9] The protection of western investments is of particular importance here. The repatriation of profits will also have to be guaranteed. So far it has been a grave problem in the CIS that western investors have been worried about political and economic stability, and thus have postponed investment decisions, in ironic contrast to the stability and predictability under Communism.

Finally, DG XVII has proposed applying the competition legislation also to the upstream oil and gas sector. This means that rules governing concessions must be harmonized and that national companies cannot enjoy a privileged position. However, the control over depletion levels and other resource management issues is to remain with the member state.[10] This directive proposal also met considerable resistance, and was adopted in a modified form only. The energy-producing countries were sceptical about the intentions of the Commission, fearing that the aim was to control national energy resources. The Commission argued that market actors must have non-discriminatory access to participation in the exploration and production phases of energy trade, and that national companies could not retain a privileged position. Also Norway, at the time a prospective EU member, protested vigorously against the original draft directive.

The common energy policy proposals

Some energy policy proposals primarily aim at creating a freer energy market while others entail concentrating more policy power at the European level. The distinction refers both to the contents of the policy and the consequences for EU actors, especially the Commission. A market-liberalizing proposal would not necessarily increase the Commission's powers, but, for instance, to adopt open access for gas implies that the Commission might become the regulator monitoring the implementation of rules and setting transport tariffs. Although the IEM may aim at more market-responsive rules in general, the political power to set the rules may be moved from the level of the state to that of the EU. I shall return to this issue in the concluding section.

The CEP proposals imply that a given policy proposal could not be implemented unless the EU institutions acquire an explicit or implicit transfer of power. The proposal itself may also aim at a freer market, but at the same time presumes a greater integration of energy policy. The best example of this is the Energy Charter. Its aim is *inter alia* to extend the IEM (or at least market-based operations) across Europe, and it allots a major policy-making and coordinating role to the EU, which provides the secretariat for the treaty.

The more radical CEP proposals involve a more explicit EU competence in energy policy. In 1990 the Commission sought to apply for membership in the IEA, implying that the EU would speak with one voice in that organization. This proposal was not accepted; instead the EU has observer status in IEA, with the Commission representing the EU. Another proposal, presented shortly after Iraq's invasion of Kuwait, would have given a much greater role than before to the Commission in deciding when to use emergency oil stocks.[11] Both these proposals would have meant a more central role for the Commission in energy policy, and both were rejected in their original form by the Energy Council in May 1990. The position on IEA membership for the Commission was not, however, negative, but could not be clarified without agreement on a clear demarcation line between the EU and the member states themselves. It is significant that both proposals had been 'hastily put together at the start of the Gulf crisis since Cardoso e Cunha, then energy commissioner, saw the political opportunity a Gulf crisis might offer' and signified that the Commission 'was requesting sweeping powers'.[12] The Commission tried to use such 'windows of opportunity' to enhance its role as an actor in international energy policy.

Obviously the whole story would have been different if energy policy had been included in the Treaty of Rome (EEC). Energy policy has instead been based on the ECSC and the Euratom treaties, general policy provisions, and more recently the TEU provision on TENs. During 1989 and 1990 various suggestions had been made about the need for a legal basis for energy policy. Thus the Commission proposed to the Intergovernmental Conference (IGC), that 'as far as energy is concerned, the Treaties could be consolidated into a single chapter making it possible to implement a common energy policy' (Commission 1990c).

The effort made to establish the IEM had underlined the need for a more extensive common policy in the sense that in some important cases it implied decision-making power on the part of the EU which as of yet did not exist. The more far-reaching of the Commission's proposals highlighted this lacuna. The proposal to the IGC for a legal basis for energy policy would have included the IEM, a Community-wide system of supply security, closer integration of energy and environmental policy, and an 'external policy which enhances the Community's standing and influence on the world stage' (Commission 1990c). This proposal could be so far-reaching because of the political 'windows of opportunity' which presented themselves at the time. The turn of external events—again in particular the energy-environmental crisis in central and eastern Europe and the Gulf crisis—underlined the need for additional policy-making powers in the energy field, according to the Commission.

As the IEM developed further, the need for a coordinated policy for security of supply became apparent on the logic that the freer the market, the more vulnerable the importer becomes to an interruption of supply, as no state actor can secure supply for a market that extends beyond national borders. The point here is that the more internationalized the energy market becomes in terms of common rules, the less control the national government can exercise. There is thus logic in the call for EU-level policy interventions to take care of problems that the market cannot solve. It is therefore not surprising that the Commission included this point with its proposals to extend the IEM.

The CEP loomed large in the Commission's subsequent work programmes on energy. The argument was also made that the IEM activities demanded a greater Commission involvement in the field of infrastructure, to oversee and to aid the development of the electricity and gas grids. In the Work Programme for 1995 announced by the Santer Commission the energy priorities remain the completion of the IEM and the development of a CEP.

During 1994 DG XVII drafted a Green Paper on the CEP, with three pillars: security of supply, the IEM, and the environmental implications of energy policy. Each of these can be used to claim a more extensive role for the Commission, as a regulator of the IEM, a role which the Commission would like to occupy, and, in due course, as manager of an external supply policy. These are both responsibilities traditionally associated with the state.

Finally, some of the most important policy issues lie in the field of the environment. Environmental problems are very often also energy problems, stemming from 'dirty' or excessive energy use. The TEU considerably reinforces the legal basis for environmental policy. This could stimulate calls for a Community energy policy which favours some types of energy over others on environmental grounds. This is where the future battle over the CEP may well lie. The environmental consequences of energy use present a host of political problems that must be solved at the international level, from global warming to local and regional pollution stemming from emissions to air of NO_x and SO_2. There already exist important environmental regimes and conventions in this area in which the EU acts collectively. In the future it will be increasingly important for energy policy to be aligned with environmental policy. The case will be made for developing a common approach and for negotiating collectively through the EU. The Commission will certainly argue that this reinforces the need to develop a CEP.

The idea of a CEP thus directly addresses the merging of energy and environmental policy. So far the most tangible result of this has been the proposal for an energy/CO_2 tax, explicitly designed to penalize energy and carbon use. Various proposals have been floated since the first oil crisis of 1973. It would be a big step to agree on the direct levying of a Community tax. Resistance from several governments, including the British and the Spanish, has, however, put the tax proposal on the 'back burner' for the time being, but it is certain to resurface on the agenda in the future. The Commission has reached the view that eco-taxes have come to stay, and further efforts can be expected to promote agreement on EU-level environmental taxes. The emphasis of European policy on direct regulation as an instrument of environmental policy may prove hard to sustain in a modern market economy. As energy markets become more international, market actors will also increasingly demand common environmental rules and regulations to ensure the 'level playing-field'. This demand has already started to make itself felt.

Another stimulus for a CEP, with an environmental dimension, is driven by developments in central and eastern Europe. Already environmental and energy-saving measures are being linked to much financial assistance to the area. The Charter contains clauses on the environment and allows for transitional arrangements for this region. In this development phase the policy that emerges will often be initiated by the EU as well as funded by it (Matlary 1993: 2). In the economic development of these countries the challenge will be how to achieve economic growth without taxing the environment unduly. The influence of EU policy in this region could be critical and much will depend on the linkages between an intra-EU common policy and the projection of that policy eastwards.

The Commission has, as we have seen, produced a large number of policy ideas and proposals on both the IEM and the CEP in the period since 1989. While some have been rejected, many have been negotiated and adopted in some modified form. Importantly, the need for a CEP has been persistently advocated throughout this period by the Commission, which has seized opportunities to advance by small steps. The Commission might well propose a more explicit competence for energy policy to the 1996 IGC, building on the White Paper scheduled to be published in 1995.

In the 'national' interest?

Energy policy is an area where it still makes much sense to talk about national interests. They not only exist, but they are also largely 'givens'. In policy discussions on energy issues member governments have fought to assert their national interests, yet they have been willing to change the organization of their domestic energy sectors to some extent. At the outset all member governments were keen on protecting what they saw as the 'givens', but the debate on deregulation and the presumed benefits of a common energy market have none the less been significant. To some extent perceptions of how national interests can best be promoted have altered. An example of such a change is the UK, which deregulated its domestic energy sector before the IEM debate and despite entrenched national producer interests. The UK became the most willing 'deregulator' in the EU: it has supported the IEM proposals, but balked at all attempts to make a CEP, while France and Germany have been resistant to too much open access in the gas and electricity sectors. Italy and the 'cohesion countries' have been more ready to accept both the IEM and the CEP.

Table 10.3. Policy responses to the IEM and CEP

	France	Germany	Italy	UK
IEM, stage 1	yes	no	yes	yes
IEM, stage 2	no	no	no	no
Charter	yes	yes	yes	yes
Treaty revision	yes	no	yes	yes

Table 10.3 shows the views of the few larger member states on the various energy-policy proposals and is based on an extensive empirical study of their role between 1985 and 1994 (Matlary, forthcoming). Among the member states there are at least two blocs of interests: the importers and the exporters. The importers are in a majority and favour more of a common energy policy than do the exporters, which include the UK, the Netherlands, and Denmark. As Black pointed out in 1977, the UK has resisted all attempts at creating a common energy policy in the EU. However, the UK has supported a freer energy market and thus most of the internal energy-market proposals. Other major energy importers have different interests—Germany still subsidizes its domestic coal industry, while France strongly favours the expansion of nuclear energy. Finally, the 'cohesion' countries want a more common energy policy, as this may ensure additional funds for infrastructural and other developments in the energy sector.

National interests vary between countries, between sectors, and between the richer north and the less rich south. In the south the energy sectors are not very well developed in terms of infrastructure, while in the north new types of energy-related problems come to the fore, notably those related to the environment. By all measures national interests in energy policy remain strong, and thus the interesting issue is really to what extent they may be transcended by the increasing need for a 'level playing-field' and by interest-group pressure for this.

The role of interest groups

Interest groups do not form a homogeneous group, but can roughly be classified as energy producers, energy transmitters (transporters), and energy consumers. In addition, environmental groups are increasingly important to the energy-policy process. Energy producers sometimes favour a freer market in energy in Europe, but are wary of the implications of moving away from long-term contracts that typify trade in certain products such as natural gas. Energy transmitters, especially for gas, dislike the prospects for open access to gas infrastructure, because this will subject their business to politically imposed rules and tariffs. Energy consumers typically favour a more open market, with transparent pricing, as this will lead to lower prices, whereas environmental groups are sceptical about the free-market proposals exactly because lower prices would lead to increased use of energy.

Interest groups play a major role in the policy process at all stages, perhaps especially in the initiating stage. They are consulted, and many have developed their own European umbrella organizations with secretariats in Brussels. National interest groups lobby their own governments at home, while at the same time they often form part of the European organizations. Interest groups have been very important in providing DG XVII with policy input and ideas in the initial stages of policy formulation, as they possess great expertise. On the often technical issues that concern energy policy, they play an important role. The Commission has a close relationship with these groups. However, they have not been able to change the basic internal energy-market concept which is 'nested' firmly in the internal market mandate.

The role of the Commission and competition policy

In general the Commission has pursued a two-pronged strategy. Knowing the extent of member governments' opposition to many proposals, DG XVII has gone slowly, favouring the negotiated approach and 'directive route', even in cases where the competition legislation could have been applied directly. DG IV has had a relatively minor role so far in the energy field, but there is evidence that this role is being stepped up. The particular nature of the Commission's powers in this is very important.

A key strength of the Commission lies in its strong legal powers in operating the competition rules. In the energy area these rules had not been invoked; it was only with the advent of the single market in 1985 that governments began to accord the Commission a wider legitimacy in the field. Since this period the member governments have found it harder to judge either the Commission or its CEP proposals. Our view is that competition legislation can be successfully used by the Commission only in a period where its role as policy-maker is accorded general legitimacy, as was the case during the internal market period.

When Sir Leon Brittan was the Commissioner responsible for competition, he made the application of the competition rules more rigorous, including in the energy sector, an approach that has been continued by his successor, Karel van Miert. Brittan, for example, sharpened the definition of permissible monopolies: 'they can only be allowed if they are absolutely necessary for the provision of a public good which the market is unable to provide' (Matlary 1993). In effect, this implies that the gas and electricity monopolies must be dismantled, in terms of both import and export monopolies and transportation. The Commission then asked all member states which had such monopolies to abolish them or to explain why they were deemed necessary, arguing that they were unlawful under Article 37 (EEC) (*Agence Europe*, 22 Mar. 1991). This move was the first step in a legal procedure to force the member states to dismantle the monopolies. In late 1994 the Commission took these states to the ECJ over this issue.

In his statement in the annual report on competition policy in 1990, Brittan stressed that a major aim was to achieve competition in previously shielded sectors, characterized by national monopolies. Only minimal norms would apply in sectors where there arguably existed a public service function such as gas, electricity, postal service, transport, banks, and audiovisual services (*Agence Europe*, 24 June 1991: 11). DG IV has been very insistent that the energy sector also be subjected to competition legislation. DG XVII, however, has been wary of such direct intervention, as this might be a very costly political strategy. In the long run the Commission relies on the continued support of the member states in the sense that they have to consider the Commission's work legitimate. However, DG IV has consistently used the internal market mandate in sectors where exercise of the rules had been dormant or never applied. It has applied it strategically, starting with the 'easier' sectors like telecommunications, and moving to notoriously 'difficult' sectors with heavy monopoly traditions, like energy.

Utilizing 'windows of opportunity'

The CEP proposals, however, demanded a different strategy. The Commission launched a host of proposals that related to energy in central and eastern Europe, energy and the environment, energy supply, and so on. These proposals were aided by *external events*, which provided DG XVII with 'windows of opportunity'. Commission officials sought to provide policy to fill a policy vacuum—the Energy Charter is a good example—also hoping to enhance the Commission's role as a policy actor. In the absence of a formal competence the Commission needs to substantiate the functionalist argument that there is a *need* for policy solutions at the European level. This is a plausible conclusion in areas where individual member states alone cannot deal with problems, such as: the issue of nuclear energy in the CIS and eastern Europe; or regional and global environmental problems. There is a clear rationale for making policy at the international level, though it does not follow that the EU is the obvious level. Some of the proposals, such as the Charter, clearly filled such a 'policy need'. Others have not convinced member states, notably the CO_2/tax proposal. The acceptability of the Commission's policy has depended greatly on the current political climate.

Summing up, the process of creating both an IEM and a CEP has been slow, and proposals have mostly had to be modified to win agreement. The Commission has, however, drawn strength from the competition rules and the possibility of recourse to the ECJ. Its role as an *agenda-setter* has been of crucial importance: interest groups as well as member governments turned to the EU as an arena for policy-making. The role of member governments is formally very important, yet a policy process which is so much based on non-political, technical, and factual arguments can shape decisions. National representatives have to master the complexities of the policy problems and to 'translate' their interests during negotiations through technical arguments. Majority voting puts more pressure on governments to make their arguments count and not simply to rely on their apparent political weight. It is too early to pass judgement on the success of the Commission's persistent advocacy of a broader energy policy, but we may at least conclude that it has succeeded in putting this at the centre of the agenda for Europe's energy policy-makers in the span of only five years.

Once it becomes accepted that policy actors must relate to the agenda in Brussels, the Commission can play the role of 'policy entrepreneur', a role that it is uniquely positioned to play (Majone 1994*b*: 22). This is due both to the institutional prerogative of the Commission as the 'drafter of the texts' and to its ability to engage a wide range of policy-makers, from national experts to interest groups.

The Commission as regulator

As deregulation has gradually replaced public and state ownership in the economy, the tasks of the state have become increasingly related to regulation (Majone 1994*b*). A regulator needs a strong legal basis for intervention, with watch-dog functions rather than large budgets and power in the traditional sense. The Commission is thus well positioned to make an impact as a regulator, given the institutional relationship between the Commission and the ECJ and the very strong legal basis in competition policy. Another important factor is that interest groups may well prefer a European regulator to national supervision, in that the need for a 'level playing-field' across Europe is a prime concern for multinational corporations. 'Another important element in an explanation of the growth of Community regulation is the interest of multinational, export-oriented industries in *avoiding inconsistent* (my emphasis) and progressively more stringent regulations in various EU and non-EU countries. Community regulation can eliminate or at least reduce this risk' (Majone 1994*b*: 17).

This is exactly what has happened in the energy field. Although interest groups were at first mostly opposed to the IEM and any Commission involvement in the energy field, they gradually accepted that the IEM debate was not going to go away, and therefore sought to influence its development. The concern for a 'level playing-field' was the foremost reason for the involvement of the European electricity industry, which initially had sided with their national governments in keeping national privileges. In late 1994, however, Eurelectric, the association of European electricity companies, changed its former position completely, arguing that the changes towards deregulation that take place at the national levels in Europe are so important that some European regulation is needed to avoid market distortions between countries. 'Substantial changes in the electricity sector have been introduced or are being considered in almost all member states. It is important to ensure a sufficient degree of convergence between these different structures' (*EC Energy Monthly*, 21 Oct. 1994: 16). Thus, there is scope for the Commission to play an increasingly important role in ensuring that rules are followed, as well as in shaping them.

Conclusion

The policy process in energy is thus marked by the importance of embedded national interests, but also by the force of ideas and by the changing character of the market. It is not surprising that national interests figure greatly in the political calculus of member governments, or that interest groups have sought to define the agenda so that their particular interests can carry the day. The balance of the argument shifted to favour an approach to energy policy that would be more European as the single-market philosophy became established. Only a few years earlier, it had been assumed that energy policy was a national concern and that too much market freedom would undermine the role of the state. Today there is much more pluralism in thinking about energy policy by both national and interest-group actors. We argue that this is the beginning of a paradigm shift.

Also the process itself has been important in the sense that market actors have started to anticipate that energy policy decisions will increasingly be made at the EU level and that they should therefore direct their attention there. This has become something of a self-fulfilling prophecy: actors turn to the EU, and as some rules become shifted to the European level, pressures develop to follow with more European regulations. Time plays in favour of both an IEM and a CEP.

In terms of substance the policy process has been slow. There is still not much of a common policy at the EU level. Yet this picture is not so very different from other policy areas. The Commission's goals have been ambitious; proposals have had to be modified and sometimes abandoned. The big achievements have been 'progress' in applying competition legislation and the increasing recognition of the way to pull energy and environment issues together. The Commission has struck a balance between satisfying national interests, and proposing policy innovation as a clever policy entrepreneur.

Notes

This chapter draws substantially on the author's book *Energy Policy in the European Union*, to be published by Macmillan Press. The relevant sections of the chapter are reproduced here by permission of the publishers.

1. Glen Toner (1987) has analysed the policy process of the IEA.
2. *The List of Obstacles to be Overcome to Complete the Single Energy Market: An Initial Commission Analysis*; Supplement to *EEC Energy Policy and the Single Market of 1993* (Brussels: Prometheus, 1988/9).
3. *Transparency of Consumer Energy Prices*, COM(89) 123, *Draft Directive on Electricity Transportation*, COM(89) 336, and *Draft Directive on Natural Gas Transportation*, COM(89) 334, Commission, Brussels.
4. A so-called Declaration of European Interest has been proposed for networks. The Commission can grant a special status to projects in the transport or network sector, which in turn is expected to facilitate both the funding of and the choice of a given project (*Agence Europe*, 25 Mar. 1992: 9; and 26 Mar. 1992: 10).
5. A list of forty-three energy projects was adopted by the Council of Ministers on 29 Nov. 1994. These include electricity and gas networks in cohesion countries, for the whole of Portugal, large parts of Spain, gas lines from Algeria to Spain, and further to France. Also included is a new gas link from Russia/Belarus to Poland/EU (*Europe Energy*, 25 Nov. 1994: 8). The question of financing of the TENs is controversial. DG XVII can provide three kinds of aid: feasibility studies, loan guarantees, and interest-rate subsidies. The amount available for the period 1995–9 is, however, small, only 105 million ecus. Energy TENs receive the smallest amount of the funds; transport e.g. has been allocated 1,868 million ecus in the same period (*EC Energy Monthly*, 11 Apr. 1994: 4). However, the intention for the energy projects is that TENs-status will help in attracting private funding.
6. The harmonization of the VAT encompasses a range from 14.5 to 22 per cent.
7. *Treaty of the Energy Charter*, text for adoption, 14 Dec. 1994, Conference on the European Energy Charter.
8. A proposed text for the legally binding Basic Agreement was presented by the Commission after consultation with all member states in early Apr. 1992 (*Agence Europe*, 3 Apr. 1992).
9. 'Progress on European Energy Charter raises hope for December signing' (*Energy Monthly*, Nov. 1991). This article is a detailed summary of the Charter's contents.
10. This communication was presented in main outline on 25 Mar. 1992, but had not been published in full at the time of writing (*Agence Europe*, 26 Mar. 1992).
11. These draft directives are discussed in 'Council agrees approach for Commission to coordinate oil crisis management measures' (*Energy Monthly*, Nov. 1991: 3)
12. Ibid. 3.

Further Reading

For the pre-history see Black (1977). On the recent evolution of policy and for a theoretical appraisal see Padgett (1992). For a critical analysis of both energy policy and broader regulatory issues see McGowan (1993). For a more detailed overview see J. Haaland Matlary (forthcoming).

Black, R. A., Jnr. (1977), '*Plus ça change, plus c'est la même chose*: Nine Governments in Search of a Common Energy Policy', in H. Wallace, W. Wallace, and C. Webb (eds.), *Policy-Making in the European Community* (Chichester: John Wiley and Sons, 1977), 165–96.

Matlary, J. Haaland (forthcoming), *Energy Policy in the European Union* (London: Macmillan).

McGowan, F. (1993), *The Struggle for Power in Europe: Competition and Regulation in the Electricity Industry* (London: Royal Institute of International Affairs).

Padgett, S. (1992), 'The Single European Energy Market: The Politics of Realization', *Journal of Common Market Studies*, 30/1: 53–75.

CHAPTER ELEVEN

ECONOMIC AND MONETARY UNION: THE PRIMACY OF HIGH POLITICS

Loukas Tsoukalis

Political, rather than economic, arguments have driven demands for economic and monetary union (EMU). Thus, from the first calls at the Hague Summit in 1969 for a common currency to the provisions for EMU put down in the TEU of 1991, politicians have demanded closer economic cooperation. However, the history of failed attempts shows the difficulty of linking divergent economies and the asymmetry of economic burdens borne by the member states. EMS for a time provided a half-way house. Political coalition-building at Maastricht won élite agreement on the principle of EMU, in spite of persistent scepticism from some economists. The agreement did not win ready support from the citizens of Europe, for whom national currency is an important symbol of national sovereignty and national identity, leaving questions of legitimacy unanswered.

Introduction

Economic and monetary union (EMU) has occupied a prominent place on the European agenda for many years. It has had a chequered history, and it remains a subject of considerable controversy among policy-makers and academic economists. There is a rapidly growing literature on the pros and cons of EMU, the necessary preconditions to be fulfilled before the adoption of a single currency, and the requirements for its successful operation. Much of it has built on the early literature dealing with the advantages and disadvantages of fixed exchange rates. Perhaps inevitably, most writings on the subject are of a highly technical nature. Thus the initiated are separated by high fences from the ordinary folk and political debate is rendered difficult— difficult, though of course not futile, given the political significance and the symbolism which money carries in relation to national sovereignty, not to mention its wider economic ramifications.

EMU is an important subject in itself but not only for economic reasons. Money has also been frequently seen as an instrument for the achievement of wider political objectives. It may, therefore, be rather unfortunate that markets and economic fundamentals do not always oblige by adjusting themselves to the exigencies of high politics. The history of European monetary integration can be seen, using a slightly old-fashioned terminology, as a dialectical process between wider political objectives and market realities.

EMU constitutes the most important and also concrete part of the Maastricht revision of the treaty as the Treaty on European Union (TEU). Following the old political maxim that difficult decisions are best left for later, it adopts a back-loaded approach: a relatively long transitional period, with minor changes envisaged for the early stages, while the crucial transfer of power from the national to the European level has been put off until the end of the century. It appears highly unlikely that any serious modifications will be made to the text during the Intergovernmental Conference (IGC) of 1996.

In this chapter I shall examine briefly the history of European monetary integration, concentrating on the more recent period. I shall attempt to throw some light on the way in which decisions have been taken and how different monetary 'regimes' have been administered. I shall also examine the interaction between economics and politics in an area where political imperatives often clash with the logic of international markets. With respect to EMU, the most important decisions are yet to be implemented; their implications will be discussed as well as some of the difficulties that may lie ahead. EMU is certainly not a typical case-study of decision-making in the European

Union, because it has direct implications for so many other policies. After all, the establishment of a full EMU would radically transform the whole process of European integration.

Much ado about nothing: early plans for EMU

The first twenty years after the establishment of the EEC in 1958 were characterized by much talk and little action. The debate did, however, help to prepare the ground for the more substantial phase of monetary integration which started with the setting up of the European Monetary System (EMS) in 1979.

The Treaty of Rome contained very little in terms of binding constraints in the field of macro-economic policy: some wishful thinking in terms of coordination of policies; provisions for balance of payments assistance; and considerable caution with respect to the elimination of exchange controls. There was absolutely no reference to the creation of a regional currency bloc, nor could there realistically be such reference at the time. The international monetary system was based on the dollar standard and European countries were still plagued by the idea of dollar shortage (the external convertibility of EEC currencies was introduced only in 1958). Moreover, Keynesianism was still at its peak. This meant that national governments were zealous in retaining the independence of their monetary and fiscal policies for the pursuit of domestic economic objectives, and a heavy armoury of capital controls was considered as an acceptable price to pay for this independence (Tsoukalis 1977).

Interest in monetary integration grew during the 1960s, largely in response to the increasing instability in the international system and the perceived need to insulate Europe from the vagaries of the US dollar. The fact that this need was perceived more by the French than by the other Europeans, who were not keen on a confrontation with the United States, complicated matters and helped to confine notions of a European interest to a rather general debate which remained far short of any kind of serious joint action by the Six. The latter also lived for several years under the illusion of a *de facto* monetary union, an illusion created by a relatively long period of unchanged exchange rates and further strengthened by the setting up of the common agricultural policy (a typical example of trying to put the cart before the horse).

The situation seemed to change in 1969 when the political leaders of the Six adopted for the first time the target of a complete EMU. It was a political decision reached at the highest level, and it was directly linked to the first enlargement of the Community and the further deepening of integration. It was also the first important example of a Franco-German initiative, involving President Pompidou and Chancellor Brandt, although the common basis for this initiative proved subsequently to be very fragile. The negotiations which followed revealed the existence of a broad agreement about the contents of the final stage, but little agreement about how to get there. The Werner Report (Werner *et al.* 1970), issued by a high-level committee of central bankers and high-level national finance officials, chaired by the Prime Minister of Luxemburg, made a valiant attempt to conceal those differences behind the thin cloak of their so-called strategy of parallelism.

Solemn decisions were then taken at the highest level. They did not, however, prove strong enough to survive the adverse economic conditions of the 1970s. EMU became the biggest non-event of the decade. The collapse of the Bretton Woods system brought down with it the fragile European edifice. Fixed exchange rates, with narrow margins of fluctuation, which had been seen as the most concrete manifestation of the first stage of EMU, proved incompatible with increasingly divergent economic policies and inflation rates. Thus, political commitments, taken at the very top and usually not translated into the appropriate economic policies, finally gave way under market pressure.

After 1974, what was left of the ambitious plan for EMU was only a mutilated snake (the term used for the EC system of exchange rates), wriggling its way in the chaotic zoo of international exchange markets. The majority of EC currencies had been forced to leave the system, thus turning a Community arrangement into a *de facto* German bloc, in which the currencies of some small European countries (including non-Community members) were tied to the Deutschmark (DM), and so were their monetary policies. This was also the first concrete manifestation of the growing importance of Germany in the Community and Europe more generally, and not only economically.

Coordination and asymmetry in the EMS

The setting up of the EMS in 1979 marked a turning-point. It was a renewed attempt to establish a system of fixed and periodically adjustable exchange rates between EC currencies, operating within relatively narrow margins of fluctuation. Unlike many academic economists who have long remained relatively agnostic about the costs of a floating system, the large majority of European policy-makers and businesspeople have always stressed the advantages of fixed (but not necessarily irrevocably fixed) exchange rates. The experience of the 1970s was read as a confirmation of this long-held belief which had been temporarily shaken by new ideas about the alleged efficiency and stability of financial and exchange markets, ideas which were mainly imported from the other side of the Atlantic.

Thus the EMS should be seen first of all as an attempt to bring some stability to European exchange rates and thus provide a more solid foundation to the common market which remained at the centre of European integration. One of the novelties of the system was the European currency unit (ecu), consisting of fixed amounts of each EC currency, which in turn had a central rate defined in ecus. Central rates in ecus were used to establish a grid of bilateral exchange rates, with 2.25 per cent margins of fluctuation around those bilateral rates, except for the lira and the punt, which began with margins of fluctuation of 6 per cent, and sterling which stayed out. Central bank interventions were compulsory and unlimited, when currencies reached the limit of their permitted margins of fluctuation. Central rates could be changed by common consent. Provisions were also made for large credit facilities in order to permit intervention in EC currencies.

The adoption of the divergence indicator, another important novelty of the EMS, was meant to guarantee some form of symmetry in the adjustment burden between strong and weak currencies, surplus and deficit countries. It was an original approach to a very old problem which had been apparent in other monetary systems. In simple words, the introduction of the divergence indicator implied that average behaviour should constitute good behaviour, although this did not square well with another implicit feature of the EMS agreement, namely a more general alignment to Germany's anti-inflation strategy. The contradiction was soon to become apparent, and the result was that the divergence indicator was never put into effect.

Chancellor Schmidt and President Giscard d'Estaing, the architects of what was perhaps the most important decision of the 1970s, had much more than exchange rates in mind. Money was once again directly linked to European unity and the strengthening of the EC construction. In the case of Germany, it was a decision imposed from the top upon a recalcitrant, yet constitutionally independent, central bank. The other countries followed, except for the UK, where the majority of the political establishment shared neither the economic nor the political goals associated with the EMS.

Stability up to a point

The history of the EMS was one of increasing stability of exchange rates based on a progressive convergence of inflation rates downwards (Gros and

Thygesen 1992; Tsoukalis 1993); that is, until hell was let loose in 1992. This is shown in Fig. 11.1. In this respect, three groups of countries can be distinguished; and there is a close correlation between inflation rates and participation in the exchange-rate mechanism (ERM). The first group comprises the seven original members of the narrow band of the ERM (2.25 per cent margins of fluctuation), which also experienced the highest degree of convergence and the lowest rates of inflation. The second group consists of those countries which had long remained in the wider band of the ERM (6 per cent margins) or completely outside it, namely Italy, Spain, and the UK. They had higher inflation rates for most of the period and they were the first to suffer from the crisis which hit exchange markets in 1992. The remaining two countries, Portugal and Greece, constitute a category of their own, with significantly higher inflation rates and, in the case of Portugal (Greece never joined the ERM), an unstable and relatively short participation in the exchange-rate mechanism.

Exchange-rate stability relied basically on the convergence of inflation rates and the main instrument used for that purpose was monetary policy. Participation in the snake and even more so in the ERM acted as an external constraint on domestic monetary policy. This was, at least in some cases and most notably in Italy, joyfully accepted by central bankers, who thus increased their independence *vis-à-vis* national politicians in the pursuit of anti-inflation policies. There was close cooperation among central bankers in terms of exchange-market interventions and the everyday running of the system in general (Goodman 1992). This cooperation relied more on informal networks and less on established EC institutions and committees.

On the other hand, there is precious little evidence of convergence in the case of budgetary policies. To the extent that such convergence did take place, it was the result of autonomous decisions leading to the reduction of public deficits in several member countries (for a large part of the 1980s) and not the product of an effective coordination of national fiscal policies within the EC. In fact, the mechanism set up for the coordination of policies never worked properly: it was strong on procedure and weak on implementation (Mortensen 1990). Multilateral surveillance has acquired new teeth with the procedure adopted at Maastricht and is addressed essentially towards profligate members. A whole range of measures will now be available to the Ecofin (Council of Economic and Finance Ministers), ranging from public recommendations to the imposition of fines. The use to be made of those measures remains to be seen, but it is no longer a question of only 'soft' coordination.

While the everyday running of the EMS was left to central bankers, politicians came into the scene in times of crisis and when realignments of central

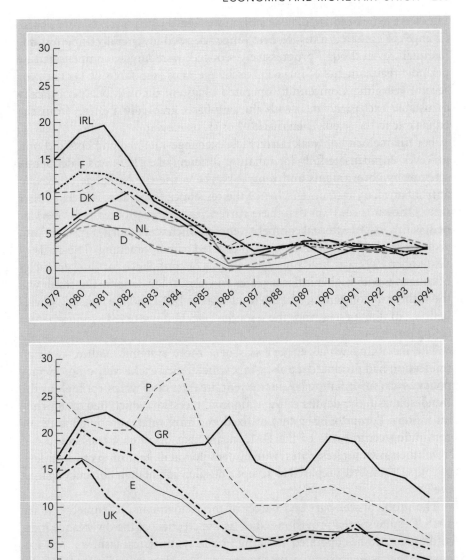

Fig. 11.1 Price Deflator Private Consumption in EC12, 1979–1994

Note: unified Germany since 1992.

Figures for 1994 are European Commission estimates from Dec. 1994.

Source: European Commission, *European Economy*, n. 58 1994, for 1979–1986; *European Economy Supplement A*, Dec. 1994, for 1987–1994.

rates were to be negotiated, invariably at weekends when markets were closed. Changes in exchange rates were no longer decided unilaterally. Inflation differentials, even though progressively reduced, were largely compensated by periodic realignments of intra-EC exchange rates (see Table 11.1). Thus for several years the Community operated a system of fixed but periodically adjustable exchange rates, with the emphasis gradually shifting from the adjustable to the fixed. Then the 1992 crisis broke out.

For the majority of weak currencies, exchange-rate realignments did not, however, compensate fully for inflation differentials. There was a deliberate attempt by governments and central banks to use the exchange rate as an anti-inflationary instrument; hence the resistance to devaluation. The other side of the same coin was that these currencies became progressively overvalued, with a loss of external competitiveness, which was in turn translated into growing trade deficits. In the end, this acted as a boomerang. The crisis of September 1992 was fundamentally a crisis of confidence for the currencies of countries with higher inflation rates and large trade deficits. The result was the withdrawal of sterling and the lira from the ERM and the repeated devaluations of the peseta and the escudo (De Grauwe 1994; Johnson and Collignon 1994).

The instability which ensued was of a more systemic nature. German unification had produced a policy-mix which relied exclusively on monetary policy as an anti-inflationary instrument. Interest rates were kept high by the Bundesbank in the depths of the economic recession which had meanwhile hit western Europe. The system of fixed exchange rates, combined with the continuing central role of the DM, meant that other countries could not lower their own interest rates, while unemployment kept on rising—unless, of course, they were prepared to accept a devaluation of their currency against the DM.

The problem was part and parcel of the fundamental asymmetry in the EMS. Despite special provisions, such as the creation of the divergence indicator, the EMS has operated all along in an asymmetrical fashion, thus following the earlier example of the snake. The degree and nature of the asymmetry have changed over time and so has the assessment of its effects. This asymmetry relates to the central role of the Deutschmark, which is explicitly referred to as the 'anchor' of the system (Delors et al. 1989: 12).

The source of the asymmetry is dual: the German low propensity to inflate (the period following unification has proved both rather short and exceptional) and the international role of the DM. The former, combined with the economic weight of the country and the priority attached by other EC partners to exchange-rate stability and the fight against inflation, has enabled Germany to set the monetary standard for the other countries.

Table 11.1. EMS realignments: changes in central rates (% change: minus sign (−) denotes a devaluation)

Currency	24 Sept. 1979	30 Nov. 1979	23 Mar. 1981	5 Oct. 1981	22 Feb. 1982	14 June 1982	21 Mar. 1983	22 July 1985	7 Apr. 1986	4 Aug. 1986	12 Jan. 1987	8 Jan.[a] 1990	14 Sep. 1992	16 Sept.[b] 1992	23 Nov. 1992	1 Feb. 1993	14 May 1993	6 May 1995
Deutschmark	2			5.5		4.3	5.5	2	3		3		3.5					
French franc				−3		−6	−3	2	−3				3.5					
Netherlands guilder				5.5		4.3	3.5	2	3		3		3.5					
Belgian and Luxemburg franc					−9		1.5	2	1		2		3.5					
Italian lira			−6	−3		−3	−3	−6				−3	−4					
Danish krone	−2.9	−4.8			−3		2.5	2	1				3.5					
Irish punt										−8			3.5			−10		
Spanish peseta[c]													3.5	−5	−6		−8	−7
Pound sterling[d]													3.5					
Portuguese escudo[e]													3.5		−6		−6.5	−3.5

[a] The lira reduced its fluctuation bands from ±6% around its central rate to ±2.25%.
[b] Participation of the pound in the ERM was suspended and the Italian authorities temporarily abstained from intervening in the exchange-rate markets.
[c] The peseta entered the ERM on 19 June 1989 with a fluctuation band of ±6% around its central rates.
[d] The pound entered the ERM on 8 Oct. 1990 with a fluctuation band of ±6% around its central rates.
[e] The escudo entered the ERM on 6 April 1992 with a fluctuation band of ±6% around its central rates.
Source: Eurostat and Bank for International Settlements.

No realignment of ERM central parities has ever involved a depreciation of the DM in relation to any other currency. Fig. 11.2 gives some indication of the margin of manœuvre used by the other members of the ERM *vis-à-vis* Germany, ranging from the Netherlands on the one extreme, with an almost complete alignment on German monetary policy, to Italy, which had repeatedly resorted to devaluation of its currency, even before the crisis of 1992.

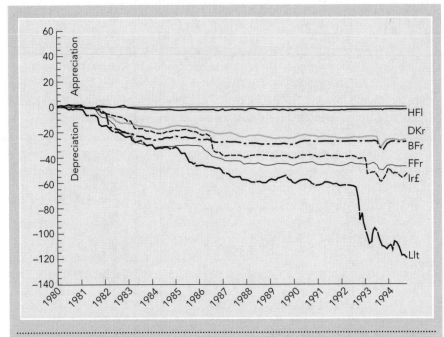

Fig. 11.2. Bilateral DM rates of 6 ERM currencies, 1980–1994 (first quarter), showing % change in relation to Jan. 1980 (price of DM in terms of each of the 6 ERM currencies; the exchange rates are monthly averages)

Source: Deutsche Bundesbank, *Devisenkurs-statistik*, Nov. 1994.

Asymmetry in a system of fixed (even if periodically adjustable) exchange rates is linked to the unequal distribution of the burden of intervention and adjustment and also of influence in the setting of policy priorities. The divergence indicator which had been designed to work against such asymmetry was never allowed to function. The burden of supporting the ERM fell disproportionately on the shoulders of the countries with the weaker currencies. They were also the ones which were forced to adjust their interest rates as a means of defending the exchange rate. In practice, only Germany set its monetary policy independently. The other countries had either to adjust or to leave the system.

The disadvantages of asymmetry have usually been magnified in times of economic recession and growing unemployment, when the link with the DM was considered by economists and lesser mortals as imposing an undesirable deflationary bias. During those periods, internal resistance to the continuation of the link with the DM grew in several countries, while markets began to doubt the sustainability of existing policies, thus creating a vicious circle which often ended in speculative crises. The experience of the early 1990s is a good example.

In September 1992 some currencies were forced to leave the system and others to adjust. But the instability continued as market operators bet on a more general realignment of currencies, at least as long as German interest rates remained unchanged. The resistance to a new realignment of central rates (basically a euphemism for a general devaluation against the DM) could be justified in terms of the so-called economic fundamentals and the political commitment to anti-inflationary policies. But there was much more at stake, namely the prestige and credibility of governments; and this was, perhaps, most strongly felt in France.

Given the size of private funds shifting across national borders and currencies and the relatively recent dismantlement of the old and increasingly ineffective system of capital controls (capital liberalization being an integral part of the internal-market programme), European governments were finally forced to give in in August 1993. To save face, it was announced that the margins of fluctuation of the ERM were to be widened to 15 per cent, which would be little short of floating.

The adoption of wider margins has not fundamentally changed the asymmetrical nature of the system. The nine currencies left in the ERM (subsequently joined by the Austrian schilling) have so far made rather limited use of the wider margins, which seem to be generally considered as a protective shield against the speculative force of the market to be used only in times of extreme danger.

In the exchange crisis of 1992–3, markets won a decisive victory over governments in a game in which the latter had already been deprived of most of their weapons. Markets won because they had become more powerful and also because governments had shown themselves unable to make use of the flexibility offered by the system they had themselves designed. During the crisis, intergovernmental coordination at the European level showed its limits.

Economists are sceptical . . .

Although EMU has been on the table of European Councils for many years, it has not generated much enthusiasm among academic economists, who have been, arguably in their majority, sceptical about the whole project. True, several economists, among the most reputable and also among the most prolific on the subject, have argued in its favour, but the arguments put forward were often political rather than economic (for representative views, see Krugman 1990; De Grauwe 1992; Gros and Thygesen 1992; Padoa-Schioppa 1994; Eichengreen and Frieden 1994). European institutions, and the Commission in particular, have invested much money and mental energy in order to influence informed opinion; and the effect has not been insignificant.

To put it in simple terms, the opposition to EMU among professional economists is based on the following reasoning: western Europe is not yet an optimum currency area where the gains from a common currency can outweigh the losses resulting from the abandonment of the exchange rate as a policy instrument for the correction of external imbalances. Economic integration in Europe has not yet reached the level where capital and especially labour mobility can act as near substitutes for changes in the exchange rate; or that wages and price movements in different countries correspond to changes in productivity rates so as to make exchange-rate realignments redundant; or even that the European economy is sufficiently homogeneous so that different countries and regions are not frequently subject to asymmetric external shocks. But is this true of the United States or even Italy? It is largely a question of degree.

Inter-regional budgetary transfers play a major role in reducing economic disparities within all existing federations. Despite the remarkable growth of structural funds since 1989, there is hardly any prospect of a similar development occurring inside the EU. Can we expect a much greater flexibility of European labour markets as a means of absorbing asymmetrical shocks in the future, or would still higher rates of unemployment (and wider income disparities) become the necessary price to pay for EMU?

In the discussion on EMU, some differentiation is needed with respect to individual countries. The effectiveness of the exchange rate as a policy instrument is closely related to the openness of an economy to international trade, while the relative importance of intra-regional trade is also an important factor to be taken into consideration before entering a monetary union. The UK is not the same as Luxemburg, or even Belgium. Countries with small and open economies can use the exchange rate only sparingly as an instrument of adjustment; a devaluation, for example, has a strong effect on domestic inflation, which in turn helps to undo very quickly the initial positive effects on the external competitiveness of the economy. It is not, therefore, surprising that the Benelux countries, Denmark, and also Ireland have shown in practice greater attachment to the system of fixed exchange rates than other countries which have also tried their luck with the same system.

The scepticism with respect to EMU shown by many academic economists (and central bankers, until they began to discover with considerable delight that participation in the EMS usually went hand in hand with greater independence for themselves *vis-à-vis* national governments), does not necessarily mean that they are in favour of floating. The experience of the 1970s helped to dissipate any enthusiasm that might have existed. Exchange markets, and all financial markets, have shown repeatedly their inherent instability. Economic fundamentals play little role in the short term, and the herd instinct leads to the so-called overshooting. The real question then becomes: what is the price of market instability and how much confidence should we have in the ability of governments to manage?

Many economists would be in favour of a system of fixed but adjustable exchange rates for a group of countries with a high degree of economic interdependence, such as the members of the EU. But they would also be wary of the apparent tendency of governments to forget progressively the adjustable part of the arrangement. The experience of the EMS is very indicative. The combination of fixed exchange rates and free capital movements means that governments can no longer use monetary policy for the achievement of domestic objectives; and this has proved a very unpleasant political conclusion which governments often refuse to draw.

. . . But money is highly political

With the notable exception of the British, especially during the Thatcher years when the magic of the market-place was almost unchallenged, the large majority of European policy-makers have shown strong attachment to exchange-rate stability. The common market, now elevated to the higher stage of the internal market, is generally considered to be incompatible with widely fluctuating exchange rates. Indeed, the direct cost of foreign exchange transactions, to which the cost of exchange-rate uncertainty should be added, has been treated as yet another barrier to be eliminated in the construction of the internal market (Commission 1990e). Thus EMU has been presented as the logical consequence of the 1992 programme.

The caution of economists has usually been overcompensated by the enthusiasm shown by some political leaders. EMU has been seen as an economic means to a political end, rather than an end in itself. This was true of the 1960s, when plans for European monetary integration were used mainly as a means of rallying the other Europeans behind the French challenge of the dollar standard. It was even more true of the ill-fated attempt to establish an EMU in the 1970s. Together with European Political Cooperation (EPC), it was then chosen as the main weapon to fight against a possible dilution of the Community after the entry of the British 'Trojan horse'. It was also to serve as the basis for Franco-German cooperation and as 'la voie royale vers l'union politique'.

The same story was repeated once again with the establishment of the EMS in 1979: it was the peak of Franco-German cooperation and the apotheosis of high politics, with money being used as the instrument of high politics *par excellence* and with political leaders imposing their will on recalcitrant and 'narrow-minded' technocrats (Ludlow 1982). With the benefit of hindsight, it can be argued that the setting up of the EMS was indeed a major turning-point in the process of European integration, after a long period of crisis which threatened to undo the very foundations of the European edifice. Giscard d'Estaing and Schmidt proved capable of longer vision than several of their colleagues, not to mention those in charge of monetary policy and the exchange rate.

The 1991 intergovernmental Conference followed the same pattern. EMU was seen as the main instrument for the strengthening of the EC, which was in turn linked directly to German unification and the creation of a new political and economic order in the East. Monetary union was the preferred way of integrating the German giant more tightly into the Community system. Money was thus part and parcel of the European balance of power, as it, perhaps, ought to be.

Earlier initiatives in the field of European monetary integration had been largely motivated by external preoccupations; the instability of the dollar and US policies of 'unbenign neglect' had served as powerful federalizing factors in Europe. This was not true of the various initiatives which finally led to the new treaty provisions adopted at Maastricht for the establishment of EMU— or, at least, not to the same extent as in the past. True, the reform of the international monetary system was not on the cards and the lack of unity among European countries remained an important factor behind the continuing asymmetry in the international system. But this asymmetry was now less evident, the US Administration did not adopt the aggressive stance of its predecessors, and intra-ERM exchange rates appeared at the time less vulnerable to the gyrations of the dollar. Perhaps less preoccupation with external factors was also a sign of the new collective confidence of the Europeans.

The driving force for the latest relaunching of monetary union came from Brussels and Paris, with EMU representing the flagship of the European strategy of both the Commission (Jacques Delors in particular) and the French government. There was also strong support from Belgium, Italy, and, with some qualification, Spain. Initially, Germany showed little enthusiasm: the government and the central bank were happy with the *status quo* and any move towards monetary union was perceived, quite rightly, as leading to an erosion of Germany's independence in the monetary field. In purely economic terms, there was in fact precious little for the Germans in a monetary union. What later tipped the balance was the perceived need to reaffirm the country's commitment to European integration in the wake of German unification. This is how the matter was presented in Brussels and Paris. Thus the German decision (Helmut Kohl's to be precise) to proceed with EMU was highly political, as it was indeed in most other member countries.

Once a Franco-German agreement had been reached on the subject of EMU, the process appeared almost unstoppable, thus becoming a repetition of earlier patterns of European decision-making. The Dutch shared much of the economic scepticism of the Germans, but their margin of manœuvre was extremely limited. The main concern of the small and less developed economies was to link EMU to more substantial budgetary transfers and also to avoid an institutionalization of two or more tiers in the Community. As for Britain, it remained the only country to question in public the desirability and feasibility of EMU, on both economic and political grounds. The situation had apparently changed little since 1979. Realizing its isolation, the Conservative government made a conscious effort to remain in the negotiating game. In the end, it reconciled itself with an 'opt-out' provision in the TEU for the final stage of EMU.

In terms of decision-making, the negotiation on EMU during the Maastricht IGC bears considerable resemblance to earlier European initiatives and especially the one which had led previously to the adoption of the internal-market programme. The gradual build-up of momentum, the steady expansion of the political support base through coalition-building, and the isolation of opponents were combined with an effective marketing campaign orchestrated by the Commission and addressed primarily towards opinion leaders and the business community. Central bankers were closely involved early on, notably through their participation in the Delors Committee which produced the report on EMU. Later, they played an active role in the drafting of the relevant articles of the treaty. Functional spill-over was also successfully mixed with high politics and the appeal to 'Euro-sentiment'—a recipe which had proved quite powerful in the past.

Some fundamentals: the problems of turning ambition into reality

Money and macro-economic policy in general remained in the exclusive domain of national governments for a long time; the external constraints imposed on their freedom of action were much more the result of international economic interdependence than European integration. The Community started to be considered as a possible focus of attention in response to instability in international markets and the desire to assert Europe's role. External factors progressively lost in importance as money became more and more directly linked with the internal construction of Europe. Geopolitical considerations, European ideology, and symbolism have characterized much of the debate on EMU conducted at the highest political level.

Initiatives came from the very top, and the role of personalities was absolutely crucial. The role of European institutions was essentially supportive, except for the period of the Delors presidency, when the Commission often succeeded in entering the top league. One can hardly detect an important role played by pressure groups. As for public opinion, it provided at best some kind of permissive consensus. Monetary union has been a subject for the select few: difficult to penetrate because of its technical nature and also rather distant as a goal. Plans for EMU have always included long transitional periods, with the real transfer of power being postponed for the final stage.

Things have, however, started to change as the D-day approaches. The German public shows little enthusiasm at the prospect of the DM being replaced by a 'softer' European currency, while the champions of national sovereignty seem to touch a sensitive nerve of the British public. Furthermore, EMU is becoming increasingly identified in the public eye with austerity policies: stabilization measures are frequently justified by governments with reference to Maastricht and the need to prepare for entry into the final phase of EMU. Austerity measures are never popular, even less so in times of recession and high unemployment. Thus public opinion could become a much more important factor in relation to EMU, and not necessarily a positive one.

While the main parameters of the European debate on EMU have been set at the highest political level, the negotiations of specific arrangements and the everyday running of monetary 'regimes' have been entrusted to small numbers of technocrats and most notably central bankers, who often had more in common with each other than with their national political masters, to the extent that this term can be used in relation to more or less independent central banks. Informal transnational alliances have developed in this area.

Collective management of European monetary 'regimes' has not been so far accompanied by the development of common institutions. The setting up of the European Monetary Institute (EMI) is the first step leading to the creation of the European Central Bank (ECB), but which will happen only in the final stage. This will be the moment when power over the conduct of monetary policy is transferred from the national to the European level.

Economic globalization, especially as it is manifested in capital markets, has substantially narrowed the autonomy of national governments in terms of economic policy. Time and time again, member governments have experienced major set-backs on the road to monetary union, when the particular combination of national policies was deemed by markets to be incompatible with fixed exchange rates. Political agreements reached in European Councils are not by themselves sufficient to secure the stability of exchange rates, and even less so to guarantee a workable monetary union.

The part of the TEU devoted to EMU is a typical Community compromise. The French secured a clear commitment enshrined in the treaty, as well as a specific date. The Germans made sure that the date would be distant, with little happening in between, and that the new European model would be as close as possible to their own. The British secured an 'opt-out' for themselves, followed later by the Danes, while the poor countries obtained a more or less explicit link with redistribution. Unfortunately, political compromises do not always stand the test of time and markets.

The conditions for admission to the final stage of EMU, otherwise known as convergence criteria, are quite explicit and they concentrate exclusively on monetary variables. The first refers to a sustainable price performance, defined as a rate of inflation which should not exceed that of the three best-performing member countries by more than 1.5 percentage points. The second relates to national budgets: actual or planned deficit should not be above 3 per cent of GDP, while the accumulated public debt should not be above 60 percent of GDP. With respect to this criterion, however, the treaty leaves some room for interpretation.

Exchange-rate stability is the third criterion: the national currency must have remained within the 'normal' fluctuation margins of the ERM for at least two years; after August 1993 'normal' seems to be defined as 15 per cent, which makes this criterion easy to meet. The fourth criterion is meant to ensure that the exchange-rate stability is not based on excessively high interest rates; thus the average nominal interest rate on long-term government bonds should not exceed that of the three best-performing member states by more than two percentage points.

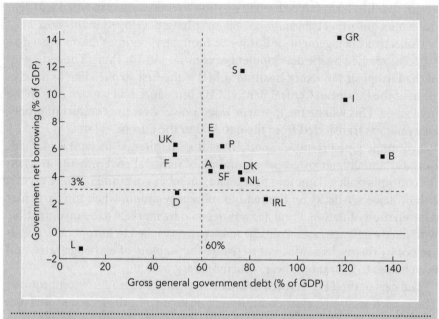

Fig. 11.3. Gross general government debt and government net
borrowing for 1994 in the EU 15
Source: European Commission.

The convergence criteria can be criticized on many grounds: they are
mechanistic, some of them are arbitrary and, perhaps, also superfluous (see
also De Grauwe 1994). They do, however, reflect the strong influence of
Germany as well as prevalent economic orthodoxy; and the two are closely
interrelated. The same applies to the whole approach towards macro-economic policy in the context of EMU. Monetary policy will be run by independent technocrats in the ECB, the Community budget will remain very
small, and the coordination of fiscal policies will be almost exclusively
addressed towards profligate members. But who will set macro-economic
priorities for the EU as a whole? And will the system prove flexible enough to
accommodate a possible change of the economic paradigm in the future?

The TEU said that before the end of 1996 the Council would decide, by
qualified majority and on the basis of reports from the Commission and the
EMI, whether the majority of member countries met the four convergence
criteria and would thus give the green light for the beginning of the third and
final stage of EMU. A qualified majority decision would be reached about
whether there is a majority fit to enjoy the fruits of paradise (and EMU): this
would be a rather uncommon case in international and even domestic

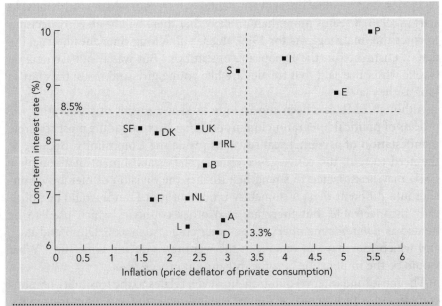

Fig. 11.4. Inflation and long-term interest rates for 1994 in the EU 15

Note: Long-term interest rates for Greece n.a..
Source: European Commission.

politics. Under French pressure the authors of the treaty went further. It was stipulated that the third stage would start on 1 January 1999 at the latest, irrespective of how many members are found then to fulfil the necessary conditions, again the Council deciding on each case and on the basis of qualified majority.

According to the latest statistics available for EU15 (see Figs. 11.3 and 11.4), only Germany and Luxemburg would fulfil all four convergence criteria in 1994 plus Ireland, if we were to adopt a loose definition of the public-debt criterion. Those countries would make an unusual monetary union. However, since most countries fail to qualify in terms of their public debt and current deficits, an improvement of the economic conjuncture, combined with a generous interpretation of the debt criterion, would allow more countries to join the privileged group sooner rather than later. They include Austria, Denmark, Finland, the Netherlands, and perhaps also France and the UK.

It is therefore just about conceivable that the majority of member countries would meet all criteria by 1997 (Denmark and the UK may exclude themselves because of their 'opt-outs', negotiated for political reasons). But given the relatively recent experience of major turbulence in exchange markets, the weak public support which EMU currently enjoys in several European countries, and the new Intergovernmental Conference which is scheduled to start in 1996, it seems more than unlikely that there will be an early decision to enter the final stage. As for 1999, this is still a long time ahead. True, the date is enshrined in the European constitution. But was it not General de Gaulle who once said that treaties are like young girls and roses: they last as long as they last?

Although EMU has been consistently used as a means of accelerating the process of political integration in Europe, it has been so far the most concrete manifestation of asymmetrical relations inside the Community. Because of the need to fulfil the convergence criteria, the establishment of a complete EMU may be expected to strengthen further the division of member countries into different tiers. A monetary union without France would be politically inconceivable; but there are also other countries which used to be treated as *bona fide* members of the inner circle, such as Belgium and Italy, and which may have to remain in the waiting-room for some time. What would be the implications for the integration process?

The most fundamental question, however, relates to the feasibility of monetary union without political union, and indirectly to the issue of legitimacy. Can we have a central bank with absolute authority in terms of monetary policy at the European level, whose decisions will have a direct impact on unemployment in Andalusia or Attica, without a corresponding political authority and the political system that it implies? The last Intergovernmental Conference failed to deliver the political system which would match EMU. It remains to be seen whether the next one will be more successful.[1]

Note

1. I should like to thank Outi Jääskeläinen and Andrea Rossi for research assistance and especially for the preparation of graphs and tables.

Further Reading

For a comprehensive analysis of the EMS and monetary integration in general see Gros and Thygesen (1992) and Padoa-Schioppa (1994). For the exchange-rate crisis of 1992–3 see De Grauwe (1994) and Johnson and Collignon (1994). The Delors Report (1989) is also recommended as the official document that laid the basis for economic and monetary union as negotiated at Maastricht in the TEU.

De Grauwe, P. (1994), 'Towards European Monetary Union without the EMS', *Economic Policy*, 18 (Apr.): 147–85.

Delors, J., *et al.* (1989), Committee for the Study of Economic and Monetary Union, *Report on Economic and Monetary Union in the European Community* (Luxemburg: Office for Official Publications of the European Communities).

Gros, D., and Thygesen, N. (1992), *European Monetary Integration: From the European Monetary System towards Monetary Union* (London: Longman).

Johnson, C., and Collignon, S. (1994) (eds.), *The Monetary Economics of Europe* (London: Pinter).

Padoa-Schioppa, T. (1994), *The Road to Monetary Union in Europe* (Oxford: Clarendon Press).

CHAPTER TWELVE

EU POLICY IN THE URUGUAY ROUND

Stephen Woolcock and Michael Hodges

The European Union's growing role in world affairs, coupled with the increasingly political nature of commercial policy, has forced the EU to move away from its traditional technocratic approach to commercial policy. As commercial policy has become more politicized, decision-making has become more complex. As the history of the GATT Uruguay Round demonstrates, internal tensions—between the member states, within the Commission, and between the Commission and Council—can hinder the EC's bargaining position in multilateral negotiations. This was notable in discussions on agriculture; but in other areas the EC suffered no greater internal difficulties than other trading entities. Indeed the EC was able to play a shaping role as the Round drew to a close. Questions of the trade-off between accountability and efficiency remain to be resolved.

Introduction

This chapter addresses the complex process of EU commercial policy-making in the Uruguay Round (UR) of multilateral trade negotiations within the General Agreement on Tariffs and Trade (GATT). Where space allows there is reference to the positions of the various EU member states, interest groups, and trading partners, and we have touched on the economic and political environment within the EU against which decisions were taken. Those interested in a full picture of the EU in the Uruguay Round will therefore have to read more widely.[1] The aim here is to provide an overall view of how the EU, then still the European Community (EC), operated in the important area of multilateral trade negotiations and to offer some insights into what went on 'in Brussels' during the Uruguay Round.

As in most areas of policy, EU commercial policy is characterized by two areas of tension. The first concerns the pendulum of national versus Community competence, the second the balance between efficiency and accountability. The EC experience in the UR shows how globalization of the world economy is pushing more and more issues on to the agenda of commercial policy. In response to this sustained pressure, the member states of the EU have recognized the benefit of policy collaboration at a European level. A single voice is considered more effective, especially in a global trading system devoid of benign leadership from the United States and the rules of which continue to be shaped and interpreted with a good measure of power-politics. As in previous multilateral trade rounds, member governments preferred to stop short of granting the EC exclusive competence, but accepted the need for the EC to negotiate with one voice in areas of mixed competence. In most cases member governments also accepted the requirement of cohesion and of maintaining Community discipline which followed this commitment. Thus the EC had, if anything, fewer problems in areas of mixed competence, such as services and intellectual property, than in areas of exclusive EC competence, such as agriculture. If one adopts a wider definition of competence in the sense of who controls EC commercial policy, the picture is different. Indeed the UR was characterized by repeated tensions between the Commission and Council over who was in control.

The second most important area of tension was between efficiency and accountability in EC decision-making (see Fig. 12.1). The EC policy process in commercial policy remains essentially technocratic. In a formal sense political accountability is to national parliaments through national civil servants and ministers, but the EC is characterized by a lack of effective scrutiny of the national and Commission technocrats by national and European par-

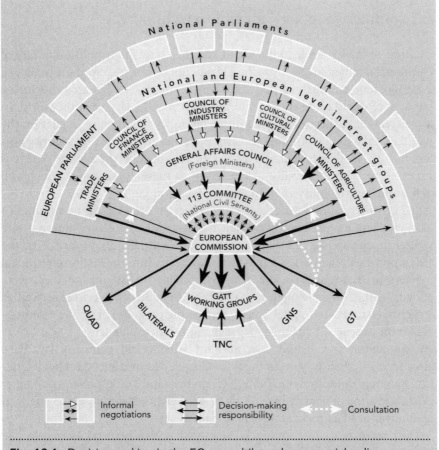

Fig. 12.1. Decision-making in the EC on multilateral commercial policy
Note: The British, French, German, and Italian governments also participate directly in the G7.

liaments. This system may promote efficiency: the officials have a grasp of detail and an understanding of what can be 'sold' to their domestic political constituencies, as well as those of their negotiating partners, both inside and outside the EC. In general the EU was effective in the UR when these technocrats were in control. But there are questions concerning the accountability of EC commercial policy decision-making. The UR also showed the difficulties the EU can get into when political decisions are needed, such as in agriculture. Where commercial policy is politicized, technocratic approaches will tend to be ineffective as well as to raise issues of legitimacy.

EC commercial policy: an end to technocracy?

Changes in the EU's role in the world and the increasingly political nature of commercial policy suggest that the EU will not be able to afford the luxury of a technocratic commercial policy for much longer. The EU accounts for 40 per cent of world trade and its internal market is the largest in the world. In a very real sense therefore EU commercial policy has a major impact on the world and the EU's world image. Unlike the case in foreign and security policy, the EU is already a major player in commercial diplomacy. Much is expected of it and its ability to match up to these expectations which will determine its credibility as an international actor.

EC commercial policy decisions also affect an ever increasing number of 'domestic' policy preferences. When the theories of customs unions were written in the 1950s, the trade agenda was essentially about tariffs. In the 1970s non-tariff barriers (such as subsidies, anti-dumping actions, technical barriers to trade, and preferential government purchasing policies) were added. In the 1980s regulatory barriers were added in services, as well as structural barriers to market access, such as the existence of public monopolies or the absence of effective national competition/anti-trust policies. Now environmental law, labour law, and investment and company law have already found their way on to the post-Uruguay Round agenda. The distinction between 'domestic' (including EU) policies and 'trade' policies no longer exists and a much wider range of domestic constituencies must now be seen as endogenous factors in commercial policy.

Decision-making in commercial matters has also been politicized by the blurring of another distinction, that between foreign/security policy and commercial diplomacy. After the cold war Europe's security interests have less to do with the deterrence of a potential military aggressor and more to do with promoting economic prosperity and social stability in neighbouring countries. In this battle commercial policy plays a central role. Outside Europe the proliferation of regional economic agreements, such as the North American Free Trade Agreement (NAFTA), Asia Pacific Economic Cooperation (APEC), the Free Trade Agreement for the Americas (FTAA), and the Common Market of the Southern Cone (Mercosur), is further evidence that commercial policy will play an ever greater role in international relations.

These changes mean that the balance between efficiency and accountability in European commercial policy, established on the basis of treaty provision by years of political dialogue, practice, and the rulings of the European Court of Justice (ECJ), must be continuously revised, if the EU is to be a credible and effective player in commercial policy.[2]

Box 12.1. The EC commercial policy-making process

....

The Negotiating Mandate. Articles 110–16 of the EEC Treaty had established a clear competence for the EC in implementing a common commercial policy, with the Commission making proposals for a negotiating mandate for approval by the Council by a qualified majority vote. Ultimate responsibility lies with the ministers of foreign affairs sitting as the General Affairs Council, although other Councils play a role (notably the Agricultural Council in the Uruguay Round GATT negotiations) (see Fig. 12.1). The fact that there is no regularly convened Council of Trade Ministers has meant that the trade ministers have played a less influential role than their counterparts in agriculture.

Conduct of Negotiations. During negotiations the Commission is the sole spokesperson and negotiator for the EC, acting on the basis of the Council's mandate. The mandate provides the Commission with varying degrees of discretion and flexibility, depending on the issue involved: the Uruguay Round negotiations sometimes aroused fears in the Council that the Commission was stretching or even exceeding its mandate in order to reach an agreement with its negotiating partners. In its negotiations the Commission is advised by the Article 113 Committee, a collegial body of experienced senior officials from member states, who confirm modifications to negotiating mandates on issues considered to be of a technical or minor nature. When politically sensitive or significant changes to mandates are necessary, these are referred to the Committee of Permanent Representatives (Coreper) and thence to the Council. The composition of the 113 Committee varies according to topic.

Competence. Article 113 grants the EC exclusive competence for commercial policy but makes no definition of commercial policy. The European Court has confirmed EC exclusive competence in areas of non-tariff as well as tariff barriers. In each round national governments have accepted the Commission as sole negotiator for issues of mixed competence, but without prejudice to the debate on competence. In this way the Commission was sole negotiator for the EC on all issues in the Uruguay Round. The competence question was then subsequently decided by the European Court in November 1994.

Consultations. There are no formal procedures for consultation with interest groups, but the Commission has informal contacts. Sectors also lobby their national governments. There are consultation procedures with the European Parliament (EP) and in practice both the Commission negotiators and Council presidents brief the EP. Scrutiny at the national level is up to each country, but in practice there has been little effective scrutiny.

Endorsing an Agreement. The results of the negotiations are endorsed by the General Affairs Council on behalf of the member states. Ratification by the national parliaments was also more of a rubber-stamping exercise, except in France. In Britain there was no vote. The EP was asked to give its 'assent' to the results, because they touched on issues of EP competence (under Articles 40, 56, 110a, 238 et al. of the Treaty), but not under the main Treaty provision of Article 113.

Launching the Uruguay Round: from Punta del Este to the Mid-Term Review (MTR)

Early in the 1980s the United States had suggested a new round of GATT negotiations, but the idea found little support in Europe, which was in both the depths of economic recession and the grip of Eurosclerosis. The developing country Contracting Parties (CPs) to the GATT were also cool on the idea, particularly since some of the 'new issues' raised at the GATT Ministerial meeting in November 1982 (trade in services, trade-related investment measures, trade in high technology) were ones which were perceived as most advantageous to the industrialized countries, while the latter still resisted trade liberalization in textiles and agricultural products (Golt 1982).

The Reagan Administration's determination to start a new round was in part out of fear that without demonstrable progress in multilateral negotiations, the US Congress could not be prevented from adopting protectionist legislation in response to the record trade deficit of $140 billion in 1983. The EC initially resisted US calls for a new round: although Britain and Germany favoured commencing negotiations, France resisted, arguing that a new multilateral round would do nothing to reduce the US trade deficit and could raise false expectations of the multilateral system, which when not fulfilled would lead to more rather than less protectionism. The French government was also a leading proponent of the view that parallel action had to be taken in the field of international monetary policy if rapid changes in exchange rates were not to negate the effects of any trade liberalization. The fact that the USA was the *demandeur* of a new round and the EC the only entity able to resist US pressure gave the negotiations a strong US–EU dimension from the start.

In March 1985 EC trade ministers, meeting within the framework of the General Affairs Council, decided, on the basis of a paper by the European Commission, to support a new round—despite French misgivings (*Agence Europe*, 20 Mar. 1985). This decision was taken against a background of economic recovery from the recession of the early 1980s and renewed political optimism in Europe. For example, the Milan European Council in June 1985 was soon to endorse the EC's '1992' internal-market programme, which envisaged measures in areas of technical and regulatory barriers to trade that were also on the UR agenda.

From March 1985 the Commission had been working on position papers covering the potential agenda. These provided the basis for negotiations between Commission and national officials in the Article 113 Committee. During the early months of 1985 there were also two guideline debates on the EC's position in the Council. By the end of May 1985 the Commission produced a position paper (*Agence Europe*, 29 May 1985) and in June 1986 the Council adopted, by consensus, a negotiating mandate for the round, setting out overall objectives.

The European Parliament (EP) adopted a comprehensive position on the

EC's negotiating objectives, but arguably too late to shape the mandate, since its report was not debated until September 1986. National parliaments had even less input. Indeed, many parliaments, including the British, were not informed of the EC's negotiating mandate even after it had been adopted, because it was thought that open political debate would prejudice the EC's negotiating objectives (House of Lords 1985). The negotiating mandate was consequently not a public document, but the broad outlines of the European position are summarized in Box 12.2.

Box 12.2. European objectives in the Uruguay Round

....

Reasonable expectations. Trade negotiations should not be expected to solve imbalances and problems with origins that lie outside the trade field, such as erratic exchange-rate changes. The EC therefore sought a *multi-polar* approach with parallel results in the field of monetary and fiscal policy. This was a relatively downbeat assessment of the effects of a trade round and contrasted with what many in the EC saw as the US administration's 'over-selling' of the benefits of the round to the US Congress.

Support for a **strong multilateral system** based on a balance of rights and obligations (a reference to the need for Japan to do more to open its economy) and able to demonstrate its continuing vitality compared with bilateral and plurilateral deals (a reference to a perceived US bilateralism—the USA was negotiating a bilateral semi-conductor agreement with Japan at the time—and the US threats to pursue plurilateral approaches (such as within the Organization for Economic Cooperation and Development (OECD)) if no multilateral round was launched. Existing *grandfather* provisions were targeted: exemptions from GATT discipline which pre-date the GATT, such as the 'waiver' which exempted US agricultural support schemes from GATT discipline. Also the special advantages accruing to *federal states*, notably the US, since state-level non-tariff barriers, such as local technical regulations or 'buy-America' provisions in public purchasing, were not covered by the GATT provisions negotiated during the Tokyo Round in the 1970s.

Differential and more favourable treatment for developing countries. The EC continued to support this approach, which was incorporated in Part IV of the GATT added during the Kennedy Round, but envisaged a 'graduation', with the developing countries accepting more obligations as their economies progressed. This position placed the EC between the hawkish US position and that of the so-called hard-line developing countries, such as Brazil and India, which sought to use the retention of Part IV as a means of avoiding any new obligations under GATT.

Globality, which meant, in effect, that there should be no agreement on the Round as a whole until there was agreement on all elements. This was based on past experience and a concern that the USA would single out specific areas, such as agricultural export subsidies.

EC acceptance of the Punta del Este declaration launching the round in September 1986 provided an example of problems to come in EC decision-making. Agriculture was an issue from the start, with the French delegation seeking to avoid the EC's position being prejudiced from the outset by reference in the declaration to the 'removal of export subsidies' or even 'phased

Box 12.3. Issues in the Uruguay Round at a glance

....

MARKET-ACCESS ISSUES

Tariffs. Still some high tariffs, especially in developing world. In Europe customs union created uniform tariff reductions, in USA still some tariff peaks. EC objective 'graduation' among developing countries to accept more tariff bindings and reduce tariff peaks.

Non-tariff barriers. Quotas and quantitative restrictions (QRs). Commission supports removal of QRs as part of internal market initiative. Sensitivities in some member states.

Textiles and clothing. LDCs and NICs exporters seek removal of QRs in OECD countries under Multi-Fibre Arrangement (MFA) as *quid pro quo* for accepting new multilateral disciplines in services and intellectual property. In EC Portugal and Italy resist phasing-out of MFA, which is otherwise broadly accepted.

Tropical products. Trade liberalization (zero tariffs and removal of QRs and tax barriers for products such as coffee, cocoa, tea, etc.) seen as a way of showing LDCs they have something to gain from the round. Asia, Caribbean, and Pacific (ACP) countries concerned about loss of preference with the EC.

Agriculture. USA and Cairns Group (led by Australia and New Zealand but with support from countries such as Argentina, Thailand, etc.) press for elimination or reduction of export subsidies to reduce competition from subsidized European exports. As CAP is based on price support EC resists this and calls for gradual reduction in Aggregate Measure of Support of all subsidies including domestic support used in USA and other countries. USA seeks to tie EC down to commitments in each area of subsidy: EC seeks flexibility to 'rebalance', i.e. redistribute support across sectors and forms of support.

....

STRENGTHENING THE MULTILATERAL RULES

Dispute settlement. GATT failed effectively to enforce rules agreed in Tokyo Round in the 1970s. US initially strong on more adjudication in dispute settlement. EC as well as Canada and large number of NICs joined to support stronger dispute settlement, in part, in an effort to ensure multilateral discipline of US unilateral trade laws. Issues centred on removing 'veto' right which had been used by large CPs to block the establishment or adoption of a dispute settlement panel report.

Subsidies and countervailing duties. Continuation of Tokyo Round debate with US seeking tighter GATT controls of subsidy programmes. EU accepting tighter controls but with green light for such things as regional and R&D subsidies.

Safeguards. Also continuation of Tokyo Round, which failed to establish a functioning safeguard provision. EC argued this required an element of 'selectivity' but was opposed by developing countries, NIC exporters and GATT purists violently opposed to undermining of principle of non-discrimination.

GATT Articles. Included issues such as revision of anti-dumping code. Japan and NIC exporters seeking tighter discipline, USA with EU backing opposing multilateral controls on scope for imposing anti-dumping duties.

Government Purchasing Agreement (GPAs). Negotiations on extension of 1979 GPA (not formally part of the round). EC seeking comprehensive extension to cover utilities (whether public or privately owned) and local and regional governments. USA and Japan wish to draw the line at central (federal) government and publicly owned utilities.

Box 12.3. (*cont.*)

....

NEW ISSUES

General Agreement on Trade in Services (GATS). Pushed on to the agenda by US private sector interests, but adopted by the EC as a main objective. US concern to avoid outcome which sets rules (which have the effect of opening the US market) but does not produce concrete liberalization elsewhere, hence back-peddling when round failed to produce hoped-for results. EC sees USA as establishing a set of rules and first stage of liberalization process. Developing countries, especially India and Brazil, resist pressure for multilateral rules on services.

Intellectual property rights. Another issue pushed on to the agenda by the US private sector against resistance of developing countries. EC adopts middle position then shifts to support the US position as European multinational companies also appreciate the benefits.

Investment. Initially USA sought a comprehensive agreement but was forced to moderate demands when developing countries opposed multilateral discipline on their domestic investment policies and EC adopted a lukewarm position.

reductions' in agricultural subsidies. The joint Colombian–Swiss draft containing these references was opposed by France, with support from Ireland, Greece, Spain, Denmark, and Belgium, while the British, then holding the Council presidency, Germany, and the Netherlands wanted to accept the text in order to get negotiations started. The declaration ultimately adopted referred in more general terms to reductions in agricultural support. This early division in the EC, like others that followed, provided an irresistible target for point-scoring by the chief US negotiator in Geneva, who asked 'why the rest of the world should suffer because the EEC countries were unable to stick together' (*Financial Times*, 1 Aug. 1986).

On most of the fifteen negotiating groups established in Punta del Este, cooperation between the Commission and national trade experts in the Article 113 Committee enabled the EC to develop positions at an early stage of the Round. In September 1987, for example, the EC submitted proposals on tropical products, an issue of critical importance to some developing countries. During 1987 the Commission, impelled by the momentum of its own internal-market programme, also sought to take the lead in rolling back existing quantitative barriers to trade. The EC led the OECD countries in tabling proposals to phase out the Multi-Fibre Agreement (MFA) regulating trade in textiles and clothing. The EC showed an innovative capacity, for example, in its December 1987 paper on trade in services, which set out how the negotiations could advance and was complementary to the earlier (and very broad-brush) US proposals. Likewise the EC paper of February 1988 on intellectual property rights (IPR) was one of the most comprehensive and detailed contributions thus far in that negotiating group. On most of these 'technical' issues the General Affairs Council merely endorsed the position agreed in negotiations between the Commission and 113 Committee.

To satisfy domestic concern to see results, the USA pressed for an 'early harvest' in areas in which it wanted progress. But the overall mood in the USA was one of qualified support for multilateralism. In 1987, for example, the US Congress threatened the US trading partners with the Gephardt amendment (which envisaged sector-by-sector trade-balancing in cases of bilateral trade deficits), stronger US provisions on Section 301 (the main instrument of US aggressive unilateralism), and changes to US laws on anti-dumping and countervailing duty provisions that would clamp down on 'unfair' import competition. On textiles and clothing, tropical products, and roll-back, the US Congress would not countenance any US offers.

On agriculture, in contrast, the US proposed a radical 'zero 2000' in July 1987 to eliminate all agricultural subsidies. This had the intended effect of raising the political stakes on agriculture, but also the unintended effect of allowing the EC the luxury of avoiding serious discussion on a 'credible' level of subsidy reductions. The EC simply responded by suggesting that subsidies should be frozen at existing levels. In July 1988 the Cairns Group of agricultural exporters, led by Australia (and named after the Australian resort in which it had been formed in 1985), came forward with a broadly intermediate proposal, calling for 'substantial and progressive' reductions of all forms of agricultural subsidies. In November 1988 the EC's Agriculture Council took firm control of the EC position and 'recommended' the General Affairs Council to adopt a tough line in the Mid-Term-Review meeting to be held in Montreal in November 1988. The French minister of agriculture, Nallet, presented a detailed justification for this tough position and was supported by the German, Irish, Spanish, and Greek ministers for agriculture (*Agence Europe*, 19 Nov. 1988). Britain and the Netherlands tried, without success, to get the EC to adopt a more moderate position closer to that of the Cairns Group and thus to isolate the United States with its extreme position.

With irreconcilable differences over agriculture the Montreal meeting collapsed and the Cairns Group sided with the USA, leaving the EC isolated. There were problems other than agriculture, such as a north–south divide over the liberalization of textiles and clothing, safeguards (where the EC held to the goal of selectivity), and IPR, where the developing countries opposed the idea of GATT standards for IPR protection. The EC shifted its position on agriculture in the early months of 1989, when Arthur Dunkel, the GATT Director-General, negotiated compromise wording calling for 'substantial, progressive reductions (of agricultural subsidies) over a specified period of time'—which was close to the wording proposed by the Cairns Group. In permitting the

Agriculture Council to dictate an uncompromising position the EC allowed itself to be put in an isolated and inflexible position on agriculture, which it retained, at great diplomatic expense, for much of the rest of the round.

The MTR also provided an early example of EC difficulties concerning the interpretation of Council mandates. In April 1989 the Commission negotiated a conclusion of the MTR in consultation with the Article 113 Committee, but at the eleventh hour Portugal (referring to 'vital national interests'—the code for a threat to invoke the Luxemburg Compromise) and Italy objected to the wording on textiles and clothing; they claimed that the Commission had exceeded its negotiating mandate. The issue was resolved, but only after the whole multilateral process had been held up while the EC resolved internal differences. In subsequent Council meetings the Portuguese complained bitterly that there had been inadequate communication between the Commission, Article 113 Committee, and the Council. The Italian trade minister, Renato Ruggiero, called for the Council Presidency to sit beside the Commission in future negotiations to avoid such problems, a move which was blocked by the Commissioner responsible, Willy De Clercq, on the grounds that it would undermine the Commission's credibility and flexibility. A compromise was reached on the need for more effective communication between the Commission, Article 113 Committee, and Council, but similar problems recurred at later stages.[3]

The road to Brussels

The round had been scheduled to be completed in Brussels in December 1990, but after the MTR there was a lull in the negotiations. Momentous political events in the shape of the fall of the Berlin Wall, German unification, and the end of the cold war were the focus of political attention. Against such events the GATT negotiations understandably were not given priority by many politicians. But 'technical' work progressed in Geneva in which the EC played a positive role, exhibiting considerable flexibility and bringing its expertise on market integration to bear. Under the guidance of the Article 113 Committee, the Commission championed the strengthening of the multilateral institutions, by tabling proposals, jointly with Canada, for the creation of a new Multilateral Trade Organization (or World Trade Organization (WTO), as it became known later). This reflected a desire to discipline US unilateralism, but it also reflected an EC willingness to accept multilateral discipline over its own policies (Commission 1990b). This willingness was also reflected in the March 1990 EC paper on improved multilateral dispute-settlement procedures. This represented a major shift in EC policy, which had hitherto been sceptical of multilateral adjudication of disputes. The shift was, in part, due to the greater acceptance of rules-based approaches within Europe and in part aimed at disciplining US unilateralism. Paradoxically, it came at a time when the USA, which had previously been the great champion of rules-based approaches, was moving in the opposite direction, in order to placate Congressional fears of a loss of sovereignty to multilateral institutions.

As prior to the MTR, the EC maintained a coherent position in most negotiating groups. In textiles and clothing (T&C) it continued to favour a phasing-out of the MFA. In contrast, the USA was unable to submit a position paper on T&C until February 1990 when it merely sought global quotas to replace the existing bilateral quotas negotiated under the MFA, an approach which removed discrimination but at the price of increased protection. In a reverse of the situation on agriculture, the EC was able to maintain a coherent policy to phase out the MFA and thus gain support for its position from developing and middle-income countries; it was the US that was isolated.

On the road to Brussels, as in other phases of the negotiations, work in the other negotiating groups, covering issues which accounted for 89 per cent of world trade, was overshadowed by agriculture (11 per cent of world trade). There are a number of interlinked reasons for the EC's failure of commercial diplomacy in agriculture in 1989/90. Perhaps the most important was the lack of any political focus on the need for a reform of the Common Agricultural Policy (CAP) in order to enable the EC to have a coherent and sustainable position in the negotiations. The 1986 EC mandate for the round did not help. This specifically excluded any concession which would undermine the fundamental mechanisms of the CAP, which EC agricultural interests interpreted as a mandate to oppose any CAP reform.

The only body that could bring political focus on the need for reform and override the Agriculture Council was the General Affairs Council. But foreign ministers were occupied with other things. Following the fall of the Berlin Wall in November 1989, foreign ministers were concerned with how to respond to German unification and to build a post-cold-war European order, which included the launching of negotiations on European political and monetary union. Against this background foreign ministers found issues such as 'tarification' of quantitative import restrictions on dairy products or rebalancing of concessions between agricultural products less than compelling. They left the 'technical' issues to the Agriculture Council and thus effectively left the farm ministers with a veto over the EC's position in the Uruguay Round. The agriculture ministers developed a coherent but restrictive position in the—what proved to be the mistaken—belief that the US would, as in previous rounds, accept the conclusion of a round without agriculture.

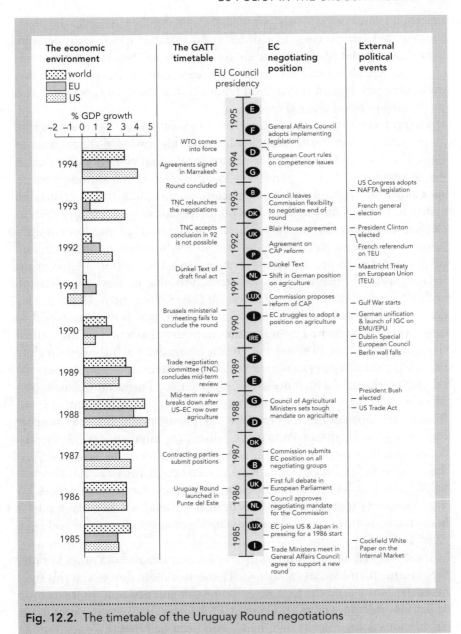

Fig. 12.2. The timetable of the Uruguay Round negotiations

The presidency of the Council played a role in the treatment of agriculture (see Fig. 12.2). As in other policy areas, the presidency cannot dictate policy, but it can influence the timing and agendas of Council business. At the crucial period in the negotiations in 1989 and 1990 Spain, France, and then

Ireland, all countries strongly opposed to changing the CAP, held the presidency of the Council. These presidencies were therefore happy for agriculture to be dealt with by agriculture ministers alone. As will be shown later, it was only when the Dutch and British Presidencies enhanced the role of informal trade ministers' meetings that this preponderance of agriculture in the EC's negotiating position was redressed.

The problem of getting foreign ministers to focus on GATT and agriculture in particular was exacerbated by a scheduling nightmare in the autumn of 1990. This was made worse by the failure, for the first and only time during the negotiations, of the Commission to agree on a draft position paper on agriculture in September 1990. Agricultural ministers therefore had an excuse to procrastinate and foreign ministers had no opportunity even to consider the 'wider Community interest'. 'Jumbo Council' meetings of agriculture and foreign ministers were held to try to break the deadlock and resolve scheduling problems, but France, with the support of Germany, Ireland, and the Mediterranean countries, was able to block any decision until after the special European Council in Rome at the end of October. This delay was welcomed by the German government, which neither wanted to upset its farmers in the run-up to the first general election in the unified Germany in October 1990 nor the French government at a time when close Franco-German cooperation was needed to steer the agenda for the Intergovernmental Conference (IGC) through the European Council. Mrs Thatcher tried to push for a coherent position on agriculture, but she and Britain were isolated in October 1990 and, paradoxically, her only support came from the European Parliament, which was impotent on commercial policy. The 'Jumbo Council' finally adopted a mandate on 15 November, just three weeks before the planned final conference in Brussels, but this was tightly drawn and left the Commission little scope for negotiation. As US negotiators wryly observed, 'the Commission would not negotiate until it had a mandate from the Council and could not negotiate once it had been given one'.

Trade and agricultural ministers were present in Brussels in order to ratify the agreement and to ensure that the Commission did not exceed this tight mandate. This placed the Commission in a difficult position. In the run-up to Brussels it had called on the other CPs to show flexibility in order to get negotiations going, but its own hands had been tied by the Council. In an effort to rescue the situation and break the log-jam, the Commission acted independently and made informal offers to the effect that the EC would make (unspecified) commitments on reduction of agricultural subsidies, but that to reach agreement the EC needed concessions from the others (especially the USA) on services, intellectual property, and market access. This move provoked another crisis over the definition of the mandate. The national ministers present in Brussels, who had nothing to do but wait for consultation meetings with the Commission, learned of what was afoot from the Australian delegation. France protested that the Commission had exceeded its mandate, and the Council as a whole reprimanded the Commission and threatened to place a representative, from the Italian presidency, in all negotiating groups. This effectively undermined the Commission's credibility, with the result that the Commission's informal approaches—to the effect that the EC might be able to accept commitments to reduce subsidies in each area (the concession the USA was looking for to continue negotiating)—were not taken seriously by its negotiating partners, who simply walked away. The negotiations collapsed, with the EC taking the blame for the collapse and national ministers of the EC adding to the chorus of criticism, even though they were ultimately responsible.[4]

The EC presses for completion: the limits of negotiating flexibility

After the Brussels debacle, the EC negotiating position improved. First, the Commission was given greater negotiating flexibility. Second, the Council of Trade Ministers was given a greater, albeit informal role, which helped counterbalance the excessive influence of the ministers of agriculture. Finally, there was a shift in the position of some member states, with Germany, in particular, moving to accept concessions on agriculture as a means of concluding the round, in order to ensure continued market access for German manufactured products. In the autumn of 1991 Chancellor Kohl felt able to weigh into the internal debate in Germany and tip the scales in favour of concessions in agriculture even if this meant breaking ranks with France. The shift in Germany's position meant that there was no longer a potential blocking minority (with France and Ireland) in the Council against concessions in agriculture if the issue came to a vote. But the experience of this 1991–2 phase of the round also shows there are clear limits to the ability of the European Commission, even with backing from some member states, to press ahead in critical negotiations without a consensus of the member states.

These changes from 1990 ultimately enabled the Commission to present a more coherent position, avoiding the isolation it had experienced at the Brussels meeting, but this was not without further delay. In the crisis meetings in October 1990 it had become clear that agricultural interests would have to be 'bought off' by offering income support for smaller farmers affected by reform of CAP price support. This was the approach proposed by the Commission in early 1991 and elaborated in July 1991.[5] The fact that the EC was considering reform helped, but it took until May 1992 to work out the details and get the package of reforms through the Council. This is to be expected with such a major reform of the CAP, but it meant that vital time was lost in the Uruguay Round negotiations.

During 1991 the Luxemburg and, even more, the Dutch Council presidencies sought to involve trade ministers more in policy decisions. This also helped to offset the influence of agricultural interests in the EC's consideration of preparations for the Draft Final Act (or Dunkel Text, as it became known) in the autumn of 1991. The Dunkel Text was completed on 20 December 1991 and represented a draft agreement for each negotiating group, but not an agreed text. Nevertheless, it consolidated texts which had been conditionally agreed at the Brussels ministerial meeting in December 1990.

The initial responses to the Dunkel text were mixed. The Council found itself unable to make a formal EC response because of French and Irish opposition to the provisions on agriculture, which proposed reductions of domestic and export subsidies, and the tarification of import restraints. Britain and Germany supported acceptance of the text. In March 1992, under pressure to respond to the text from other GATT CPs, the Commission responded that the Dunkel text provided the basis for a final agreement, but that this was without prejudice to the final position of the EC (*Agence Europe*, 4 Mar. 1992). This move, supported by Britain and Germany, was a break from the consensual approach followed thus far in the round and it enabled the Commission to begin to present detailed proposals on the outstanding issues in the round. But France, now close to being isolated, responded by arguing that the Commission was exceeding its mandate.

The EC's flexibility appeared to pay off in so far as other CPs could no longer use the EC as a scapegoat, but were obliged, in March 1992, to submit their detailed schedules on the market-access negotiations. These showed that it was not the EC alone that was facing difficulties with the Dunkel text. The USA confirmed its retreat from its previous enthusiasm for liberalization in the services sector, withdrawing various sectors (such as shipping, potentially telecommunications, and financial services) from its offer and seeking derogations from the general principle of most-favoured-nation (MFN) status in the event that its requests for market-opening in certain other countries (mainly Japan and other east Asian countries) were not satisfied. The Bush administration was also concerned that the proposals for a WTO would create problems with ratification in the US Congress.

In June 1992 outside political events again intervened. Following the

Danish 'no' vote in its first referendum on the Maastricht Treaty on European Union (TEU), President Mitterrand called a referendum on the TEU in France for September 1992. Mitterrand claimed that forcing France to accept what was presented as concessions for the Americans would result in a 'no' vote in the referendum, and indeed the final vote in favour of accepting the TEU was very close. The outcome for the Uruguay Round was, however, a further four months of relative inaction.

In October 1992 negotiators in the outgoing Bush administration offered concessions in an attempt to conclude a deal. On services it said it would accept the principle of MFN, if offers from the Asian CP were adequate. On tariffs it offered to negotiate some reductions in tariff peaks, and on public procurement it was able to offer the coverage of federal and at least some of the state purchasing contracts, following an agreement by state governors in some of the key states that they would renounce state 'buy American' legislation. At the same time the USA adopted unilateral retaliation in the oil-seeds case in order to try to force the hand of the EC on agriculture.[6] It is not clear what the American motivations were in taking such action. Perhaps the aim was to keep the EC under fire on agriculture and thus to reduce the danger of the US being exposed on one of its vulnerable issues in the round. The effect, however, was to make acceptance of the Dunkel text harder, because France was now being asked to make two concessions under US pressure: and all this just a few months before a French general election in March 1993.

In November 1992, shortly before the US presidential election, US and EC negotiators met in Chicago in an unsuccessful attempt to reach agreement on agriculture and the oil-seeds dispute (Commission 1992b). The Commission negotiated in Chicago, although with the British minister for agriculture, John Gummer, then president-in-office, hovering in the corridor outside. Again differences over interpretation of the mandate occurred, this time within the Commission. The Agriculture Commissioner, Ray McSharry, and Trade Commissioner, Frans Andriessen, wanted to accept the deal on the table and take their chances when it came to getting it through the Council, but Jacques Delors, the Commission President, argued that the deal would not be acceptable to all member states.

The Clinton victory in the 1992 US Presidential election (and fears that the Democratic congress and administration would be tougher on trade issues than the Bush administration) prompted the Commission, with the encouragement of the British Council Presidency, to return to the negotiating table. Meeting in Blair House—across the street from the White House—an agreement was reached in late November. The Blair House (pre)agreement contained an EC agreement to limit its oil-seed plantings, guaranteed minimum levels of oil-seed imports and the essentials of a deal to reduce farm subsidies

over six years in return for a US 'peace clause' guaranteeing no further agricultural trade challenges under US trade law. The Commission negotiated the
deal without a consensus in the Council, leaving an opening for member governments to attack it. This is what France duly did, indicating, for example,
at the General Affairs Council meeting on 3 March (just weeks before the
French general election) that it would oppose approval of Blair House 'by all
means', including the Luxemburg Compromise. In the event the Danish
Presidency avoided putting the issue to a vote.

By May the need to confirm the Blair House agreement was becoming
pressing. The absence of Council approval was creating uncertainties in
Washington (where the change in administration had in any case made US
policy less predictable) and undermining the EC's credibility in negotiations.
The Trade Commissioner, now Sir Leon Brittan, pressed for a decision, but
the member governments were not prepared to push the incoming Balladur
government in France into a vote on such a sensitive issue. Blair House therefore illustrates the risks inherent in a strategy in which the Commission 'gets
out ahead' of a consensus among the member states. Such a strategy may
make for flexibility, but when major member states are not 'implicated' in the
deal they are free to attack it and thus undermine EC credibility.

The endgame: the EC becomes proactive

By spring 1993 the political and economic climate had improved; Europe was
following the US out of recession, the Maastricht ratification process was
coming to an end, and there were no general elections pending in any of the
major member states. This provided a fairer wind for an agreement, but there
was still much at stake politically in how the deal concluded (OECD 1993a).
In fact the UR negotiations in 1992 and 1993 were more about political pre-
sentation than substance.

1993 saw new negotiators in both the United States and the EC.
Immediately on acquiring the Commission trade portfolio, Sir Leon Brittan
established links with Mickey Kantor, the US Trade Representative. Brittan
stressed from the outset that the Commission's policy was to stick as closely
as possible to the Dunkel Text and that the Blair House '(pre)agreement' was
not subject to renegotiation (as France wanted). This was to head off problems within the EC, but also from the new US administration, which wanted
to reopen parts of the Dunkel Text covering anti-dumping, textiles and clothing, dispute settlement, services, and the WTO.

There was considerable uncertainty about US trade policy in the initial months of the Clinton administration, which took a number of trade actions that again inflamed tensions with the EC. Sweeping countervailing and anti-dumping duties were introduced against European steel exports to the US. 'Retaliation' against perceived unfair trade practices in procurement under Section 301 was announced, and President Clinton made a number of speeches suggesting that he wanted to the revoke US–EC agreement to limit levels of subsidies for civil airframe manufacturers that had been agreed in 1992. However, the Clinton administration eventually gave notice that it would seek from Congress a one-year extension of fast-track negotiating authority, which had run out in March 1993, thus effectively setting a (US) deadline for completion of the round on 15 December 1993.

In Europe Prime Minister Balladur produced a memorandum on the Uruguay Round (*Agence Europe*, 5 May 1993) which summed up the new French government's position. Critical of the manner in which the outgoing administration had boxed the French into a corner, Balladur sought to shift the focus of the negotiations away from agriculture and on to issues, such as the creation of a WTO and trade in services, in which the EC had a coherent position. Nevertheless the Balladur memorandum also sought to reopen the Blair House pre-agreement—an option that Sir Leon Brittan strongly opposed. Balladur's attempt to win support for his view on Blair House was largely unavailing, even after the USA made a rather heavy-handed attempt to attack EC protectionism in audio-visual trade.

In July the General Affairs Council approved the oil-seeds part of Blair House, but not the part on general agricultural subsidies. To sign off on agriculture before the end of the global negotiations would have weakened the EC's overall position. France now pressed for another 'Jumbo Council' of foreign and agriculture ministers to 'reopen discussion' on Blair House. The Belgian presidency, at pains to avoid a rerun of the 1990 experience when giving agricultural ministers too much of a say had resulted in deadlock, first consulted widely on how to proceed. In these meetings it established the principle, accepted also by the Commission, that agreement on the final package would be by consensus and not by qualified majority vote as envisaged in Article 113 of the EEC Treaty, for issues within EC competence. The Presidency then convened a 'Jumbo Council' on 20 September, but finessed the issue by treating the meeting as a non-voting 'discussion' of the EC's position. This avoided tying the hands of the Commission or any member state being out-voted (Devuyst 1995).

This was followed by repeated endorsement of the Commission's negotiating position by the Council, which, together with a more sustainable position on agriculture, gave the Commission greater credibility in bilateral and multilateral negotiations. In Washington the Clinton administration's willingness to conclude was reflected in concessions and talks in Washington in October and November 1993 resulted in progress on all the issues raised by the Council in September, including 'clarifications' of issues in the Blair House agreement. The outcome of these bilateral negotiations was endorsed in the General Affairs Council before the focus of negotiations shifted to the multilateral round in Geneva. Having squared the agricultural issue with the USA, the Commission was free to pursue a more aggressive approach in the multilateral negotiations. It pushed for the conclusion of market-access negotiations, calling for maximum coverage of services and across-the-board tariff reductions. It also pressed for wording in the WTO agreement which would oblige CPs to amend any national legislation in conflict with multilateral discipline. This 'proactive' strategy paid off in the sense that the EC was not isolated as it had been in 1990 and 1991 and succeeded in getting concessions from its negotiating partners (Commission 1993c).

At an extraordinary meeting of the General Affairs Council on 2 December to consider the EC's final negotiating tactics, the member states again endorsed the Commission's position. Unlike at Brussels in December 1990, the national ministers felt no need to disown the negotiations or to blame the EC or the Commission for failure. Unlike in Chicago or Blair House, all member states were associated with the final package. The General Affairs Council met again on 8 and 13 December to consider final concessions, and on 16 December to decide unanimously to accept the final package. Support from France was conditional upon a decision to strengthen the EC's commercial policy instrument, which was agreed by qualified majority, with Germany and the Netherlands voting against. Side-payments were also made to Portugal, in the form of subsidies and grants for trade-adjustment assistance for the textile sector.

Subsequently some national governments, in particular France, had to justify this decision to their parliaments, but at that stage there was little that any national parliament could do to change the substance of the final agreement. The main impediment to ratification of the agreement by the EC was a dispute over competence between the Commission and the Council. At the outset member states had allowed the Commission to negotiate for the EC on all issues, but without prejudice to the question of competence. The Commission claimed that the results of the round had to be seen as a package to be ratified as a whole by the EC. Member states rejected this bid for increased EC competence on issues which they considered to be mixed competence, such as services and intellectual property. In November 1994 the ECJ ruled broadly in favour of the Council, stating that non-cross-border provision of services and intellectual property were mixed or national competence.[7] But this ruling is unlikely to resolve the issues of competence, and continued clashes can be expected on the representation of the EU in the WTO.

Conclusion: the trade-off between efficiency and legitimacy

The picture behind the headlines was of EC decision-making machinery working with reasonable efficiency when the experts or technocrats in the Commission and Article 113 Committee were able to develop common positions and make non-controversial changes to the negotiating mandate (Commission 1994e). In this way the EC was able to make a number of important, but low-profile, contributions to the Round. The Commission worked generally well with a small staff both to prepare position papers and to represent the EC in negotiations. On the vast majority of issues the General Affairs Council simply rubber-stamped the position worked out between the Commission and Article 113 Committee. The real problems—as was evident during the Montreal Mid-Term Review and at the abortive Brussels negotiations—came when ministers in specialist Councils, in particular the Agriculture Council, involved themselves in the detail of tactics and content, because of the importance or sensitivity of the issue. The Agriculture Council exercised a disproportionate (and, in terms of the negotiations, negative) influence, because the General Affairs Council lacked the expertise and focus to inject the necessary sense of the overall Community interest. The European Council could do little more than make general statements about the desirability of completing the round, which had no impact. The absence of a formal Council of Trade Ministers meant there was no counterbalance to the influence of the Agriculture Council.

The Uruguay Round negotiations therefore provide further evidence of a paradox in EC decision-making: the most efficient and effective parts of the process are primarily technocratic and non-transparent, lacking democratic oversight and consequently legitimacy. Although the European Parliament was reasonably well informed about developments in the Round, it had negligible influence over the negotiations and national parliaments had even less; only the power of the Council to control the Commission's negotiating mandate could be said to provide legitimacy and transparency to the negotiating process. Tensions between the Commission and Council over how the latter should exercise this control function created problems at important stages in the negotiations, inhibiting the Commission's ability to be flexible and proactive. The Commission then had to decide when to risk 'getting ahead' of the consensus in the Council. As Blair House showed, a member government not implicated in a deal can attack it from the outside. But not to have taken such risks would have condemned the EC to a purely reactive negotiating position.

Having said this, the Community was more effective when it maintained a clear, credible policy (often based on its own experience and expertise in market liberalization, much of which was adaptable to a multilateral forum) and developed it, consistently but flexibly, in the dynamics of negotiation, as in its push for multilateral solutions on a range of issues in 1993. 'Jumbo Councils' do not lend themselves to fancy footwork, and when the EC attempted this its negotiating partners (witness the US in agriculture) were able to out-manœuvre the EC and leave it wrong-footed and isolated. This contributed to the negative image of the EC in the Uruguay Round negotiations—an image that was less than just if one examines the Community's actual contribution to the successful conclusion of the round, including elements that strengthen multilateralism.

The experience of the Uruguay Round shows that the EC finds it difficult to deal with issues once they have become politicized. Real, as opposed to formal, democratic accountability was not much in evidence during the round. The main legitimizing body is the General Affairs Council, but this was not able to focus sufficiently on the detail of the negotiations to control the negotiations effectively. In future the General Affairs Council will have to become either more effectively engaged in the detail or to enlist the support of trade ministers on a more systematic and formalized basis in order to do so. The fact that the distinction between commercial policy and foreign and security policy is becoming more blurred would seem to argue against delegating ultimate power to trade ministers. The EP, despite its limited or non-existent powers, made a good effort at providing some scrutiny of the negotiations. The EP was arguably more effective as a scrutiny body than national parliaments (with the exception of the French Assemblée Générale), which appear even less willing or capable of getting to grips with the detail needed for effective scrutiny than their national ministers, who were too often willing to limit themselves to general reports of progress combined with rhetorical support for the importance of concluding the round.

The dog that did not bark, at least during the negotiations, was the issue of national versus Community competence. National governments accepted the need for a single voice in negotiations, although they did attempt to exercise especially close control of the Commission when it was negotiating on mixed-competence issues. This could have had some negative effects in reducing EC negotiating flexibility, but there is little evidence of this from the services and IPR negotiations, where there was broad support for the policies pursued.

Finally, it is not possible to close without some words on the substance of the EC positions. First of all, the EC was able to maintain coherent positions on most issues. In fourteen of the fifteen negotiating groups the EC performed on a par with, for example, the USA, if not better, in terms of presenting coherent consistent positions. The fact that the EC had no sustainable, coherent position in agriculture proved very costly as it provided its trading partners with a means of shifting the pressure and blame for lack of progress in the round on to the EC.

With regard to delivery, that is the representation of the EC mandate in negotiations, the Commission performed well. When it had difficulties, such as in Brussels, this was to do with differences with the Council (MTR and Brussels), and sometimes within the Commission (Chicago), over the interpretation of the mandate.

Did the EC contribute to multilateralism, or is it, as some American commentators seem to believe, destined to undermine multilateralism? There is clear evidence that the EC contributed to a strengthening of multilateral rules in the round. This was the case in its support for the WTO, for stronger multilateral dispute-settlement provisions, for its unqualified support for multilateral rules for services, investment, and IPR, and even in its support for comprehensive multilateral rules on a range of non-tariff barrier issues such as technical barriers to trade, government procurement, and a more grudging acceptance of tighter rules on subsidies and safeguards. The main qualification that needs to be made concerns EC resistance to tighter discipline over anti-dumping actions and the persistent rearguard action against multilateral disciplines in agriculture. This record compares reasonably favourably with that of the USA, which proved ultimately to be sceptical about the benefits of the WTO and genuine multilateral rules for services, and even seemed to adopt a guarded view on the new WTO dispute-settlement procedures, because of the impact on US sovereignty.

Even with regard to concrete liberalization on agriculture the EC record is not as bad as the headline debate would have us believe. It did accept multilateral discipline of agriculture and the phasing out of the MFA quotas on textiles and clothing, albeit with the inevitable strings attached. It even accepted tougher multilateral discipline on 'voluntary export restraints'. Again the negative side of the equation must include retention of extensive discretionary power to impose anti-dumping duties. The point here is not that the EC is any better than any other trading entity when it comes to the substance of trade negotiations, but that (despite the difficulties developing common positions) it is no worse, and that in some cases there may be more checks and balances against protectionism than within an individual country.

Notes

1. For more detail on the Uruguay Round itself, see GATT (1994). Assessments of the outcome can be found in OECD (1993a) and Schott (1994). For a general outline of the politics of the negotiations see Woolcock (1993), Murphy (1990), and May (1994). For more coverage of the roles of the member states in negotiations see Hayes (1993).
2. For a summary of the EC's procedures and substantial policy issues see the GATT's trade policy review for the EC (GATT 1993). For a discussion of issues in current debate see Maresceau (1994).
3. On this phase of the negotiations but also more generally on decision-making in the EU, see Murphy (1990).
4. See e.g. Ruggiero (1991).
5. See above, Ch. 4 on the domestic pressures on the CAP and Ch. 3 for the budgetary context.
6. The dispute on oil-seeds had been simmering since 1987. The USA initiated two GATT dispute procedures against the EC's policy on imports and a subsidy programme for crushing oil-seeds. When it lost the second the EC opted to compensate the USA rather than change the scheme, but no agreement could be reached on the level of compensation. Frustrated with the delays in resolving the dispute, the USA decided to set the level of compensation unilaterally on the basis of Section 301 of US trade legislation.
7. See Prise de Position de la Cour, European Court of Justice, Nov. 1994.

Further Reading

For a general treatment of EU policy and decision-making see GATT (1993). On the issue of the legal competence of the EU and broad trade policy issues, see Maresceau (1994) and on the economic context see Tsoukalis (1993), especially chapter 9. On the EU role in the Uruguay Round see Devuyst (1995).

Devuyst, Y. (1995), 'The European Community and the Conclusion of the Uruguay Round', in C. Rhodes and S. Mazey (eds.), *The State of the European Community*, iii (Boulder, Col.: Lynne Riener, 1995).

GATT (1993), *General Agreement on Tariffs and Trade, The European Communities: 1993*, 2 vols. (Geneva: GATT).

Maresceau, M. (1994) (ed.), *The European Commercial Policy after 1992: The Legal Dimension* (Dordrecht: Nijhoff).

Tsoukalis, L. (1993), *The New European Economy: The Politics and Economics of Integration*, 2nd edn. (Oxford: Oxford University Press).

CHAPTER THIRTEEN

EU POLICY FOR THE BANANA MARKET: THE EXTERNAL IMPACT OF INTERNAL POLICIES

Christopher Stevens

The external implications of internal EU policies were highlighted in the early 1990s by the unlikely issue of bananas. Implementation of the single-market legislation necessitated a rationalization of the EU's import regulations for bananas. The attempt to move from specific arrangements for individual EU members to common rules led to testing negotiations not only with traditional exporting states—Latin America, Africa, and the Caribbean—but also with the US and in the GATT. Differences in national customs and tastes persist, linked also with overall European development policy, as do conflicting attitudes to preferential arrangements versus free trade. Thus European policy-making veers between internal and external constraints.

Introduction

During the completion of the single European market (SEM) Europeans were wont to counter the expressions of concern and requests for consultations of third parties with the argument that it was a purely internal affair. Always false (a trading bloc as large as the European Union (EU) cannot change its internal arrangements substantially without producing ripples throughout the world trading system), the external implications of domestic policy change were highlighted, somewhat bizarrely, by the case of bananas.

The EU found itself not only with a political problem but also with a legal one. Both had external and internal dimensions. The political problem arose from the potential impact of SEM-induced changes to the European banana market on the economies of a group of favoured developing countries, and on the differing concerns of the member states and of interest groups within them. Whilst often portrayed in terms of a simple clash of 'liberal traders vs. protectionists', the range of interests underlying the conflicting positions was more complex. The legal problem arose from a clash of commitments: between the Community's treaty obligations under the Lomé Convention, its obligations under the General Agreement on Tariffs and Trade (GATT), and its internal-market rules. The humble banana became a symbol of the external consequences of domestic European changes, of the tension between preferential and generalized trade agreements, and of the balance between national custom and tastes on the one hand and Community rules on the other.

In this chapter I analyse the nature of the problem within the context of broader changes in EU relations with developing countries. Decision-makers have attempted to balance a complex set of interests. These have their roots in the trade-policy regime that grew up for bananas over the past forty years, and the pattern of bilateral trade that it reflects. This set of national policies was incompatible with the SEM and so had to be replaced by the end of 1992. But it had to be done in such a way as to maintain the EU's treaty obligations both to preferred groups of banana suppliers and to the wider international community through the GATT, as well as to minimize disruption to the interests of traders and consumers in the member states.

The chapter also explains the underlying tensions in 'Euro-South' policy that added an extra political charge to the negotiation of a new banana regime. This derived from the widening disparity between the focus of political relations with developing countries and Europe's underlying economic interests. European attempts to square the circle and to find a new set of instruments that would be compatible both with the SEM and with its treaty obligations are analysed in the penultimate section of the chapter, where the sources of resistance to the new arrangements are also described.

The banana problem

The completion of the SEM involved changes to the market for bananas because they were one of the few agricultural products not yet covered by Community rules. Not only were bananas not subject to the common agricultural policy (CAP), they were also not subject, in effect, to the common commercial policy (CCP). This anomaly was both a cause and a result of differing national practices with respect to the various sources of supply. France, Italy, and the UK all had favoured sources which were protected from the full force of competition from other suppliers through a variety of measures. The challenge posed by the SEM was how to create a uniform system within the Community that would still provide special privileges to the favoured suppliers.

The regulatory pattern

The three regimes   Traditionally about half of the EU's consumption of bananas is supplied by the African, Caribbean, and Pacific (ACP) states and by the Community itself (Guadeloupe, Martinique, Crete, and the Canary Islands), while the other half has consisted of 'dollar' bananas, mostly from Latin America. Hence in 1988 Latin America supplied 55 per cent, the DOM/TOM 23 per cent, the Caribbean 12 per cent, and Africa 6 per cent of EU imports (See Fig. 13.1).[1] By 1992 the Latin American share had risen to 63 per cent—an increase that gave rise to post-1992 problems.

These competing sources of supply operated through different policy and institutional regimes, each of which had created a set of vested interests over the years. The DOM/ACP bananas enter the Community under special arrangements designed to preserve traditional markets, institutionalized since 1975 under the Lomé Convention. Thus France provided a guaranteed market for bananas from its overseas départements and from Cameroon and the Ivory Coast, as did Italy and Britain for Somalia and for the English-speaking Caribbean plus Surinam respectively.

This colonial-based distinction between the member states was not the only one that complicated the banana market. In addition, the treatment of the commodity in Germany was different from that in Belgium, the

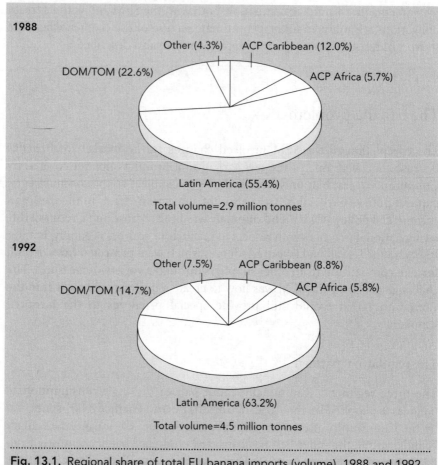

Fig. 13.1. Regional share of total EU banana imports (volume), 1988 and 1992
Source: Derived from Eurostat Comext database.

Netherlands, Luxemburg, Denmark, and Ireland. The CCP imposed a 20 per cent most-favoured-nation (MFN) tariff on bananas, but in practice it was effective only in Benelux, Denmark, and Ireland. Imports from the DOM were, of course, duty free, as were imports from the ACP under the preferential trade regime of the Lomé Convention. In consequence, the preferred imports into the UK, France, and Italy were exempted from duty. Germany had a duty-free quota, negotiated as part of the Treaty of Rome, which was about the size of total consumption, so that, in practice, it also imported bananas duty-free.

In summary, there were three tariff regimes operating within the Community:

- duty-free imports by virtue of preferences for specific exporters (in France, UK, and Italy);
- duty-free imports by virtue of a special derogation for the importer (Germany);
- and 20 per cent duty-paid imports in the Benelux countries, Ireland, and Denmark.

Their origin and underpinning   This complex arrangement had developed in an *ad hoc* fashion as the Community had grown, and was founded in a mixture of law, regulation, institutional responsibility, and commercial practice. The effect of the SEM was to remove a corner-stone from the rambling edifice that had been created, threatening to bring the whole lot down.

The special links of France and Italy were acknowledged in the Treaty of Rome, as was the German duty-free quota, agreed as part of the trade-off of benefits between the Six.[2] When the UK joined the Community in 1973 it negotiated a continuation of its Commonwealth preferences for bananas from the ex-colonies in the Caribbean (which were insufficient to satisfy more than a part of total demand). This combined the 20 per cent MFN tariff on non-Commonwealth imports with a system of import licensing that limited the volume of non-Commonwealth imports in order to ensure an adequate market for the Commonwealth products which were imported duty-free. The licensing was administered by two UK ministries and an 'advisory committee' which included commercial representatives (See Fig. 13.2).

The import preferences of France and Italy were combined with those of the UK in the first Lomé Convention (1975–80), which incorporated a special protocol on bananas, maintained in subsequent Conventions up to and including Lomé IV (1990–2000). This contains the explicit promise that: 'in respect of its banana exports to the Community markets no ACP State shall be placed, as regards access to its traditional markets and its advantages on those markets, in a less favourable situation than in the past or at present'.[3] Greece, Portugal, and Spain all joined the Community with heavily protected markets that favoured domestic production (in Crete, Madeira, and the Canary Islands respectively). In Spain such domestic production continues to meet virtually all demand, although in Greece and Portugal there are some additional imports (largely from Latin America).

The problem for the Community has been that the instruments adopted to give effect to Community preference and to its Lomé commitments were incompatible with a single market, but the creation of new instruments which would be compatible required changes that were difficult for both internal and external parties to accept. The fundamental difficulty (explained

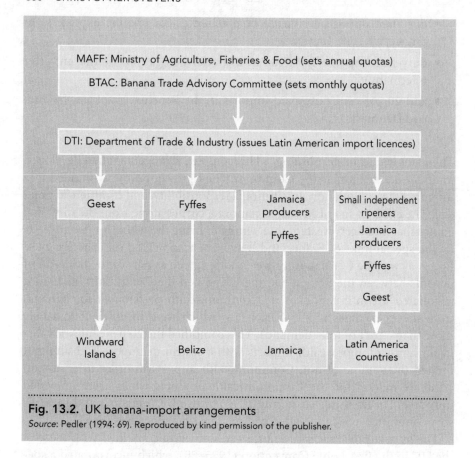

Fig. 13.2. UK banana-import arrangements
Source: Pedler (1994: 69). Reproduced by kind permission of the publisher.

below) is that the varied regimes resulted in markedly different prices in national markets. To prevent sales from cheap to expensive markets undermining the system, its guarantees were enforced through controls on intra-EU trade to prevent dollar fruit being re-exported from, say, Germany to France. These were legitimized by Article 115 of the Treaty of Rome, which allowed individual member states to maintain some national trade measures, subject to Community conditions. With the completion of the SEM, Article 115 became inoperative.

Interest groups   The pre-SEM banana market in Europe was a complex mixture of national regimes that had been in place for many years and which involved close links between importing governments, exporting governments, and commercial interests. It is not surprising, therefore, that it spawned an intricate set of interest groups. In consequence, the number of parties active in the negotiation of a successor regime was substantial. Table 13.1 lists some of the more important, and their principal concerns.

Table 13.1. Banana interest groups

Main types	Sub-divisions	Interests
European Commission	DG VI DG VIII DG I	Support for EU producers Support for ACP Support for GATT negotiations
European Parliament	Development and External Affairs Committees	Support for ACP
Member states	UK, France, Italy	Continuation of preferential links
	Belgium, Germany	Unrestricted imports (most via Antwerp)
	Netherlands	As Belgium/Germany, but balanced with development concerns for Caribbean and Surinam
Companies	Geest, Fyffes	Protection for high-cost trade (plus access to other markets)
	United Fruit, Standard Fruit, EU importers and ripeners	Unrestricted imports; maintenance of dominant position in 'dollar' trade
Supplying states	Caribbean	*Strong* protection in traditional markets
	Africa	Protection in traditional markets
	Latin America	Unrestricted imports

The European Commission played a central role but did not have an entirely consistent position. The Directorate General for Agriculture (DG VI) was primarily responsible for the formulation of a new regime but, because the pre-SEM competences were largely national, its leadership role was not as marked as is normally the case for agricultural products. The Directorate General for Development (DG VIII), as custodian of the Lomé Convention, had responsibility for ensuring effective implementation of the Banana Protocol under any new regime. In addition, the Directorate General for External Relations (DG I) had an interest in the banana debate both because its North-South Directorate was responsible for relations with Latin America and, more importantly, because of its responsibilities for the GATT and the concurrent Uruguay Round negotiations.

The European Parliament (EP) played a role in the banana story, even though it had no formal right of initiative. Both the Development Committee and the External Relations Committee prepared reports on the subject that were generally favourable towards ACP interests.

Those member states with strong interests in the matter tended to act in two main groups but with some movement between them which, as explained in the penultimate section, turned out to be of critical importance in obtaining a final agreement. The UK, France, and Italy all consistently supported a regime that would give clear guarantees to their preferential sources of supply. Germany was equally clearly in the opposite camp, demanding a continuation of its duty-free access to bananas from any source. Bananas acquired a symbolic importance in Germany, probably accentuated by their association in the eastern part of the country with the change in politico-economic regime.

Germany was supported strongly by Belgium, which, in addition to supplying its domestic market with dollar fruit, had an interest in the role of Antwerp as a port of entry for much of Germany's consumption. The Netherlands was in a somewhat ambivalent position: its consumption of dollar bananas, as well as support for liberal trade and successful completion of the Uruguay Round, counselled support for the German position; on the other hand, the Netherlands's traditional interest in development matters, together with its close ties to Surinam, tended in the other direction.

The companies involved in transporting, marketing, and, in some cases, producing the fruit, fell into two similar categories, but once again with some cross-cutting interests. The two European companies most closely involved with the import of fruit from the Caribbean, Geest and Fyffes, were active in pushing for a new regime that would defend their traditional markets. Geest, in particular, is believed to have worked in close alliance with Caribbean producers lobbying in support of the Banana Protocol (Pedler 1994: 70). On the other side, the US companies involved in plantation production in central America and in transporting dollar fruit—United Fruit (Dole) and Standard Fruit (Chiquita)—together with importers and ripeners based in Belgium, the Netherlands, and Germany, wanted a removal of restrictions on the banana trade.

As will be seen from the riposte to the new banana regime, described below, such 'protectionist' and 'free trade' stances were not quite as clear as they might seem. What both sets of companies really wanted was to fend off competition in their traditional markets whilst opening up possibilities elsewhere in the EU.

The supplying states can be roughly categorized into the same two groups—protectionist suppliers of EU and Lomé bananas and liberal suppli-

ers of dollar fruit—but here again these broad distinctions hide subtle variations. Among the ACP there was a latent tension between the Caribbean and
African banana exporters. As is explained in the next section, the former are
the highest-cost external producers and so need absolute guarantees in their
traditional market and would not benefit from improved access to other
European markets. African producers, by contrast, are medium cost and
might be able to offset losses in their traditional market against gains in the
other European markets. The Latin American states favoured a liberal trade
regime, although each had their own interests (depending in part on whether
production was indigenously controlled or undertaken by a US company, or
both), resulting in policy differences.

The pattern of trade

The four sources of supply   Under this variegated system of regulation a
trade developed in which there are clear bilateral patterns. There were four
main sources of supply—domestic, Caribbean, African, and Latin
American—and within the three external sources a limited number of supplying states. In 1992 five Caribbean states supplied over 90 per cent of the
region's exports; they were three Windward Islands, Jamaica, and Surinam
(see Fig. 13.3). In the same year, two African states supplied an equivalent
share of imports from that region. They were Ivory Coast and Cameroon, and
in earlier years they had been joined by Somalia before the collapse in banana
production consequent upon the civil war there (in 1988 Somalia had
accounted for 32 per cent of the region's supplies to Europe). Similarly, in the
case of Latin America five states—Ecuador, Colombia, Costa Rica, Panama,
and Honduras—supplied 95 per cent of the total.

The producer-trader-import nexus   These four groups of supplying countries tend to be linked to particular importing countries, via trade dominated
in each case by a small number of companies (because transporting bananas
is capital-intensive and specialist). In the case of the UK, for example, the
Windward Islands and Jamaica accounted for just over 50 per cent of 1992
imports (see Fig. 13.4). The trade was handled largely by two companies—
Geest and Fyffes. In the case of France, the DOM/TOM of Martinique and
Guadeloupe accounted for almost 60 per cent of 1992 imports, with Ivory
Coast and Cameroon filling virtually the whole of the remainder of the market. In the cases of Germany and the Benelux states (as well as Ireland and
Denmark), the five main Latin American exporters held virtually the whole
of the market, with trade being dominated by a small group of essentially US
companies.

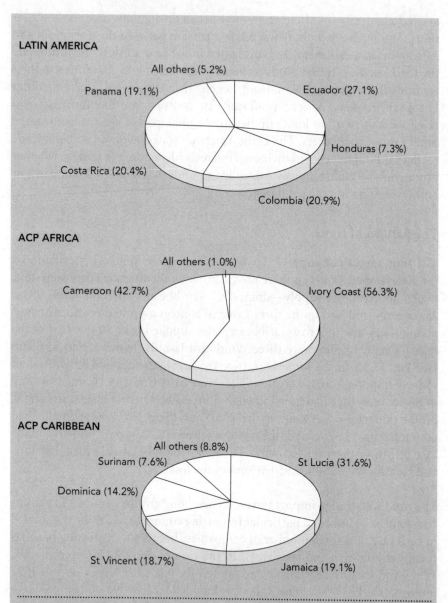

Fig. 13.3. The main sources of banana supply to the EU market (by volume), 1992
Source: Derived from Eurostat Comext database.

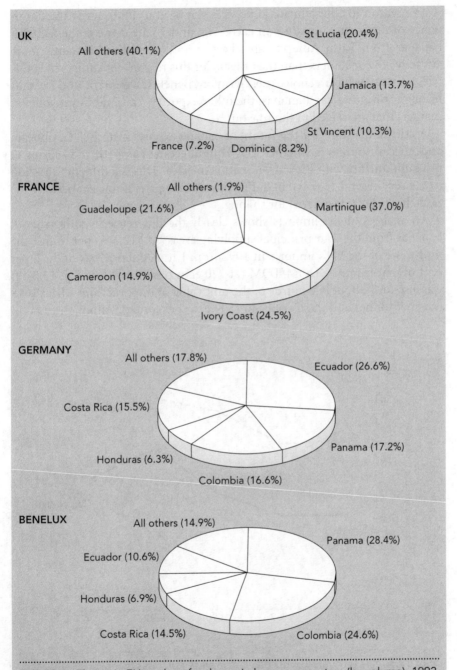

Fig. 13.4. The main EU markets for the main banana exporters (by volume), 1992
Source: Derived from Eurostat Comext database.

The reason for these bilateral flows is not hard to find, nor is the dominance of different companies in each. The underlying cause is the fact that bananas from Latin America are cheaper and, according to many, more attractive to consumers than the others. For this reason, consumers in countries where they have a choice (i.e. Germany, Benelux, Denmark, and Ireland) favour Latin American bananas; the other exporters can find buyers only in markets that are rigged in their favour.

A critical question for the future of banana exports from EU, Caribbean, and African sources is whether they could become competitive in terms of price and quality with those from Latin America. This is a difficult question to answer since the true extent of the difference in price is unclear because the trade has been so regulated for so long.

An analysis of EU imports shows clearly the difference in unit value of bananas from the four principal supplying areas. At 412 ecus per tonne, the unit value of the EU's imports in 1992 from Latin America was only 57 per cent of that from the DOM/TOM (at 720 ecus per tonne) (see Fig. 13.5). It was also only about two-thirds of the unit value of imports from Africa (619 ecus per tonne) and the Caribbean (599 ecus per tonne). Although the unit

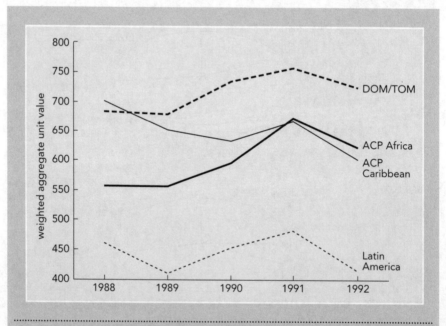

Fig. 13.5. The unit value of EU banana imports from main suppliers, 1988–92 (ecu per tonne)

Source: Derived from Eurostat Comext database.

value of Caribbean fruit fell over the period 1988–92 (due in part to the appreciation of the ecu against the US dollar, the dominant currency for the exporting states), bananas from the three 'preferred sources' clearly have much higher unit values than those from the non-preferred source. An analysis made by the UK Ministry of Agriculture for 1988 and 1989 indicates the distribution of costs for ACP and dollar bananas (see Fig. 13.6). This suggests that production costs in ACP states are some 50 per cent higher than those in Latin America.

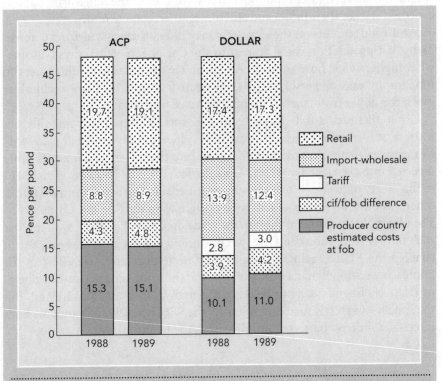

Fig. 13.6. Annual gross mark-ups for ACP and dollar bananas in the UK market (1988 and 1989)

Note: Average of monthly figures weighted by volume of imports, ACP or dollar; 1988 January to December; 1989 January to June.
Source: UK Ministry of Agriculture.

The problem arises in judging how far these market prices indicate real costs of production and how far they simply reflect the distortions in the market. On the one hand, the market rigging that has occurred in the UK and France may have allowed production costs in the preferred suppliers to rise above the levels at which they would otherwise have stuck. If true, this would mean than the cost of Caribbean, African, and DOM/TOM exports *could* be reduced if competitive pressures forced through increased efficiency. On the other hand, the non-preferred suppliers, too, benefit from market rigging in the sense that consumer prices in the protected markets are higher than they otherwise would be. Hence, whilst these countries are unable to supply as many bananas as they might wish to the UK market, those that are permitted entry may be sold at a price that is higher than would pertain in an unrestricted market. If this is the case, then the disparity between production costs is understated by figures on import unit values, since the dollar fruit suppliers could reduce their prices and still make an acceptable return.

Whilst judgements on these matters may differ, there is widespread agreement that production costs, at least in the Caribbean, will always be somewhat higher than those in Latin America. This is because of differences in topography and ownership. The Caribbean fruit is grown on smallholder plots; the dollar fruit is grown mainly on large plantations.

Given this structural disparity in the cost of imports from different sources, it is easy to see why the pre-SEM system could be held in place only by measures that prevented dollar imports into the non-rigged markets being diverted into the rigged markets. As explained, this was based on Article 115 of the Treaty of Rome, which allowed member states to restrict imports from other Community members of goods originating outside the Community. The oligopolistic nature of the banana trade made it relatively easy to impose such border restrictions and the system worked quite well in the sense that it enabled the EU to maintain Community preference and give effect to the Lomé Convention Banana Protocol, while at the same time permitting liberal or relatively liberal access to Benelux, Germany, Denmark, and Ireland. But the system was clearly incompatible with the SEM objective of removing both direct and indirect barriers to trade within the Community.

The wider context of development policy

As if the specific problems of bananas were not enough, the issue was sucked into a much wider debate on future relations between the EU and the developing world. The ACP banana exporters feared, with justification, that the erosion of preferences on this one item was the (not so) thin end of a large wedge that was being forced into the EU's traditional relationship with the ex-colonies.

Tensions between formal policy and effective interests

At the heart of the wider problem is a change in the structure of Europe's economy and a growing disparity between the focus of the EU's formal development policies and its economic and political interests in the South. Over the past decades there has been a change in the relative importance of various sources for European growth, with non-traded services and intra-developed country trade increasing in relative significance. The distortions caused by the CAP have simply accentuated a trend away from the traditional colonial trade pattern of importing raw materials from the South and exporting manufactures to it. In its place, a trade has developed with parts of the South that emphasizes a two-way flow of manufactures and services. But the ex-colonial banana exporters are not well represented in this new trade pattern.

The leaders of the new pattern of trade have been, on the European side, the states with relatively weak colonial ties (notably Germany) and, in the South, the countries of east and south-east Asia. By contrast, formal development policy has been fashioned largely by the major ex-colonial states (France and the UK) and has focused on the recent colonies, particularly the ACP.

A tension has developed between the focus of formal policy towards the South and the focus of the EU's immediate economic interests. The tension has been defused partially up to now because each EU member state has retained control over many of the most potent commercial policy instruments. Export credits, investment promotion, and debt rescheduling remain member-state responsibilities; Germany, for example, may use them to promote its interests in south-east Asia regardless of the Community focus on the ACP. Indeed, it may prefer the Community to concentrate on the ACP so as not to queer its pitch in Asia. But as powers are transferred increasingly from national to Community level this capacity to run an independent shadow policy withers; the emphasis of Community-level policy acquires a direct importance for national interests. This is the broader significance of the SEM for Euro-South relations.

A similar tension has developed on the political front. It is not that the interests of Africa and the Caribbean will be deliberately downgraded in the EU. Rather, the problems of the Mediterranean and eastern Europe will force themselves to the top of the pile, leaving less time for others, especially those relating to areas of declining economic importance for the EU.

The shift to new policy instruments

Such changes come at a time when the foundations of the Community's relationship with the South are trembling. European officials have managed over the years to fashion with some skill a quasi-foreign policy based on the limited range of Community-level instruments. Trade preferences bulk large in the relationship with the South. The Banana Protocol is just one example. But the value of trade preferences to the beneficiary is related inversely to the level of protectionism (at least if the matter is viewed only in a short-term, static perspective). With the completion of the SEM and the GATT Round, there is a reasonable chance that the 1990s will be a decade of liberalization, reducing the vitality of any preferences.

Hence the whole edifice built up over the years by the EU is likely to subside gently as its foundations are weakened by liberalization. Since this is happening at a time when the pace of European integration is quite fast (despite periodic reverses), we may expect that a new edifice will be thrown up to replace the old. The Community institutions will acquire a wider range of powers. Among them, no doubt, will be instruments that are of value to the South and may be used to construct a new relationship.

The implications for the ACP

Why should the new ways not be just as favourable to the ACP as the old? Why should the new instruments not be used to offset the effect on the ACP of the withering of the old? The answer is that perhaps they will, but that full replacement of the old by the new is unlikely. There are two reasons for this pessimism. The first is that current EU policy reflects the interests of the past; the new policies and practice are more likely to reflect current interests in which the ACP are less prominent. The second is that the ACP states, and especially the Caribbean countries most likely to be adversely affected by a change in banana policy, appear to be ill prepared to take advantage of Europe's new methods and instruments.

The cost for the Caribbean of losing its privileges in the UK would be particularly great, because they are more valuable than those enjoyed by the generality of ACP. In terms of the foreign exchange involved, the guaranteed market access that the Banana Protocol confers on the Caribbean (plus Ivory Coast and Cameroon) is one of the most important parts of the Lomé trade regime, and is buttressed by the Sugar Protocol, which also provides very substantial foreign exchange benefits to the Caribbean and a few African and Pacific states. The potential benefits of the SEM are dimmer because the Caribbean has been less successful even than Africa in the past in exploiting the dynamic potential of the Lomé trade preferences.

Negotiating a solution

Given the importance of the Banana Protocol, the creation of a new set of instruments to give effect to its promises was a highly charged affair. Even so, the full extent of the fallout—which put a question mark over the future of the Lomé Convention and threatened a trade war between the EU and the USA—was unexpected. It arose because although it was possible to identify a number of mechanisms that would have underpinned the *status quo* on bananas without contravening the aims of the SEM, each of them suffered from political problems and some were economically less desirable than others.

Proposed solutions

In the debate preceding the 1992 deadline, four types of solution were discussed widely. These were the use of tariffs, deficiency payments, quotas, and of palliatives to ease the pain. Since the dust blown aloft by the EU's chosen option has not yet settled, it is worth while examining each, since one or more may need to be reconsidered in the future.

A free market with tariff   Under this proposal the various national regimes would be replaced by a single Community market with a standard tariff. Depending upon the level of the tariff, prices in the protected markets would tend to fall and those in Germany to rise. The problem for the Caribbean countries was that, even with a 20 per cent tariff, they would have been unlikely to be able to compete with dollar fruit. As explained above, because the banana trade in the French and UK markets has been controlled for so long, it is difficult to identify how far prices of Caribbean exports could be cut and still remain viable. However, the data in Figs. 13.5 and 13.6 suggest that quite substantial tariffs would be required to protect ACP imports from Latin American competition.

Whilst a uniform tariff on which the ACP would receive a preference was one of the simpler mechanisms being proposed, it was by no means without controversy even at the European end. Any tariff would require the German government to accept an increase in the domestic price of bananas. And even the Benelux governments indicated their unhappiness with a tariff as high as 20 per cent; their initial preference was for complete abolition of the tariff, but they indicated that they would accept a tariff 'reduced substantially from existing levels' (Benelux 1990). This was despite the fact that the Netherlands has tended to be sympathetic to Third World interests in general and has a particular relationship with Surinam.

Deficiency payments Under this proposal there would be a free market within the EU (with or without a tariff), but high-cost EU and ACP producers would receive direct income support to increase their return to an agreed level. Although this solution would approximate more closely to the existing regime than a free market, it would still tend to disfavour high-cost banana exporters such as the Windward Islands if the deficiency payment were standardized for all preferential suppliers in order to avoid a cost-plus system creating inefficiency. A payment that would be sufficient for the most efficient ACP suppliers would be inadequate for the least efficient. The principal problem with instituting such a system was its budgetary cost to the Community.

A global quota A proposal which received some attention in Community circles was to establish an overall EU quota either for dollar fruit or for all bananas. The restriction on supply would tend to result in an increase in prices in the 'free' markets and a fall in the 'protected' markets. The extent to which this system would satisfy Caribbean interests would depend critically on the overall size of the quota, any agreed growth in the quota over time, and whether it was supported by ancillary measures such as a tariff and/or deficiency payments. Once again, however, the Windward Islands would face the problem that a price level acceptable to the most efficient ACP suppliers could still leave them unable to compete.

Other palliatives These three options could be combined both with each other and with other palliatives, notably an increase in aid. Whilst financial assistance may be a useful part of any package of measures, it could never be a satisfactory alternative to an adequate trade mechanism. This is partly because the economic impact of large-scale aid would be quite different from that of banana-export earnings. It is also because the EU would be most unlikely to be willing to provide finance on a scale adequately to 'compensate' for the loss of banana exports. In the case of St Lucia, for example, Lomé aid would have had to rise in the region of twentyfold adequately to compensate for the loss of banana exports. Deficiency payments would be much cheaper.

The solution adopted

The solution finally adopted by the EU was to implement a two-tier import regime. A temporary regime was introduced to cover the first half of 1993, and a permanent system established thereafter.[4] In both cases ACP (and, of course, DOM/TOM) bananas continued to enter the Community duty-free, but dollar fruit was to be subject to a two-tier tariff. The first two million tonnes of imports from Latin America (technically known as a tariff quota) were to pay a specific duty of 100 ecus per tonne; for imports above this threshold, the duty increased to 850 ecus per tonne. To put these figures in perspective, they can be compared with the unit value of imports in order to obtain an *ad valorem* equivalent. The unit value of EU imports from Latin America in 1992 was 412 ecus per tonne, at which level the lower tariff for the first two million tonnes was equivalent to 24 per cent and the higher tariff was a penal 206 per cent.

The object of the two-tier tariff is to allow the Latin American countries to continue to supply their traditional share of the market while imposing a serious barrier to attempts to increase their market share at the expense of preferred suppliers. ACP suppliers no longer have an absolute advantage in the UK and French markets, but they have a substantial tariff advantage *provided* that the Latin American tariff quota has been set at a level that will leave space for them.

An additional feature of the new regime, which has fuelled much of the subsequent controversy, was the introduction of three types of licence for importers of the two million tonne quota for dollar fruit. Licences for 66.5 per cent of this amount were to be allocated to traditional operators involved in the dollar banana trade; these were known as A quota licences. Around 30 per cent (B quota licences) were to be allocated to traditional importers of EU and ACP bananas. The balancing amount of around 3 per cent (C quotas) were to be allocated to newcomers.

The reason for this innovation was that the new regime was expected to lead to a fall in prices in the protected markets and an increase in price in the free markets. The companies engaged in supplying the former argued, successfully, that their profit margins would be eroded seriously. The creation of B quotas was designed to allow them to make profits in the free market in order to achieve a rate of return overall that would enable them to continue to ship ACP/EU fruit. One consequence, of course, was that the traditional suppliers of dollar fruit found themselves with unwelcome new competitors in their traditional back yard.

How the decision was reached

It would appear that the square of mutually incompatible positions was circled to produce a common decision by the usual combination of chance and design. An important element in the successful outcome was the fact that the Commission's proposal was to introduce a CAP regime for bananas under Article 43 of the Treaty of Rome. This meant that the matter would be decided in the Agriculture Council and any decision would be taken by qualified majority vote (Pedler 1994: 78). The draft proposal issued on 1 August 1992 was broadly favourable to the cause of EU/ACP bananas. A quick review of the voting strengths of the main protagonists would have suggested that the 'free traders' would have sufficient votes to form a blocking minority. Germany, Benelux, Denmark, and Ireland had twenty-eight votes between them, with only twenty-three required to block the proposal. Yet the final decision reached after votes in December 1992 and February 1993 was broadly in line with the Commission's initial proposal (the main trade-related difference being that the latter proposed an absolute quota of two million tonnes from all sources).

How did this come about? The short answer is that countries did not vote entirely in accordance with their assumed position within the 'protectionist' and 'free trade' camps. That they did not do so was a result partly of persuasion, partly of accident, and partly of broader considerations. The fact that the UK was President of the Council during the second half of 1992 and lobbied hard in favour of the ACP is likely to have enhanced the persuasion element of the equation. Ironically, the assumption of the Presidency by Denmark (a 'free trader') also benefited the 'protectionist' cause, but for different reasons.

A first vote was taken at the Agricultural Council on 17 December 1992 which dealt only with the trade aspects of the new Banana Trade Regime, leaving the internal implementation to be dealt with at the February Council. Although the details of the proposal were significantly changed from the Commission's original proposal, it remained fundamentally geared to the preservation of protected markets for EU and ACP fruit. The proposal was adopted, with the line-up of member states indicated in Table 13.2.

Several states voted in a way that was contrary to what might have been expected. In particular, Belgium and the Netherlands voted for the regime (although they would have been expected to support freer trade), while Portugal (which would have been expected to support its domestic producers) voted against. Later, the Belgian and Netherlands's governments claimed that they had been misled over the exact effects of voting for the package (Pedler 1994: 85). In the case of the Netherlands, too, the fact that their representative, Agriculture Minister Bukman, had previously been Minister for Development Cooperation may have made a contribution. In the case of Portugal, the decision to vote against was taken in protest against wholly unrelated events in the EU.

When the matter came back to the next Agriculture Council, in February 1993, for the internal details to be agreed, there was again a qualified majority, although several states changed sides. Portugal reverted to expected form, as did Belgium and the Netherlands. The result was that the Council presi-

Table 13.2. Voting on the Banana Trade Regime

The first vote in Council (December 1992)

The BTR was adopted by qualified majority (needing 54 votes):

In favour	Against
Spain	Denmark
France	Germany
UK	Portugal
Ireland	
Italy	
Netherlands	
Belgium	
Greece	
Luxemburg	
58 votes	18 votes

The second vote in Council (February 1993)

The internal details were adopted by qualified majority:

In favour	Against
Spain	Germany
France	Belgium
UK	Netherlands
Ireland	
Italy	
Greece	
Portugal	
Denmark	
Luxemburg	
56 votes	20 votes

Source: Pedler 1994: 84–5. Reproduced with kind permission of the publisher.

dency, Denmark, had the casting vote and decided in the interests of EU decision-making to vote for the proposal.

The riposte

As is the fate of compromises, this one was attacked by many of the interests it was supposed to balance. In general, it found favour with the preferred suppliers, but it was opposed vociferously by German importer interests and Latin American exporters. As a result of these complaints, two third parties—the GATT and the USA—became involved, together with the European Court of Justice (ECJ).

Intra-EU complaints   The complaint of the German importers, which was taken up by the German government and backed by those of Benelux, centred on the imposition for the first time of a tariff, albeit a relatively low one of approximately 24 per cent. The German government sought a ruling from the ECJ that it was guaranteed unimpeded access to the fruit, and that the measure amounted to an expropriation and redistribution of market share. The ECJ ruled against the two petitions (in respect of the interim and the longer-term regulation) on the grounds that the new regime did not cause 'serious and irreparable damage' to Germany, since the regulation allows the Commission to raise the tariff quota in line with demand.[5]

The action of the German government in supporting opposition to a measure that was demonstrably in the interests of France has been seen by some as evidence of a new willingness by the Germans to assert their interests. Whilst this reasoning should not be taken too far, it is certainly the case that there were distinct differences in national interest. For France, the new regime meant protection of the interests of its overseas *départements*. For Germany, it meant an increase in the price of a popular and symbolic import. Moreover, the overtones of the debate (selective support for preferred countries versus multilateralism, and protection versus free trade) reflected different perceptions in the two countries. At the very least, students of intra-European politics can derive one example from the banana history: a reiteration of the power of the ECJ in providing legal solutions to problems that are essentially commercial and political.

Latin American complaints   The dollar-fruit exporters were concerned both by the imposition for the first time of a tariff on exports to Germany and, more particularly, by the size of the quota to which the low tariff applied and the punitive nature of the high tariff. They claimed that the *tariff quota* of two million tonnes was insufficient. It was certainly smaller than the volume of Latin America's banana exports to the EU in 1992 of 2.8 million tonnes, but was higher than the volumes that had been supplied before 1990 (see Table 13.3). Hence, a continuation in 1993 and beyond of this level of exports would have resulted in part of a shipment paying the punitive 206 per cent rate. The Commission's response was that two million tonnes represented the historic level of Latin American exports to the EU and that the increases in the years leading up to 1992, evident from Table 13.3, had been part of a deliberate strategy by the dollar exporters to increase their market share with a view to the completion of the SEM.

The reaction of GATT   Five of the aggrieved Latin American exporters lodged two complaints in the GATT (one in relation to each of the EU's reg-

Table 13.3. Latin American banana[a] exports to the EU by country, 1980 and 1988–1992 (tonnes)

Country	1980	1988	1989	1990	1991	1992
Argentina	—	—	—	4	—	—
Bolivia	—	—	—	17	—	—
Brazil	1	16	41	9	22	4
Chile	—	63	63	75	27	74
Colombia	312,858	343,449	331,305	401,910	511,197	592,773
Costa Rica	296,602	341,299	449,119	548,519	569,078	577,179
Dominican Republic	15	344	855	3,829	10,201	38,489
Ecuador	246,165	320,492	273,899	352,146	600,880	767,306
El Salvador	—	—	8	—	86	—
Guatemala	54,200	34,634	61,827	9,370	13,274	63,830
Haiti	3	—	20	1	—	20
Honduras	165,970	188,811	148,846	123,381	138,396	207,872
Mexico	—	1	19	41	39	17,464
Nicaragua	—	34,726	29,037	47,600	65,291	28,423
Panama	232,188	339,827	400,497	527,507	483,217	540,116
Peru	—	—	—	6	—	—
Uruguay	—	—	11	39	21	11
Venezuela	—	203	179	50	41	45
Latin American total	1,308,002	1,603,865	1,695,726	2,014,504	2,391,770	2,833,606

[a] The volumes for 1980 are for Nimexe code 080131 ('fresh bananas'); those for 1988–92 are for Combined Nomenclature code 08030010 ('fresh bananas, including plantains').
Source: Eurostat Comext database.

ulations). They were supported in their action by the USA, which was concerned, at a critical time in the final negotiations of the Uruguay Round of trade liberalization, not to permit the EU to move in the opposite direction by increasing tariffs.

The GATT panel on both complaints ruled in the Latin American states' favour. In these rulings, the panel moved beyond the specific case of bananas to comment on the GATT-compatibility of the entire Lomé Convention. This brought the spotlight of international trade law into the rather murky area of EU policy-making. The EU claims that its hierarchy of trade preferences in favour of developing countries is justified under Article XXIV and Part IV of the General Agreement. Article XXIV permits, *inter alia*, derogations from the MFN principle to countries engaged on the creation of a customs union or free-trade area. Part IV establishes what has become known as 'special and differential treatment' for developing countries, which permits developed countries to treat their exports on terms that are better than those established under the MFN principle.

The GATT panel argument was that neither of these justifications could support the Banana Protocol. It could not be supported under Article XXIV because there was no provision in the Lomé Convention for the eventual creation of a free-trade area or customs union between the contracting parties. Indeed, one of the hallmarks of the Lomé Convention was *non-reciprocity*, that is the ACP states are not required to make any reduction in their import controls in return for their privileged access to the European market. Nor could the Lomé Convention, argued the panel, be justified under Part IV of the GATT, since it was not available to all developing countries, only to a select group of them, and effectively discriminated between one set of developing countries and another.

The EU deal with Latin America   This view was not immediately put to the test, because a compromise was hatched between the EU and the main Latin American banana exporters. Under a 'Framework Agreement on Bananas', reached in mid-1994 with four of the GATT complainants (Costa Rica, Colombia, Nicaragua, and Venezuela), the EU agreed to raise the tariff quota to 2.1 million tonnes and to grant the four Latin American countries specific quotas based on their past share of the market. In return, the four dropped their GATT complaint.

Another feature of the deal was that the four Latin American states were authorized to issue export licences, that is to determine which suppliers could take advantage of the EU's import licences. This was seen by them as enhancing their negotiating position *vis-à-vis* the US companies that have dominated the trade in dollar fruit to Europe.

This agreement stirred up, in turn, its own controversy. It was rejected by other Latin American suppliers (including Guatemala, which had been party to the GATT complaint, plus Ecuador, Honduras, Panama, Mexico, and the Dominican Republic). Guatemala indicated that it would continue its action in GATT.

The reaction of the US companies   The agreement was also opposed by some US banana companies that felt discriminated against by the overall tariff quota, by the country quotas for the four Latin American signatories, and by increased competition with other companies. The decision to establish B quotas that would allocate part of the 'free market' in Europe to Geest and Fyffes, together with the Latin American agreement that linked the allocation of A licences to the possession of an export certificate issued by the Latin American state concerned, restricted the traditional activities of United Fruit and Standard Fruit. The companies sought action from the US government, which responded by launching an investigation under Section 301 of the US Trade Act. As of early 1995, the US authorities had let it be known that preliminary investigations had found that the banana regime did adversely affect US interests, and had made threats of retaliatory action against the exports to the USA of both the EU and the Latin American signatories of the Framework Agreement.

Conclusions

The disputes arising from the EU's change in its banana-import policy had not been settled by early 1995. Some Latin American banana exporters were still in dispute with the EU, as was the USA. The question mark over the legality of the Lomé Convention had not been resolved. And the ECJ had still to consider one petition from German importers. This petition was also taken to the German Constitutional Court, which ruled in January 1995 that the EU regime damaged the interests of German importers. All of these disputes had their origin in the clash between the requirements of the SEM and existing external obligations.

One lesson to be derived from the episode is that when a major trading bloc such as the EU changes its internal rules it is likely to have external repercussions. These were particularly severe in the case of bananas, both because the change in internal regulations was substantial (from a very un-common market to a unified one) and because the *status quo ante* had substantial, differential effects on a small number of states. However, similar effects are likely to have occurred as a result of other aspects of the SEM (and of other EU internal market changes), albeit less visibly and, probably, less dramatically.

Given the likelihood of external consequences of internal changes, a second lesson to be drawn is that these effects need not be unidirectional. It is a commonplace that proposed changes to trade policy tend to provoke squeals of anguish from those who expect to lose, which are not offset by positive reactions from those who will gain. In the case of bananas, by contrast, both gainers and losers were vocal, so that the differential impact of change can be documented more easily than is usually possible. The *status quo ante* favoured some countries and restricted others; the new regime affects this balance of interests and has brought a response from all concerned. Only some of these responses have been covered in this chapter. In addition, the Caribbean states are lobbying the USA to oppose its action against the banana regime and, no doubt, so too are some Latin American states with the opposite intent.

A third lesson that flows from this is that the validity of EU 'internal' changes may be challenged in three types of fora: those that are wholly within the EU domain (such as the ECJ or the courts of the member states); those to which the EU is a party, but not a dominating one (such as the GATT); and those in which the EU has no standing at all (such as the bilateral negotiations between the Caribbean and the US). Whilst the first category of fora may be organized, as in this case, on legal principles, the others are likely to have a strong political dimension. Hence changes to EU domestic law are open to political challenge elsewhere.

Finally, the banana episode emphasizes a characteristic of an organization such as the EU that has evolved in a crab-like fashion: it can easily find itself with mutually incompatible obligations. Given time, it may be able to resolve such differences. There is a distinct possibility, for instance, that the problem over the legal status of the Lomé Convention will be resolved by its being replaced with a new set of trade policies towards developing countries that would be genuinely compatible with either Article XXIV (because a free-trade area is envisaged) or Part IV (because they are available to all developing countries, albeit with different tranches available to countries at different income levels). But this could not happen before the next century without the EU breaking its existing commitments under the Lomé Convention. Like Eartha Kitt's Englishman, the EU needs time!

Notes

1. Analysis of EU banana trade is complicated a little by the fact that the relevant category recorded in the EU's import statistics includes both bananas and plantains. It is unlikely that the inclusion of plantains in the figures cited in Fig. 13.1 and elsewhere in this chapter affects significantly the conclusions drawn from the analysis.
2. Protocol on the Tariff Quota for Import of Bananas (ex. 08.01 of the Brussels nomenclature).
3. Protocol 5, Article 1, Lomé IV.
4. The regime imposed from 1 July 1993 is set out in Council Regulation (EEC) No. 404/93 of 13 Feb. 1993 on the common organization of the market in bananas (*Official Journal* No. L47).
5. A third petition (in respect of a 1994 revision—see below) was still pending at the time of writing.

Further Reading

On the general background of EU relations with the Third World see Stevens (1992) and especially with the Caribbean see Stevens (1991). To follow the banana story in more detail see Borrell and Yank (1992), Fitzpatrick *et al.* (1993), and Read (1994). For the contrasting cases of sugar see Webb (1977) and Stevens and Webb (1983), and of fish see Shackleton (1983).

Borrell, B., and Yank, M. (1992), *EC Bananarama 1992: The Sequel,* Working Paper 958, International Economics Department (Washington DC: The World Bank).

Fitzpatrick, J., *et al.* (1993), *Trade Policy and the EC Banana Market* (Dublin: Economic Consultants).

Read, R. A. (1994), 'The EC Internal Banana Market: The Issues and the Dilemma', *World Economy,* 17/2: 219–35.

Shackleton, M. (1983), 'Fishing for a policy? The Common Fisheries Policy of the Community', in H. Wallace, W. Wallace, and C. Webb (eds.), *Policy-Making in the European Community,* 2nd edn. (Chichester: John Wiley and Sons, 1983), 349–72.

Stevens, C. (1991), 'The Caribbean and Europe 1992: Endgame?', *Development Policy Review,* 9: 265–83.

Stevens, C. (1992), 'The EC and the Third World', in D. Dyker (ed.), *The European Economy* (London: Longman, 1992): 211–29.

Stevens, C., and Webb, C. (1983), 'The Political Economy of Sugar: A Window on the CAP', in H. Wallace, W. Wallace, and C. Webb (eds.), *Policy-Making in the European Community,* 2nd edn. (Chichester: John Wiley and Sons, 1983), 321–48.

Webb, C. (1977), 'Mr. Cube *versus* Monsieur Beet: The Politics of Sugar in the European Communities', in H. Wallace, W. Wallace, and C. Webb (eds.), *Policy-Making in the European Community* (Chichester: John Wiley and Sons, 1977), 197–226.

CHAPTER FOURTEEN

POLICIES TOWARDS CENTRAL AND EASTERN EUROPE

Ulrich Sedelmeier and Helen Wallace

The end of the cold war necessitated a radical reorientation of EC policy towards its eastern neighbours, and an immediate response. The Phare programme, G24 assistance, and the formation of the EBRD provided elements of support for reforms in central and eastern Europe. The search for a long-term strategy began with the initial design of the Europe Agreements. EC policy-makers were caught between high external expectations and internal resistance to change, admitting the case for an eastern enlargement, while conscious of the difficulties involved. Foreign policy and security considerations were addressed partly through other fora and partly through the new EU framework. Since the Essen European Council of December 1994 the EU has been set on a 'pre-accession' strategy; this tests severely its capacity to learn by doing and to deliver the necessary internal bargains.

The search for policy coordinates

The cold-war context in which the European Community (EC) was conceived and developed ruled out any apparent need for a 'European policy'. A broader definition of 'European integration', aimed at overcoming the East–West division in Europe, may always have been implied. Yet part of what breathed life into the west European project was the goal of distinguishing and defending the West from the East, just as within western Europe what distinguished the EC members from the others was precisely the resistance of the latter to integration. In both cases the onus was on those outside to adjust. In this chapter the reverse is examined—the challenge for the EC of itself adjusting to post-1989 Europe.

The EC had developed over the years various kinds of formal relationships with the 'third' countries of western, northern, and southern Europe, as well as towards its non-European southern neighbours. However, the EC's image of itself as sharply distinct from the countries of central and eastern Europe (CEECs) and communism made redundant any mental or substantive preparation for a dramatic change in the relationship. Questions such as the geographical limits to EC integration and whether it could indeed serve as a pattern for transnational integration on a broader European scale therefore were simply not on the agenda. The only exception made was the special provision for the German Democratic Republic (GDR) under the guise of 'inner-German' relations.

The political coordinates of systemic antagonism on both sides precluded formal relations between the EC and the Council for Mutual Economic Assistance (CMEA), or its individual members, until the arrival of Gorbachev and the subsequent signing of the joint EC–CMEA declaration in June 1988 (Pinder 1991: 8–23). The generally low trade activity of the CEECs kept relations at the level of commercial exchanges very limited and mainly negative in their impact (Maresceau 1989a; J. Pinder and P. Pinder 1975). The EC subjected the 'state-trading' countries to anti-dumping procedures and other forms of protection (Jacobs 1989).

EC policy was therefore confined to the conclusion of limited sectoral agreements for agriculture, coal and steel, and textiles that restricted the imports from the CEECs in these 'sensitive' sectors. Otherwise trade policy was instrumental, loosely serving the political objective of encouraging individual countries towards independence from the Soviet Union. In the early 1970s Yugoslavia, not formally considered a state-trading country, was included within the EC's Mediterranean policy with substantial trade concessions. Paradoxically, in the light of subsequent history, Yugoslavia was to become the closest and most familiar of the CEECs—on the surface at least. For similarly political reasons a more limited trade agreement with Romania was signed in 1980.

More political overtures towards eastern Europe depended largely on individual member states' initiatives, which were not always unequivocally welcomed by the others. The German *Ostpolitik* was the most notable case in point. EC policy was to some extent modified by the experience of the Conference on Security and Cooperation in Europe (CSCE), widely viewed as a very positive experience for European Political Cooperation (EPC) (W. Wallace 1983*b*). It began to induce arguments for developing other forms of light cooperation and communication. The broader issue of German unification had never really been the subject of discussion in the EC, except indirectly through the Protocol on German internal trade, under which goods from the GDR could be imported tariff-free into the Federal Republic, and the extended definition of German citizenship appended to the Treaty of Rome. The dramatic events of 1989 in eastern Europe, about which the EC was bound to be pleased, made these old coordinates of EC policy redundant in a matter of months.

Immediate responses

The EC's immediate responses had therefore to be invented without benefit of experience, reflection, or any standard procedures. There was no choice possible other than hyperactivity and speed and no opportunity to relate short-term action to crafted future goals. Policy-makers had to work from imperfect analogues and from more or less good intention. The result was a curious mix of tradition and innovation. The first steps were to conclude swiftly 'trade and cooperation agreements', based on a roughly standard model, and to set up a new policy of technical and financial assistance through the Phare programme (Poland and Hungary: Aid for the Restructuring of Economies—later extended to other countries) and later the European Bank for Reconstruction and Development (EBRD). These instruments reflected a sense that 'something had to be done', but not a policy.[1] Indeed in both EC institutions and in the member states policy machinery had also to be constructed, as staff were rapidly redeployed from other tasks, often from teams experienced in development-assistance programmes for the Third World (Pinder 1991: 91).

None the less it seemed evident that the EC had accepted the central role in promoting systemic transformation in the CEECs, hence the early decision, prompted by a suggestion from George Bush as US president, that the EC and especially the Commission should take the coordinating lead for the advanced western countries—the G24. The easy part was to state the shared aim of helping to root the principles and practices of both political democracy and market economy, though it became clear that to operationalize these was much more demanding. The more difficult part was to contemplate the consequences for the restructuring of transnational cooperation more generally or the extension of European integration eastwards.

The broad endorsement within the EC of the initial policy response seemed straightforward. The objectives of transformation after all echoed shared values that were fundamental to the EC integration project. It was instinctive to advocate that the CEECs imitate west European policy models and rules. Acceptance of a central role seemed natural also for an EC that had apparently emerged from the cold war as a confident and stable anchor in the institutional landscape. To redefine Nato, for example, a child and champion of the cold war, seemed much more difficult in the changed environment.

The EC's central role was brought into sharper focus by the demands of the CEECs. Their goals for reforms of their societies, as well as of their political and economic systems, were stated with explicit references to the core values of integration. It seemed therefore natural that their principal foreign policy objective was to 'join Europe' by entering the EC (Kolankiewicz 1994). The US administration repeatedly encouraged this thinking and in any case was becoming a less present actor in European affairs (J. Baker 1989). Once German unification was under way it also appeared that the Soviet Union could live with an EC enlarged eastwards. But to follow through the logic would demand sustained and coherent policy.

In the interim the EC reached a broad internal consensus on fostering democracy and the market economy and linked them to forms of conditionality for the eligibility of individual countries for assistance and notions of phasing for enhanced support. This approach ran through Phare and EBRD programmes, while the guiding principle of 'positive discrimination' and differentiation between the CEECs appeared in the offers of trade agreements (Pinder 1991: 32–4), subsequently with the inclusion of an explicit suspension clause. However, despite the initial consensus the EC found it easier to devise *ad hoc* policy and to focus on the economic dimension of transformation than to design a more rounded approach.

Trade and cooperation

The EC's first measures were bilateral trade and commercial and economic cooperation agreements (see Box 14.1); these were concluded as each country was judged to be set on political and economic reform. They were extended across central and eastern Europe and to the then Soviet Union. Subsequently for the independent republics this latter was replaced by individual 'partnership and cooperation agreements', on the grounds that these countries, apart from the Baltic states, were different.[2]

The approach picked up from where EC relations rested with the then CMEA members before 1989, albeit relaxed by the new political context. The EC still viewed the CEECs as marked by the distortions of state-trading. The blueprint was set by the agreement with Hungary; it was negotiated in reaction to, and in the context of, the more open policy of the Soviet Union under Gorbachev and thus before the full extent of the changes in eastern Europe became evident. In so far as they removed the historical trade discrimination, their primary importance was symbolic. In terms of substance, the concrete content of these trade concessions and cooperation remained limited (Lequesne 1991: 364). These early agreements were thus quickly overtaken by events and were to prove inadequate and transitory. Several years on it is easy to criticize the lack of foresight—at the time the plausible outlook was hard to gauge.

Phare and G24

The Phare programme quickly had to bear the main weight of expectations on both sides of the new relationship. This became the first instrument explicitly to take account of the more dramatic changes that unfolded during 1989. When the Group of Seven (G7) met at the western economic summit in July 1989, it was judged necessary to underpin the reforms already under way in Poland and Hungary with a framework to coordinate the various bilateral and multilateral aid programmes. The European Commission was assigned the role of coordination of aid from the G24, to which other international organizations and agencies (OECD, World Bank, International Monetary Fund, and the Paris Club) were associated.

Box 14.1. Trade, commercial, and economic cooperation agreements

....

Non-preferential agreements with then state-trading countries for 10 years (renewable), except Poland (5 years), Slovenia (unlimited duration)

TRADE PROVISIONS

- commitment to MFN treatment for tariffs

- gradual (according to 'sensitivity' of products) elimination of all *specific* quantitative restrictions on imports from CEECs ('specific' = inherited discrimination against state-trading countries through their respective protocols of accession to GATT or in bilateral arrangements with member states

- coverage of all products *except*:
  —coal and steel products
  —textiles (subject to Multifibre Arrangement)
  —agriculture; though option for product concessions

- reinforced safeguard clause

....

COMMERCIAL COOPERATION

- to 'promote, expand and diversify trade on the basis of non-discrimination and reciprocity', through the exchange of commercial information, trade-promoting events, relaxation of import licences, investment, etc.

....

ECONOMIC COOPERATION

- to 'foster economic cooperation on as broad a base as possible in all fields deemed to be of mutual interest', specified in listed target sectors

....

INSTITUTIONS

- Joint Committee with partner to recommend further measures to enhance cooperation and to mediate disputes (not binding)

....

SUSPENSION CLAUSE

- added May 1992; linked to achievement of democratic principles, human rights, and market economy

....

DIFFERENCES BETWEEN THE AGREEMENTS

- date for abolition of quantitative restrictions reflects state of reforms at the time of negotiating the agreement (end 1995: Hungary, former USSR, Romania, Bulgaria; end 1994: CSFR, Poland)

- different priorities in economic cooperation for individual countries

- agreements with former USSR, CSFR, Romania, Lithuania not only with EEC, but also Euratom

(for a sample text on Hungary–EC see OJ L327, 30.11.1988)

For the Commission this was an extraordinary and unsought acknowledgement of its status as an international actor in its own right. The Commission did not limit its role to the exchange of information between the bilateral programmes, but decided to use its first such political mandate more actively. It submitted its own Action Plan, intended as a 'framework for action by the Community and as an incentive for the other members of the G24 to take similar and coordinated initiatives' (*Bulletin of the EC* 10-89: 8). This proposal for a more comprehensive aid policy included measures ranging from humanitarian emergency aid to the improvement of market access, and from the provision of macro-economic assistance to the financing of technical assistance for micro-economic restructuring (see Box 14.2).

The financing and provision of technical assistance for economic restructuring became the main element of the EC's own Phare programme which was subsequently set up by the Commission. A new budget line in the Community budget was created and a special service in the Commission's Directorate General for external relations (DG I) was set up. Assistance in the framework of the G24 stressed the need for the coordination of existing aid projects rather than an attempt to create the framework for a common policy (most of the G24 aid consisted of export credits). The Phare programme was presented as a new instrument of policy towards the CEECs.

Agreement in the EC on this policy was possible not only because of the Commission's active role and initiative, but also because the German government had strongly advocated a much more active Community policy. This was partly because it saw itself so directly affected by the changes in central and eastern Europe, and partly because this could have only a favourable effect on the rapprochement with the still surviving GDR, which was also made eligible for Phare assistance. The policy was promoted keenly by the British, French, and Italian governments and won ready support from other member states (Pinder 1991: 30). Furthermore, as the Phare programme did not exclude the possibility of bilateral projects, none of the member governments had to fear that EC policy could deprive them of the means to continue to express their own priorities in terms of target countries and aid philosophies.

None the less Phare, with its primary instrument of direct grants, turned the Commission into the role of patron *vis-à-vis* the CEECs. It became locked into direct bilateral and contractual relationships with individual recipient countries and had to invent a philosophy to sustain the programme. The result was a curious mix of ambition and caution. The Commission took it upon itself to channel wide-ranging advice about economic transformation and to impose demanding conditions on the clients of its policy. But it deliberately chose to confine its conditionality to market-developing measures. It

Box 14.2. EC assistance through the G24 and Phare

....

G24: assistance by OECD members and international financial institutions, coordinated by the Commission

Commission Action Plan: comprehensive framework for EU assistance

Phare: EC-funded technical assistance programme, with scope for co-financing by individual countries and other institutions

....

TRADE MEASURES

- for c.5 years ('while engaged in restructuring') GSP, providing for tariff-free quotas for certain imports and preferential rates for others, including textile and agricultural products, but excluding ECSC products (and fisheries for Poland) (Reg. 3896/89; OJ L 383, 30.12.89)

- immediate elimination of specific quantitative restrictions (as opposed to the gradual elimination in the trade and cooperation agreements) (Reg. 2281/89; OJ L 326, 11.11.89)

- suspension of non-specific quantitative restrictions ('positive' discrimination against other GATT members) (Reg. 3906/89; OJ L 362, 12.12.89); *exceptions*: textile, coal and steel, and most agricultural products

- subsequent ECSC protocols to remove autonomous quotas maintained by member states (for Hungary, Poland, Czechoslovakia only)

- subsequent increases of quantitative limits for certain textile products and improved cover of outward processing (e.g. Council Decision 90/508/EEC, OJ L 285, 17.10.90)

....

EXPORT PROMOTION

- export credit insurance and investment guarantees via member states

....

MACRO-ECONOMIC ASSISTANCE

- medium-term financial assistance (in cooperation with other institutions) for:
  —currency stabilization
  —balance of payments assistance

- debt relief (in cooperation with the Paris Club)

....

FINANCING OF ECONOMIC RESTRUCTURING (PHARE)

- grants for technical assistance in the following priority areas:
  private sector development, restructuring and privatization; agriculture and rural development; social development, labour, and health; infrastructure: energy, transport, telecommunications, etc.; banking and the financial sector; education and training (incl. TEMPUS and European Training Foundation), R&D; environment and nuclear safety; administrative reform; promotion of regional or transnational projects

- Phare democracy programme: promotion of democracy, human rights, the rule of law, chiefly through NGOs (set up July 1992)

- emergency humanitarian aid (important at outset)

- additional loan facilities through: EIB loans, guaranteed by EC budget, mainly for infrastructure projects; and ECSC loans for investment in coal and steel

- multi-annual projects from 1993

- extension of Phare resources from 1993 to develop infrastructure, not in excess of 15% of annual commitments (see Box 14.5)

was only later that the European Parliament (EP) insisted on building a 'democracy' line into the general budget for 1992, putting pressure on the Commission to pay more attention to politics and civil society, and thus to create within Phare a 'democracy programme' different in character from the rest of the Phare programmes (Blackman 1994). Lacking the resources of staff and expertise within the house, the Commission turned to outside bodies and groups as its advisers and intermediaries. Early on Phare came to depend on an army of consultants from western Europe under contract to the Commission.

As became clear only much later, this policy mode had a triple effect:

- Though it engaged a very wide range of policy actors, both public and private, it also made the programme diffuse and in significant respects fragmented. Initially the coordination had to depend on a small and overstretched staff in DG I, constantly in motion between the lengthening list of client countries and their Brussels offices. As delegations came to be set up in the countries more could be coordinated on site, but this in turn called for policy direction and control procedures that also had to be invented from scratch.

- A distinction emerged between programmes and policy; Phare consisted of a series of actions coordinated up to a point by country and up to a point by type of measure. But it did not become an overall and coherent policy. Indeed, although the trade concessions granted to the CEECs, as they became eligible for Phare, were far more important than the ones accorded under the trade and cooperation agreements, it was in important respects kept distinct from the developing trade, economic, and political relationship of the EC as a whole with the CEECs. Moreover, other DGs in the Commission also began to develop their own east European programmes, such as Tempus (Trans-European Mobility Programme for University Students) and the European Training Foundation, operationally distinct and not wholly consistent with

Phare. Thus the 'learning curve' for the Commission about its own new role and the lessons for subsequent policy were hard to consolidate.

- Albeit not intended at the outset, the financial management arrangements and the reliance on consultants led to a disturbing bias in actual expenditure away from the intended CEEC beneficiaries, a bias compounded by the rigidities of their financial systems and the higher level of costs for participants from western Europe.

The European Bank for Reconstruction and Development

Early on there was some optimism that other forms of financing would flow quickly from private capital and that loan facilities and direct investment would bear more and more of the weight. It was through a quite different channel that a new proposal emerged to encourage such investment. Jacques Attali, close adviser of François Mitterrand, developed the proposal for what became the EBRD. The story of the original proposal and the subsequent travails of the project and its architect are not our subject here (see Dunnet 1991; Weber 1994). Broadly, however, we note three points. First, Attali was deliberately advocating a high-profile and prestige-project approach, in sharp contrast to the palpably technocratic character of Phare. This consequently took the debate into a different realm of high politics around the appointment of Attali as its first Director and the siting of the EBRD in London. Secondly, the alternative option of using the existing European Investment Bank (EIB) for the purpose did not find much favour, even among those who had reservations about the EBRD proposal (Dunnet 1991: 571). Thus the early centrality of the EC was apparently diffused, as other western countries were conscripted as EBRD lenders and partners. Thirdly, in principle at least, the EBRD was established as a partnership between the lenders and borrowers, with CEECs and the successor states to the USSR becoming direct stakeholders, in apparent contrast to the patron/client relationships of Phare.

The EBRD was thus established to make public money available for directly productive investment (see Box 14.3). The agreement on this new institution, signed in May 1990, was possible because of the then consensus on the political and economic goals of the transition and their interlinkage, as well as on the predominantly 'European' character of the initiative (Weber 1994). Consensus about this broad set of ideas managed to overcome the diverging preferences about how these principles should be translated into concrete measures. Preferences had diverged partly at a technical level, but more importantly concerning the general design of the new institution.

Box 14.3. The European Bank for Reconstruction and Development

TARGET COUNTRIES

- USSR (and subsequent newly independent states); eastern Europe

'PARTICIPANTS'/SHAREHOLDERS

- capital: 10 billion ecus

- shares: EC, member states and EIB (51%); US (10%); Japan (8.5%); 'other countries' (incl. other OECD, Korea, Egypt, Morocco, Israel) (18.5%); 'target countries' (12%, of which 6% 'USSR')

AIM

- 'in contributing to economic progress and reconstruction to foster the transition towards open market-oriented economies and promote private and entrepreneurial initiatives in the Central and East European countries committed to and applying the principles of multi-party democracy, pluralism and market economies'.

- public money for investment in the private sector until private capital available

ORIGINALITY

- involvement of the beneficiaries in the bank's capital, management, and choice of projects

- division into merchant bank arm and 'genuine' development bank arm (November 1993: *de facto* geographical division into 'North' and 'South' bank)

- explicit political conditionality for lending and investment

- special obligation to promote 'environmentally sound and sustainable development'

INSTRUMENTS

- make or guarantee loans, primarily to the private sector (at least 60% of its annual total), but also for infrastructure and to state-controlled enterprises, being privatized or managed on competition principles

- investment in the capital of such enterprises

Controversy and criticism were to plague the EBRD, partly because of Attali's style and partly because of the disappointing performance of the EBRD in getting projects underway. Attali resigned ingloriously in June 1993 and was replaced, two months later, by Jacques de Larosière. The EBRD now has to earn its place by results rather than by its initially prestigious profile.

Weber (1994) may be right in his assessment that shared ideas did indeed play an important role in forging agreement on the establishment of the EBRD and that the process of its creation stimulated some convergence of ideas among the main actors. However, the *ad hoc* nature of the EC's immediate policy responses reveals that there were obvious limits on the extent to which shared ideas could galvanize either substantive policies or a longer-term political strategy. The Phare programme and the EBRD both reflect a perception of transformation in the CEECs as a predominantly technical problem, solvable through a transfer of expertise or financial resources, with the main issues and choices focused on the sequencing of technical reforms. Therefore, although EC policy and rhetoric put a strong emphasis on the *finalité politique* of economic cooperation and assistance, this political dimension proved hard to insert into a considered and sustained common policy. It was rather these diverse economic instruments that had to provide the mainstay and with somewhat mixed results (La Serre 1994: 24).

In search of a long-term strategy

The main obstacles to the development of a more strategic and long-term policy were partly related to the nature of the challenge and partly specific to the workings of the EC. The singularity of the transition to political democracy and market-oriented economies presented policy-makers both east and west with an important 'cognitive gap'. It was mirrored in academic disagreement or confusion over transformation strategies, requirements for external support, and the likely duration of the process, as well as the relationships between economic, political, and social reforms. Furthermore, decision-makers operated under considerable time pressure and the puzzling impact of the decomposition of the USSR as well. It is perhaps not surprising that policy responses were *ad hoc*, or that there was such uncertainty about the unfolding situation. The pace and scale of the changes made it difficult to think in the longer term. It is important here to recall just how long after 1945 it took the Americans and west Europeans to settle into a pattern of new transnational policies and institutions.

Internal reasons further complicated the task considerably. First, the EC is generally not very well equipped for the generation of major innovative and strategic decisions. It has been argued that in contrast to the EC's well-analysed problems at the decision-making and implementation stage (see Scharpf 1988), the EC level allows greater scope for being innovative in agenda-setting than an isolated national arena (Peters 1994). The fragmentation of EC policy-making, with its multiple points of access, gives policy entrepreneurs ample opportunities to push new issues on the agenda, which they can select from a broad range of options provided by the variety of national settings. However, a prerequisite for this is the existence or generation of valuable policy ideas at the member-state level, and these were sparse. As for the EC institutions, although the Commission had been originally conceived as a 'generator of ideas', it has over time 'acquired its own conservatism' (H. Wallace 1995). This stems to a large extent from an increasing administrative and policy overload (Metcalfe 1992). The Council is even less suited to policy inventiveness when operating outside conventional terrain.

Secondly, the European policy process is not well suited to the devising of coherent strategy in the face of conflicting signals and only vaguely articulated opinions from the member states (Pelkmans and Murphy 1991). The Council, composed of particular interests and with its unwieldy structure of coordination, is poorly equipped to set policy priorities and clear guidelines or to provide leadership. Neither the European Council nor EPC was able to fill this gap. Although there were few major disagreements or arguments about the need for active involvement in the transition process, there were many turf disputes and much confused proliferation of activities. On the key question of the extent to which support for the reforms in CEEC should be linked to a broader restructuring of transnational relations in Europe, mental confusion and conflicting preferences were manifest. This stemmed partly from different assessments of the implications of changes in the eastern part of the Continent and partly from different views on possible policy.

Thirdly, EC processes tend to favour the internal agenda above external demands. As soon as the CEECs had stated their objectives of pursuing integration with the EC as a corner-stone of their transformation, the EC found itself having to balance external and internal demands. The latter were, however, markedly different and contrary to the apparent needs of the CEECs. One set of issues concerned the standard EC agenda: the completion of the single market, progress towards monetary union, and applications for membership from other western and southern Europeans. Another set of issues related to the unification of Germany and the new geopolitics of Europe. We have not covered here the spectacular 'success' of the EC in endorsing the unification of Germany and accepting the transformation of the eastern Länder as an *intra*-EC task. This endorsement was delivered by an unusual and highly adaptive response from and partnership between the main institutions and the member governments (Spence 1991).

In sum, the EC's immediate policy response was largely about symbolism, for which the broad value-based consensus among member states and in the Commission was sufficient to generate initial agreement. But inexperience, shortcomings in the immediate policy responses, and limited competences militated against the definition of a more long-term strategy. Thus no clear political parameters were in place to guide the shift to a more comprehensive policy, which the formula of 'Europe association' was supposed to serve. Not surprisingly, therefore, difficulties emerged once the focus of debate shifted from the symbolic dimension of the relationship to substance and the need to deliver economic results. Eastern Germany was, however, given a wholly different treatment.

The Europe Agreements

The new formula of 'Europe association' stemmed from politics within the EC and the varied preferences of the individual member states as to whether support for transformation in the CEECs should take place outside the EC framework, or in a framework shared with the EC, or completely inside the EC framework. These disagreements reflected both different impacts of the new geopolitics and differently derived balance sheets as regards the economic costs and benefits of reinforced partnership with CEECs (Sedelmeier 1994: 7–20).

The beginnings of the debate about a more comprehensive and longer-term political framework for the relationship quickly became entangled in the terms and the vocabulary of a conventional debate about 'widening versus deepening'. Not much thought was given to whether the terms of the debate were still relevant in the context of a post-1989 EC (H. Wallace 1991). In this debate the British government and the Commission were at opposite ends of the argument. President Mitterrand attempted a nuanced approach, which, however, still separated internal from external policy. It consisted, on the one hand, of an intra-EC policy of adjusting to German unification by combining economic and monetary union with political union, and, on the other hand, of an extra-EU policy of a 'European confederation' to engage the CEECs in a parallel and distinct institutional framework (Vernet 1992). He had more success with the former than the latter. In the early phase the German government argued hard for an active policy *towards* the CEECs, but accepted that eastern enlargement would have to follow intra-EC adjustments. German policy was also inhibited by the continued presence of Soviet troops on German soil and thus did not give a firm steer in favour of an early eastern enlargement. Other member governments were in any case preoccupied with the immediate pressures for an EFTA enlargement. While the Commission broadly preferred deepening first, it was in several minds as to what signals to give the CEECs on how it hoped to sequence the development of policy. It was in response to these diverse preferences that Frans Andriessen, Commissioner responsible for external relations, floated on his own initiative the idea of a 'European Political Area' as a form of partial and staggered integration into the EC framework.[3]

Given this background it should not be surprising that the EC policy response drew on instruments in its rehearsed policy repertoire (H. Kramer 1993: 234). The first proposal to offer association agreements to those CEECs firmly on the path of economic and political reform was floated as early as November 1989 by the British government.[4] The European Council in Strasbourg one month later endorsed the devising of an appropriate form of association, asking the Commission to make suggestions, which it did in February 1990 (Commission 1990*d*). The special meeting of the European Council in April 1990, summoned for other reasons, agreed to 'create a new type of association agreement as a part of the new pattern of relationships in Europe'. The Commission presented further details on the aims and proposed contents of these Europe Agreements (EAs) in August (Commission 1990*a*), and these were to frame the subsequent negotiations. This concept of an improved form of trade-based agreement, combined with a more compre-hensive political partnership and backed by technical and financial assistance, was the common ground. It showed the limits of feasibility for a more daring policy and makes clear that EC policy-makers found it convenient to address the CEECs as an 'external' problem, separate from their internal agenda (Rollo and Wallace 1991). Thus, although the concurrent Intergovernmental Conference (IGC) in 1991 was accompanied by rhetoric about the need for an improved capability to deal with the CEECs, what that might imply was evaded.

In its results Maastricht was thus more an attempt to manage the diverse needs of the existing membership and the new weight of the unified Germany than a readjustment to the new Europe. The envisaged extension of the *acquis* for the Twelve could only mean additional hurdles for both immediate coop-eration with CEECs and for their potential accession. At one level the failure to devise a sustainable long-term strategy follows from the inability to dis-cuss, let alone agree on, the economic and political fundamentals of the EC integration model. At another level it reveals the fragmentation of the policy process within and between the European and the member-state levels of governance, as well as *vis-à-vis* other European institutions. The impact of this fragmented policy process became even clearer when the agreed formula of EAs as the mainstay of policy had to be put into concrete form.

The main features of the EAs are set out in Box 14.4, with the key innova-tion of a 'political dialogue', intended to underline their special character. The

Box 14.4. Europe Agreements

....

GENERAL CHARACTER

based on Art. 238 EEC; unlimited duration; preferential agreements; mixed competences (pending ratification of Europe Agreements, entry into force of trade provisions via Interim Agreements); for full text of the agreements see e.g. OJ L 348, 31.12.1993 (Poland–EU)

Modified by
- added clause on respect for democratic principles, human rights, and principles of market economy as the rule for all new agreements (Council statement, 11.5.92)
- conditional acceptance of eventual membership (Copenhagen European Council; see Box 14.5)

....

POLITICAL DIALOGUE

regular bilateral meetings and consultations on all topics of common interest, at highest political level and at ministerial (in Association Council); to promote convergence on foreign policy matters

Modified by the decision of the Copenhagen European Council on 'reinforced and extended multilateral dialogue' (see Box 14.5); Council conclusion on reinforcement of the political dialogues of 7 March 1994); decision by the Essen European Council on a 'structured relationship' (See Box 14.6)

....

PROVISIONS FOR THE FREE MOVEMENT OF GOODS

- progressive establishment of *free trade in industrial goods* (transition period for EC 5 years, for CEEC 10 years:
  - (a) tariffs: immediate elimination for some products, 1–5 years for most others (according to their sensitivity)
  - (b) quantitative restrictions: immediate elimination, with *exceptions* for sensitive sectors: (i) ECSC products (coal: quotas after 1 year and tariffs after 4 years, with special derogations for Germany and Spain; steel: quotas immediately and tariffs gradually within 5 years); (ii) textile products (quotas within not less than 5 years and tariffs gradually within 6 years)
- agricultural products: consolidation of previous concessions and some reciprocal concessions
- special provisions for rules of origin (requirement of at least 60% 'local content')
- safeguards: anti-dumping provisions; special and general safeguard clauses; unilateral measures possible
- consolidation of GSP benefits (and withdrawal from list of GSP beneficiaries, normally reserved for developing countries)

- removal from the scope of Reg.1765/82 and 3420/83 (since no longer considered as non-market economies)

> *Modified* by decision to accelerate market access by the Copenhagen European Council (see Box 14.5)

PROVISIONS ON THE OTHER FREEDOMS OF MOVEMENT

- movement of workers: equal treatment for workers legally established in the EC
- right of establishment: full application of national treatment for establishment and operation of new economic and professional activities; transitional periods for application by the associates; restriction of freedom of movement through limitation to 'key personnel'
- supply of services: progressive allowance of cross-border supply of services; special rules for transport
- payments and movement of capital: freedom of financial transfers for commercial transactions, provision of services and investment operations; repatriation of capital or investment benefits

APPROXIMATION OF LEGISLATION

Associates make legislation compatible with EC laws with EC technical assistance; identification of priority areas; *competition policy*: associates adapt their legislation to EC rules within 3 years; non-discriminatory public procurement; protection of intellectual, industrial and commercial property similar to that in EC by associates within 5 years

> *Modified* by decisions for the preparation of the CEECs for integration into the internal market (Essen European Council, see Box 14.6; Cannes European Council)

DIMENSIONS OF COOPERATION

- *economic cooperation*: covering all sectors of mutual interest, aimed at the associates' development
- *cultural cooperation*: extending existing cultural cooperation programmes to associates, with additional actions of mutual interest and identification of priorities
- *financial cooperation*: eligibility for grants under Phare and loans from EIB; possibility of macro-economic assistance through G24; no financial protocol

INSTITUTIONS

- Association Council: ministerial level at least once a year; supervises the agreement's implementation; possibility of binding decisions and dispute settlement
- Association Committee: assists the Association Council
- Parliamentary Committee: advisory role

formula, developed in the Commission by a different team in DG I from the Phare services, was still locked to the established paradigm of a classical trade agreement, although given a political gloss (F. Benyon 1992). The reference to the need for the CEECs to align themselves to EC rules and standards seems to have reflected the prevailing presumption that this was the route to having a similar economic and market structure for general purposes of transformation. Though the EC was at the time engaged in extending the single market to the Eftans, and though there was already some academic speculation about developing a European Economic Area (EEA) model for the CEECs, the implications of an agreement on 'regulatory convergence' were not uppermost in the minds of those involved in devising the formula. Having chosen the classical trade-led formula, with its emphases on products (essentially goods and to some extent services), combined with standard instruments of trade liberalization and protection, the EC was then locked into the typical political economy mode of trade negotiations. This is not to say that the EEA model could have been transposed, but equally the Andriessen suggestion of a functional equivalent was not developed. The classic trade-agreement formula conditioned the way in which the bargaining was conducted with each potential associate *and* with relevant interested parties on the EC side. The asymmetry of bargaining power made the shape of the outcome rather predictable.

The first negotiations opened with Czechoslovakia, Hungary, and Poland in December 1990. They brought into sharp focus the inadequacy of the EC's offer concerning the long-term political perspective from the point of view of the CEECs. It also began to become clear that the chosen formula and bargaining mode would make it hard for the EC to meet their expectations on economic substance. The gap between CEEC expectations and concrete proposals from the EC was highlighted by two periods of deadlock. Each time these difficulties were overcome only after the Council had agreed to make the Commission's negotiation mandate more flexible in order to take better account of CEEC demands.[5] Even so the final outcome of the negotiations did not satisfy the CEECs. Their main criticism was of EC reluctance to establish a clear link to future membership. The Commission had early on stated that, although the CEECs were seeking eventual membership, there was 'no link either explicit or implicit' between association and accession, and that 'membership is not excluded when the time comes', but was 'a totally separate question' (Commission 1990d). CEEC pressure, backed by the British and German governments, led to the adoption of a formula in which the EC recognized future CEEC membership as the associates' 'final objective'. However, this still fell well short of a firm commitment to membership and was interpreted quite widely as antipathy to enlargement.

The other main area of CEEC discontent was over substance. The EC had committed itself in principle to free trade in industrial products over five years. Yet special protocols and annexes covering certain 'sensitive' sectors, notably textiles, iron, coal and steel, and agriculture, offered slower and limited liberalization. As these sectors accounted for the bulk of CEEC exports and reflected their medium-term comparative advantages, the CEECs were dismayed at the EC's miserly trade concessions during the negotiations. Furthermore, even after the completion of the liberalization in these sectors, the provisions for contingent protection might still provide an instrument for EC producers to restrict CEEC exports (Hindley 1992). In addition, these provisos could well deter potential foreign investors.

The EAs were undoubtedly in their general design the most wide-ranging agreements ever concluded by the EC with third countries. Although they explicitly stressed the special nature and closeness of the relationship, their effectiveness and potential were greatly reduced by the EC's inability to deliver more comprehensive or predictable market access. The fragility of their political content reflected the obstacles discussed above to a more broadly based strategy. Their economic content, argued by the EC to be generous in going so far relative to other trade agreements, was seen by the CEECs as grudging, thus a major flaw in both the shorter and longer term (CEPR 1992).

The predicted consequences for the economic impact of the EAs were to provoke a barrage of criticism from the outside 'expert' community (Baldwin et al. 1992; Langhammer 1992; Winters 1992; Messerlin 1993; Rollo and Smith 1993). It is interesting to note that there had been an unusually intense involvement of newly expert groups of outside analysts in advising the Commission and the member governments on policy. Yet this did not emerge as an epistemic community strong enough to implant a different policy formula at that stage. However, the engagement of similar experts in advising CEEC governments may well have heightened the latters' awareness of the limitations to the EAs.

This outcome was at variance with the previous apparent consensus on how to respond to the CEECs. The explanation lies in a bargaining process in which the narrow economic interests of domestic EC sectors, threatened by CEEC competition, were able to gain primacy over political preferences for a more generous approach to the CEECs. As some of those involved on the CEEC side were able to observe, as fast as they identified issues on which they wanted to press for more open market access, they found that an EC-based lobby had beaten them to the EC negotiators. The fragmented process through which the negotiating directives were developed made it hard for the 'political' services to retain control of the negotiations over the details and there were ample opportunities for special interests to set the baseline of what would be on offer.

This conflict between political and economic preferences did not simply translate into a clear-cut confrontation of different national positions. Both individual governments and the Commission were faced with clashes between longer-term political objectives and the pressures from short-term economic problems. Within the Commission, this came out in the rivalry between DG I and DG III, while within most member governments the foreign ministry was competing with ministries of economics, industry, and agriculture. These latter were subjected to vertical pressures from producer lobbies and other sectoral interest groups.

The definition of EC preferences on the question of market access fell victim to a narrow sectoral logic. Considerations of broader economic opportunities, as well as of the political reasons for trade concessions, seem to have been subordinated. The outcome seriously contradicted the ambitious political goals of the Germans, but also the more modest aims of other member governments. The British, for example, were in principle favourable to a generous agreement, but wary of making concessions on migration and financial transfers. The French caused problems on agricultural products, as indeed did the Germans on coal and steel. The bias towards particularist economic preferences became magnified in the process of joint decision-making at the EC level. For an individual state the weight of a problem sector might be limited (textiles for Portugal are perhaps the outstanding exception), but aggregated in a Council meeting the sectoral objections accumulated. This narrowed the scope for compromise and amplified the voices of the reluctant liberalizers over the smaller number of protagonists of politically inspired concessions. The Commission's capacity to address problems of broader European interest was thus severely limited, as it had to rely on the individual member states to keep the pressure from their respective industries in check.

To Copenhagen and the prospect of eventual membership

The pressures on policy produced a reformulation at the European Council meeting in Copenhagen in June 1993. It had already been agreed at Maastricht and then in Lisbon, in June 1992, to examine the general implications of enlargement. The Lisbon Report (Commission 1992a) made a distinction between countries immediately eligible for membership (the Eftans), the southern applicants, and the central and east Europeans with aspirations to apply (Michalski and Wallace 1992). This had left hanging the question of what kind of 'partnership' to develop for the CEECS beyond Europe Association (Reinicke 1992).

The Commission therefore drafted a new communication on this, evaluating the progress made by the CEECs and proposing a reinforcement of the association formula (Commission 1992d). A discussion was planned by the British presidency at the extraordinary European Council in Birmingham in October, which in the event was only sketchy. The Edinburgh session in December was too preoccupied with other issues to discuss the Commission's report in detail; it left it to its next meeting to 'reach decisions on the various components of the Commission's report in order to prepare the associate countries for accession to the Union' (*Bulletin of the EC* 12-1992). This was the first official indication that the CEECs could eventually become members of the European Union (EU). In a further communication the Commission set out more concrete measures for the deepening of association (Commission 1993f). On 8 June 1993 the General Affairs Council cleared the way for the European Council to endorse the vast majority of the Commission's proposals in Copenhagen later that month.

The declarations at the Copenhagen European Council marked an important qualitative change in the evolution of policy, both in the endorsement of eventual membership and in the improvement of market access (see Box 14.5). This represented a fragile compromise that had been forged by a mix of internal political engineering, external pressure, and a certain 'learning process'. It left to be constructed the bridge between the conditional acceptance of eventual membership and its achievement.

After developing the Europe association formula EC policy-makers had found that they had much still to learn themselves about what its implementation required. We can observe a kind of learning process, albeit in fits and starts, in the subsequent period, a reflection of the acknowledgement on the EC side that they were in uncharted waters. The CEECs could not be treated like other partners on a 'take it or leave it' basis as far as trade was concerned. This learning process was induced by criticism from external experts and the CEECs, as well as by some inescapable evidence of shortcomings in the implementation of the EAs. Though new policy was hard to construct, the principle was established early on that policy could not stand still; the difficulty was to square special interests on the EC side. The story of Phare was somewhat different in that the problems of adaptation as the reforms moved forward had more to do with managerial problems on both sides.

Almost immediately after their signing, the agreements came under severe criticism from the outside 'expert community'. Although probably too heterogeneous to be called an 'epistemic community', it was largely unified in

Box 14.5. Decisions of the Copenhagen European Council

....

CONDITIONAL ACCEPTANCE BY THE EU OF EVENTUAL MEMBERSHIP OF THE CEECS

provided that *CEECs* have:

- stable institutions (guarantee of democracy, rule of law, human rights, minority rights)
- functioning market economy
- capacity to cope with competitive pressures inside the EC
- ability to adopt the *acquis*; accepted aims of political, economic, and monetary union

and provided that *EU* has:

- capacity to absorb new members without endangering the momentum of European integration

....

ACCELERATION OF MARKET ACCESS TO THE EC

more rapid (than originally envisaged in the EAs) opening of EC markets across products, including (although to a more limited degree) the sensitive sectors (Additional Protocols, OJ L 25, 29.1.94)

....

FRAMEWORK FOR REINFORCED AND EXTENDED MULTILATERAL POLITICAL DIALOGUE:

greater frequency; additional levels; additional forms of consultation; shift from bilateral towards multilateral dialogue

....

REORIENTATION OF PHARE ASSISTANCE

up to 15% of Phare budget available for infrastructure projects

denouncing the defensive nature and flawed substance of the EU's policy, notably concerning the failure to provide a clear political long-term perspective and the inadequacy of the trade measures. Academic criticism played an important role in reopening the general debate and provided the forces in favour of a more open policy with important arguments (Baldwin 1994; M. A. M. Smith and Wallace 1994).

A further part of this 'learning process' was a fuller appreciation of the scale and difficulties of the transformation process, accompanied by first indications of decreasing public support in the CEEC for vigorous reforms. This impression contrasted sharply with the inescapable evidence of the EC's apparent unwillingness to implement the agreements in a generous spirit by resisting calls for commercial protection instruments,[6] as well as of the

uneven developments of trade balances. The latter is economically not necessarily negative or surprising,[7] but was politically highly sensitive within the
CEECs and contributed to a less unequivocal support for integration with the
EC. Finally, the experience of the international community's impotence in
the conflict in Bosnia-Herzegovina (see Ch. 16) played a catalytic role in
demanding a larger policy investment towards other parts of eastern Europe.

Although the bargaining power of the associates as such had not increased,
their criticism exercised nevertheless considerable pressure, especially by
exposing the gap between EC rhetoric and substance (Inotai 1994; Saryusz-
Wolski 1994). The two Viségrad memoranda were important for their timing
and their cautious language, as well as for showing the glimmerings of a collective response from several of the associates (Viségrad 1992; Viségrad
1993). This considerably boosted the bargaining position of those within the
Commission and in member governments who favoured a more active and
generous policy.

Political engineering and leadership by the Commission were key factors in
the reformulation of policy at Copenhagen. The two responsible commissioners, Hans van den Broek and Leon Brittan, uneasily exercised their now
separated portfolios and the new division of DG IA from DG I. None the less
the Commission worked closely with the British and Danish presidencies of
the Council and with the German government. The two pressing issues had
for some time been identified as a firmer acceptance of membership and the
improvement of market access across the board. The challenge was to build a
sustainable coalition to support these goals. The careful formulation of the
qualitative criteria for membership provided the necessary safeguards for the
more hesitant member governments. To have denied the objectives would
have been incompatible with the rhetoric of policy and indeed the
Community's own self-image.

The Commission thus went out in front by trying to mould a more far-
reaching policy, though holding this line within the Commission was not
always easy. In the member states senior political players became more
actively engaged and had more attention to spare than had been the case during the Maastricht ratification period. Thus political preferences began to
weigh more heavily in the balance with specific economic and sectoral concerns about the imputed costs of greater generosity towards the associates.
Chancellor Kohl repeatedly stressed the importance of the issues, not so surprisingly, although German policy was not always clear-cut; interestingly, he
drew active support from Felipe González, the Spanish Prime Minister, and
Beniamino Andreatta, the Italian Foreign Minister. We should note here that
the repeated need to go the European Council level may have reflected the
symbolic importance of the policy. But it also underlined the weakness of the

General Affairs Council, which had not succeeded in setting firm enough guidelines to overcome sectoral defensiveness.

The Copenhagen package of trade concessions could be agreed only after a political decision to charge a high-level group of senior officials with the task of obtaining comprehensive concessions horizontally across the range of sectors, rather than relying on technical experts at the working-group level. This sheds some doubt on the neofunctionalist argument that limited political control and a bias towards sectoral bargaining increase the potential for transnational cooperation. Relations between the EC and the associates were characterized by a sharp asymmetry between economic dependence, with the 'low politics' of economic integration tending to produce zero-sum games, and the 'high politics' of rebuilding Europe, which seemed to provide more scope for a cooperative and expansive outcome (Sedelmeier 1994: 28).

Political and security dimensions

West European policy towards the reform democracies has been articulated not only through the EC and later the EU, our primary focus in this chapter (see Appendices A and B). The same question has been posed in each west European organization—how to adapt the role and membership to take account of the new Europe. For the Council of Europe the question has been answered relatively straightforwardly: countries have been admitted into the Council of Europe as and when they have made their first big jumps towards democratization, with membership being used as a goad and an encouragement to consolidate reform. This has conveniently removed from the EC the initial task of judge.

In the defence arena both Nato and the WEU have debated whether and when to establish systematic cooperation and membership for the reform democracies and CIS countries. One might perhaps have imagined that EC members would have agreed on an overall approach which would have been articulated consistently across the different transnational organizations. Curiously, however, policy has developed differently in each framework, marked by contradictory signals and different emphases, and not only because, for example, the Americans have been so much engaged in the Nato debate on this. The different organizations have elicited different policies from EC governments, reflecting different views among them about the division of labour between the EC (particularly as it developed into the EU), WEU, and Nato about the redefinition of their main roles. In turn this has complicated the task of giving content to the 'political dialogue' element of the Europe Association.

The Balladur proposal

The 'Pact on Stability in Europe' is one example of the effort to to set the security relationships into a new context. It resulted from an initiative of Edouard Balladur, shortly after he became French Prime Minister. Its aim was to promote preventive diplomacy that would stabilize relationships between the CEECs in a broad political and security framework, to which west Europeans would give their support. Issues to be addressed included the protection of national minorities and the inviolability of borders. The initiative was not unequivocally welcomed: some CEECs suspected it to be a diversion from the core issues of EU or Nato membership; within the EU there were mixed views, not least because it echoed the earlier proposal from President Mitterrand to create a 'European Confederation' and overlapped with the CSCE.

The initiative was grudgingly accepted at the Copenhagen European Council, but then came to provide an opportunity to breathe some life into the common foreign and security policy (CFSP). The Brussels European Council, in December 1993, announced the project as one of the first 'joint actions' under the CFSP. Negotiations of bilateral treaties were launched at the inaugural conference at Paris in May 1994 and followed by a series of regional conferences and bilateral round-table discussions. The Stability Pact was adopted in a concluding conference, again held in Paris, in March 1995, with a statement in which all the participants pledged to pursue its common aims and to use peaceful mechanisms for dispute settlement. 'Flanking measures' by the EU would include Phare-financed projects, in particular on minorities and cultural and cross-border cooperation. Some of these were new, but most attached a 'Stability Pact' label to projects already begun, for example under the Phare Democracy Programme. The Organization for Security and Cooperation in Europe (OSCE), the revised form of the CSCE, was entrusted with developing the Pact.

The 'French character' of the project reflects the eagerness of the French government to be seen to be making an active contribution to the EC policy, but also genuine French concerns about developments in the CEECs, a preoccupation with broad stability and the need for an extended notion of European security. In this sense the Stability Pact can be interpreted as complementing EC policy. But in another sense it exemplifies a fragmentation and ambiguity in west European policies towards eastern neighbours.

The Essen European Council and pre-accession

The Commission carried out an internal review of policy towards CEECs in the first half of 1994. The internal debate veered between support for 'consolidation' of existing policy and calls for its reformulation, a debate echoed in the discussion between member governments. The opportunity was, however, seized upon by those inside the Commission who favoured a reformulation of policy to maintain its momentum. Both Leon Brittan and Hans van den Broek came to agree on the need for a strategy to prepare the CEECs for accession. This started a discussion of which EC policies, such as agriculture, needed to be adapted for an eastern enlargement, or how to apply policies, such as the competition rules, to the CEECs to replace contingent protection.

Meanwhile some member governments were pressing for extensions of policy. The British–Italian initiative, launched by Douglas Hurd and Beniamino Andreatta as foreign ministers, aimed at deepening the political dialogue element of association by extending it to the second and third pillars, for CFSP and Justice and Home Affairs (*Agence Europe*, 22 Dec. 1993: 6). The German government continued to push for a more active policy. During the Greek Presidency it was broadly accepted that policy needed to be strengthened, but it also became clear that some, notably the French and Spanish governments, thought it comparably important to develop a more active Mediterranean policy, again with a strong emphasis on economic support and market access, but also tuned to the difficult geopolitics and stability issues of the region. There were some noticeable tensions between the French and German governments.[8] Some of these surfaced in the final stages of the accession negotiations with the Eftans; the response was first damage limitation and then the emergence of a strategic bargain between the French and German governments, which rested on French acceptance of eastern enlargement and German endorsement of the need for a new Mediterranean policy. This seems to have paved the way for agreement on a better defined 'pre-accession' strategy for the CEECs. It became possible for those in the Commission who favoured an enhancement of policy to find the support to steer it forward.

The European Council at Corfu in June 1994 called on the Commission 'to make specific proposals . . . for the further implementation of the European Agreements and the decisions taken by the European Council in Copenhagen' and to report on 'the strategy to be followed with a view to preparing for accession' (*Bulletin of the EU*, 6-1994). This explicit mandate gave those inside the Commission who wanted to advance policy towards CEECs a crucial advantage. The Commission was able to produce the first part of its report on 13 July because the responsible commissioners could build on active informal cooperation with sympathetic member governments, in setting out the major lines of a strategy to prepare the CEECs for accession (Commission 1994*a*). In contrast, efforts later in July to table a more concrete and sector-specific follow-up document proved much harder (Commission 1994*c*). This indicates the persistence of internal disagreement and extensive bargaining to accommodate the positions of the commissioners and DGs concerned. To the disappointment of the policy innovators the college as a whole took the cautious view that more thorough analysis was needed on a number of points, such as the recourse to trade-defence measures, elimination of subsidies for agricultural exports to CEECs, full cumulation of rules of origin, and the extent of financial transfers.

The Commission report outlined a 'pre-accession' strategy with the aim of a 'progressive integration of the political and economic systems . . . of the associated countries and the Union'. The strategy is built on two cornerstones: for the political dimension it offers the CEECs a 'structured relationship' with EU institutions; and for the economic dimension it advocates the progressive integration of the CEECs into the single market. The Commission promised a White Paper which would identify those parts of the *acquis* which it regarded as essential for the CEECs to adopt in their own legislation, as well as what legal and institutional framework would be required by them. For example, if a way could be found to apply EC rules of competition effectively to the CEECs, recourse to instruments of commercial defence 'could be' abandoned.

In July 1994, at the start of the German Presidency, the Council largely endorsed the Commission's broad approach, though it did not command universal enthusiasm and certainly did not guarantee support on the difficult details of policy. Informal meetings in September began to move the debate forward. At their bilateral meeting in Paris on 20 September 1994 the French and German governments agreed to agree in the Council on the merits of the proposed Commission White Paper. There remained none the less noticeable differences of emphasis and priority between them, the Germans being worried about the economic costs and benefits of the relationship and the French concerned about the financial and institutional costs.[9]

In early October 1994 the Council agreed in principle to the 'structured relationship' and to request a White Paper from the Commission. The structured dialogue responded to an important criticism of the excessive bilateralism of the EAs and aimed at reinforcing the opportunities for multilateral meetings at ministerial and official levels, as agreed in principle at Copenhagen. There was already some involvement of ministers from the associate countries in extended sessions of the Council, though on an *ad hoc* and inadequately prepared basis, hence with very mixed results.[10] It was, however, easier said than done to extend these meetings, and some member governments felt that the Germans were trying to give the associates a kind of back-door membership. These concerns remain a salutory reminder of the difficulties, both political and technical, of developing forms of 'partial membership'.

The German presidency had already announced its intention to invite the CEECs to a session at the European Council meeting in Essen, but for a while the invitation was not confirmed. The difficulties of giving substance to an improved relationship with the CEECs coincided with German electoral distractions and other tricky European dossiers. Discussions at the October and November sessions of the Council led to acceptance of CEEC participation in Essen and of the Commission's broad stance, but with a weakened text on issues of economic substance. The same stubborn problems recurred: the level of financial aid, the degree of trade concessions, the cumulation of rules of origin, and agriculture. It should here be noted that most governments found themselves internally divided as well, the German included. None the less the Germans successfully pressed for agreement to negotiate EAs with the Baltic states, helped by the knowledge that the new Nordic members would strongly support this.

In the run-up to the Essen European Council the evolution of the association policy at moments risked losing some of its momentum. Certainly what was agreed emerged as unglamorous compared with some of the rhetoric and with many of the expectations. Some care should, however, be taken in reading the results of Essen (see Box 14.6). The agreement on the structured relationship did shift consultations on to a usefully more multilateral basis, less 'hub-and-spoke' politics than before, and did pave the way for a White Paper setting out concrete measures.

The White Paper

The White Paper was published in May 1995 for discussion at the Cannes European Council in June 1995 (Commission 1995b). Nominally the second 'pillar' of the pre-accession strategy, after the 'structured relationship' it could turn out to be the mainstay. In essence it seeks to apply the successful model of the single market White Paper to the relationship with the CEECs: it concretely identifies achievable targets for the associates in terms of core legislation and staggered regulatory convergence; it stresses the importance of the legal and administrative infrastructure in the CEECs and not just the

Box 14.6. The 'pre-accession' strategy (Essen European Council)

....

STRUCTURED RELATIONSHIP

Promotion of mutual confidence and consideration of issues of common interest

- reaffirmation of Copenhagen decision to create a multilateral framework to complement bilateral dialogue of the EAs

- concrete decisions on issue areas and frequency of meetings connected to corresponding Council (General Affairs: semi-annual; Internal Market, Ecofin, Agriculture: annual; Transport, Telecommunications, Research, Environment: annual; JHA: semi-annual; Culture, Education: annual); annual meetings on the margins of European Council (stronger emphasis on meetings of 'the Fifteen' with CEEC counterparts, rather than Troika/Commission format)

- schedule for joint meetings at the beginning of each year in agreement between the two presidencies (more systematic framework than previous *ad hoc* meetings)

PREPARATION OF THE CEECS FOR INTEGRATION INTO THE INTERNAL MARKET

Creation of conditions to allow the internal market to function after eastern enlargement

EU action:

- identification of key *acquis* essential for the creation and maintenance of the internal market in each sector; inclusion of parts of legislation on competition; social and environmental measures 'for balanced approach'

- suggested sequencing for legal approximation; priority measures to be tackled first (but not priorities between sectors)

- specification of administrative and organizational structures for effective implementation and enforcement (formal transposition of legislation insufficient)

- suggested adaptation of Phare assistance for pre-accession strategy

- 'technical assistance information exchange office' (database on alignment with internal market; clearing house to match requests for assistance with expertise available in Commission, member states and private bodies)

- monitoring of implementation of recommendations

CEEC action:

- phased adoption of legislation and regulatory systems, standards, and certification methods compatible with EU

- establish national work programmes to identify sectoral priorities and timetables for alignment

SUPPORTING POLICIES TO PROMOTE INTEGRATION

Development of infrastructure; cooperation in TENs; intra-regional cooperation; environmental cooperation; CFSP; cooperation in JHA, culture, education, and training; supported through Phare 'transposition' of legislation; and by making so many of the issues technical and horizontal it reduces the opportunities for veto groups on the EU side to intervene and block progress. Mundane and dull though much of the detail appears, so was the single-market programme, but in both cases there have been both practical and tactical reasons for using this approach to policy development.[11]

The document also breaches an important taboo in two senses. Implicitly it suggests that some parts of the *acquis* are more important than others. It also starts to separate products from process in emphasizing those areas of legislative alignment that are needed to enable *products* to be exchanged in contrast to those that define the *processes* for making products and generating services. On the latter one can, after all, argue that comparable production processes cannot be achieved in the absence of higher levels of economic performance, and that in any case issues such as social and environmental standards should properly be the subject of accession, not pre-accession, negotiations, and where it can reasonably be argued that transitional periods might follow and not precede accession. In any case regulatory convergence can aid internationalization, irrespective of timetables for an eastern enlargement of the EU. This view will, however, be contested by some governments, as the Internal Market Council session of June 1995 revealed (*Agence Europe*, 6 June 1995: 7–8).

We should also note that the White Paper marks a shift in the centre of gravity of policy and in how it is coordinated: DG XV, responsible for the single market, is to handle the regulatory convergence activities, working with DG IA to draw the Phare programme in as a back-up and with the relevant agencies in the EU member states. DG II and DG IV seem to have supported the approach of DG XV, with the acquiescence of DG XI and criticism from DG V. A more direct link has in principle been made between the measures to be adopted by the CEECs and the way in which those same measures are managed for the existing member states, even though (as in the EEA) there is still no extension of the customs union. The west European investor or outward processor in the CEEC countries also stands to gain from greater certainty about the regulatory environment. It remains to be seen how far the implementation of this pre-accession strategy can induce a more symmetrical working partnership between the EU and the associates.

This is not to say that policy will now be conducted in every respect more smoothly, more coherently, and more consistently; DG XV had to struggle to put together a team to prepare the White Paper, let alone to implement it. The working relationships with the CEECs have to be turned into an operating system analogous to that of the EEA. EU member governments have to adjust to a different policy mode. But there is an opportunity for a change of style and of content to policy and for this to be based more on the practices of 'inclusion' than those of an external relationship—thus, for example, the significance of making the Internal Market Council responsible with DG XV for the follow-up.

Towards a sustainable policy?

The past five years or so have seen the EU under novel pressures to invent and to adapt policy in every sense from the definition of governing principles to the micro-management of operating systems. To adopt a 'European policy' in the wake of the 1989 revolutions has been to confront many paradoxes of the west European integration process as well as to respond to quite unexpected external developments. To mobilize strategic attention while also grappling with the tough internal agenda around the Maastricht process has been to test the policy process and political capabilites of the EU to the limits.

Many mistakes and inconsistencies have been evident and it has been easy for the outside observer to puzzle over what was happening and to criticize the rigidities of the EU policy process, many of which remain, not least as regards agriculture. The tussle over competing priorities has been recurrent, with issues of resources and budgetary costs still conflictual and with the parallel development of a Mediterranean policy still a large complication. Much still hangs on how far a consensus has been embedded on the link between policy *towards* the CEECs and commitment to eastern enlargement.

A pattern of constructive experimentation has been in competition with the reflexes of policy conservatism, with policy innovators in the Commission and member governments competing with the defenders of the *status quo*. The competition has been episodically resolved, sometimes with the innovators developing new policy and new instruments, sometimes with the ranks of vested interests blocking policy change. These tensions will no doubt continue. None the less it should be noted that the Commission has periodically been able to inject critical redefinitions of approach and only with active engagement by the Commission have policy ideas been turned into policy instruments. It has, however, been dependent on a rallying coalition being active in the Council and specifically on the varying extents to which the German government has pressed for policy development. Nor is there yet much evidence of an outside constituency on the EU side to underpin or to extend this European policy, though special interests do repeatedly figure as restraints on policy. Our best guess must therefore be that this push–pull will continue to induce disjointed incrementalism.

Notes

1. In the early stages the EC provided much humanitarian assistance, including food aid. Michael Emerson, as delegate of the European Commission in Moscow, was, for example, much occupied with this task (Buchan 1993: 63–4).
2. Policy towards the former Soviet Union is outside the scope of this chapter.
3. Andriessen had argued that 'creative thinking is now required to define the arrangements whereby the Community could offer the benefits of membership and the gains for stability, without wrecking its drive towards integration'. He suggested that 'affiliate membership' could give the CEECs 'political security . . . and a seat at the Council table on a par with full members in specific areas' (Andriessen 1991).
4. Mrs Thatcher referred very early and specifically to the Turkish association formula as a model (*Financial Times*, 15 Nov. 1989: 1).
5. The negotiations were deadlocked in late Mar. 1991 and again in July and Sept. 1991, when notably the Polish side refused to send a high-level delegation to the negotiations (*Agence Europe*, 28 Mar. 1991: 9; 13 July 1991: 7; *The Guardian*, 15 July 1991: 7; *Bulletin of the EC* 9-91: 45). The Council reached internal compromises that permitted the negotiations to be resumed on each occasion (*Bulletin of the EC* 4-91: 43; 9-91: 45).
6. Since the entry into force of the Interim Agreements the EC has used anti-dumping or safeguard measures in certain cases: pig-iron and ferro-silicon imports from Poland; seamless steel and iron tubes from Poland, Czechoslovakia, and Hungary; urea ammonium nitrate from Bulgaria and Poland, and so on. Among the most publicized restrictions was the temporary EC ban on live animal and dairy imports imposed in Apr. 1993 (*Euro-east*, 23 Apr. 1993: 26).
7. Much depends on the structure of the deficit; if the CEECs' imbalance reflects a valuable influx of investment goods the prognosis may be encouraging.
8. There was a diplomatic row over the remarks made in the press by François Scheer reflecting the increasingly assertive character of German European policy (*Frankfurter Allgemeine Zeitung*, 16 Mar. 1994: 1; and 18 Mar. 1994: 2; and *Le Monde*, 19 Mar. 1994: 6).
9. Klaus Kinkel, the German foreign minister, stressed the importance of lifting the standards of living in eastern Europe as a whole (*Financial Times*, 21 Sept. 1994: 2), while Alain Juppé noted that the potential costs of enlargement 'make us think again' (*Agence Europe*, 22 Sept. 1994: 4).
10. Rather few meetings had been convened. The Environment Council of 5 Sept. 1994 was said to have been disappointing, though the foreign ministers' meeting on 31 Oct., after the agreement to improve the arrangements, went better.
11. At the time of writing we incline to the view that this is a promising development in policy, though Rouam (1994) takes a more critical view.

Further Reading

For the background to policy before and in immediate response to 1989 see Pinder (1991). For subsequent developments in a broader context see Reinicke (1992). Critical appraisals of EU policy can be found in CEPR (1992), Kramer (1993), and Saryusz-Wolski (1994). On the development of economic relations between the EU and eastern Europe see Faini and Portes (1995). On the further development of policy see Baldwin (1994), Commission (1995b) and M. A. M Smith et al. (1995).

Baldwin, R. (1994), *Towards an Integrated Europe* (London: CEPR).

CEPR (1992): Centre for Economic Policy Research, 'The Association Process: Making it Work: Central Europe and the EC', CEPR Occasional Paper No. 11 (London: CEPR).

Commission (1995b), 'Preparation of the Associated Countries of Central and Eastern Europe for Integration into the Internal Market of the Union', COM (95) 163 final, 3 May.

Faini, R., and Portes, R. (1995) (eds.), *European Union Trade with Eastern Europe: Adjustment and Opportunities* (London: CEPR).

Kramer, H. (1993), 'The European Community's Response to the "New Eastern Europe" ', *Journal of Common Market Studies*, 31/2: 213–44.

Pinder, J. (1991), *The European Community and Eastern Europe* (London: Pinter).

Reinicke, W. (1992), *Building a New Europe: The Challenge of System Transformation and Systemic Reform* (Washington, DC: Brookings Institution).

Saryusz-Wolski, J. (1994), 'The Reintegration of the "Old Continent": Avoiding the Costs of "Half-Europe" ', in S. Bulmer and A. Scott (eds.), *Economic and Political Integration in Europe: International Dynamics and Global Context* (Oxford: Blackwell, 1994), 19–28.

Smith, M. A. M., Holmes, P. M., Sedelmeier, U., Smith, E., Wallace, H., and Young, A. R. (1995), *The European Union and Central and Eastern Europe: Pre-accession Strategies*, SEI Working Paper No. 15 (Falmer: Sussex European Institute).

Appendix A. Overview of EU policies towards CEECs

Country	Trade and cooperation agreements		Phare/G24 aid	Europe Agreements		
	signed	in force	eligible	signed	interim agreement in force	entry into force
Hungary	26.09.88	01.12.88	01.01.90	16.12.91	01.03.92	01.02.94
Poland	19.09.89	01.12.89	01.01.90	16.12.91	01.03.92	01.02.94
USSR	18.12.89	01.04.90	(1)	(2)		
GDR[3]	08.05.90					
CSFR	08.05.90	01.11.90	01.10.90	16.12.91	01.03.92	
Czech Rep.			01.01.93	04.10.93	(as CSFR)	01.02.95
Slovakia			01.01.93	04.10.93	(as CSFR)	01.02.95
Bulgaria	08.05.90	01.11.90	01.10.90	08.03.93	31.12.93	01.02.95
Romania	22.10.90	01.05.91	01.01.91	01.02.93	01.05.93	01.02.95
Estonia	11.05.92	01.03.93	01.01.92	12.06.95	[01.01.95][4]	
Latvia	11.05.92	01.02.93	01.01.92	12.06.95	[01.01.95][4]	
Lithuania	11.05.92	01.02.93	01.01.92	12.06.95	[01.01.95][4]	
Albania	11.05.92	01.12.92	01.01.92			
Slovenia	05.04.93	01.09.93	07.10.93	(5)		
Yugoslavia	(6)		01.10.90[7]			

1. TACIS (Technical Assistance for the Commonwealth of Independent States) programme for independent republics.
2. Partnership and Cooperation Agreements with independent republics.
3. German unification 03.10.90.
4. Free Trade Agreement.
5. Agreements initialled 15.06.95.
6. 1980 Cooperation Agreement suspended (25.11.91); trade benefits reintroduced for individual republics except Serbia/Montenegro (03.02.92).
7. Suspended 11.11.90.

Appendix B. Overview of the CEECs' involvement in other western organizations

Country	Council of Europe (membership)	North Atlantic Treaty Organization (cooperation through:)		Western European Union (cooperation through:)	
		North Atlantic Cooperation Council	Partnership for Peace	Forum of Consultation	Associate Partnership
Hungary	06.11.90	20.12.91	08.02.94	20.05.93	09.05.94
Poland	26.11.90	20.12.91	02.02.94	20.05.93	09.05.94
USSR		20.12.91[1]	(2)		
CSFR	21.02.91	20.12.91			
Czech Rep.	30.06.93	01.01.93	10.03.94	20.05.93	09.05.94
Slovakia	30.06.93	01.01.93	09.02.94	20.05.93	09.05.94
Bulgaria	07.05.92	20.12.91	14.02.94	20.05.93	09.05.94
Romania	04.10.93	20.12.91	26.01.94	20.05.93	09.05.94
Estonia	14.05.93	20.12.91	03.02.94	20.05.93	09.05.94
Latvia	06.02.95	20.12.91	14.02.94	20.05.93	09.05.94
Lithuania	14.05.93	20.12.91	27.01.94	20.05.93	09.05.94
Albania		05.06.92	23.02.94		
Slovenia	14.05.93		30.03.94		

1. Subsequently individual membership for independent republics.
2. Agreements with independent republics.

CHAPTER FIFTEEN

JUSTICE AND HOME AFFAIRS: COOPERATION WITHOUT INTEGRATION

Monica den Boer

A policy network among national enforcement agencies developed from the mid-1970s, in response to developments in trans-border terrorism, drugs traffic, and crime. This secretive 'wining and dining culture' of national officials, labelled Trevi, was coordinated by the rotating Council Presidency, with little political accountability and few links with EC institutions. The 1992 Programme and the Schengen Agreement, aiming to abolish frontiers among a smaller group, led to institutionalization in the 'third pillar' of the Maastricht TEU. Differences of national police and legal cultures, issues of sovereignty and civil liberties, and popular and parliamentary distrust for common action make this an acutely sensitive sector. They have delayed implementation of both the Schengen and the Maastricht agreements, and inhibited further institutional or policy development of the 'third pillar'.

Introduction

The Treaty on European Union (TEU) has added two new intergovernmental 'pillars', one of which deals with the cooperation between EU member states on justice and home affairs. 'Title VI', as this 'third pillar' of the TEU is officially called, has all the characteristics of a compromised, intergovernmental construction: it is heterogeneous with respect to its subject-matter, it lacks institutional integration with the other parts of the TEU, and it attaches an ambiguous status to matters of common interest. Notwithstanding these disadvantages, Title VI symbolizes a considerable progression in the effort to coordinate international criminal justice cooperation within the EU (M. Anderson, Den Boer, and Miller 1994: 115). Even in the mid-1980s, the creation of this cooperative domain under the auspices of the Community would have been inconceivable, as criminal justice matters firmly belonged to the realm of national sovereignty. Two developments were important in preparing the ground for the policy-making process and the achievement of a degree of consensus in the field of justice and home affairs cooperation: Trevi and Schengen.

Trevi was created in 1976 under the aegis of European Political Cooperation (EPC) as a forum to exchange strategic information for anti-terrorist purposes. It matured into a much wider policy-making forum that began to cover issues such as serious international and organized crime and public order. Despite the lack of a permanent secretariat, Trevi became a policy-making network with frequent meetings between senior officials from interior ministries, thereby consolidating its negotiating structure. The Schengen Agreement, on the other hand, was an advanced attempt undertaken by a group of only some EC member states to introduce a series of common measures to compensate for the abolition of internal border controls. The negotiations, which were often tiresome because of political disagreements, legal differences, and technological difficulties, have often been regarded as a testing-ground or laboratory for wider criminal justice cooperation in the EC.

The creation of Title VI was therefore far from unprecedented (Den Boer 1994a: 281). It could have been constituted without Schengen because various EC-wide agreements were already being prepared when the first Schengen Agreement was negotiated. But we should not lose sight of the considerable conversion of policy-making circles in the different fora. The effect was that the positions of the delegations had already been explored and that the roads towards political solutions seemed to have been paved.

This chapter begins with an analysis of the various rationales that contributed to the demand for more concerted international criminal justice cooperation. Then follows an assessment of the particular role that Trevi and Schengen played in the policy-making process which culminated in the establishment of justice and home affairs cooperation in the TEU framework. After analysing Trevi's remit, its organizational absorption into Title VI, and the parallels between the Schengen Agreements and Title VI, the structure and role of justice and home affairs cooperation are described in more systematic terms. Specific attention will be paid to the political and legal integration with other parts of the TEU, which will be exemplified by the establishment of the European Police Office (Europol). The chapter concludes with some anticipatory notes about the future challenges and dilemmas with which Title VI will be confronted. These include the diffusion of internal and external security, the social legitimacy of cross-border police cooperation, and the development of the EU into a regional security community.

Rationales for improved criminal justice cooperation

None of the initiatives in the field of criminal justice cooperation has been without reasons (Den Boer and Walker 1993: 8–11). Certain stimuli have had a structural, sometimes chronic character, others have been more contextual or incidental. In reconstructing policy-making processes, it is often difficult to disentangle genuine incentives from the discursive and ideological articulation of problems by politicians and civil servants (Majone 1989: 7, 8). Below, the rationales behind the development of criminal justice cooperation in the EU are analysed, as if policy-makers understood and articulated these rationales as fixed and well defined. In reality, however, there is a multiplicity of perspectives that fluctuate under the influence of national, institutional, and bureaucratic games that are played in the policy-making process (Den Boer 1994b: 178).

There are two elements in the category of 'structural rationales', namely jurisdictional fragmentation and the mobility of criminals. Both problems are inherently related to the limited ambit of national sovereignty and the criminal justice system. The inability of law-enforcement authorities to prosecute criminals beyond the national frontier is rather frustrating and has since long hindered the effectiveness and efficiency of criminal investigation, particularly with respect to criminal offences of an international nature. The mobility of criminals is said to have increased, partly as a consequence of improved means of transport, and partly because of improved technology and telecommunications, making the physical location or displacement of the criminal less important.

These two rationales, combined with increased anxiety about the international spread of anarchism, were sufficient reasons for Prince Albert of Monaco to convene a meeting in 1914. The conference, which was attended by 300 people from fifteen different countries, laid the basis for the creation of the international police organization—Interpol. The decision was not implemented until 1923 because of the First World War.

New, contextual, incentives were provided in the 1970s, when the problem of drug production, trafficking, and consumption assumed global proportions, and when terrorism soared in most west European societies (M. Anderson 1994: 9). Drugs and terrorism constituted security problems on a transnational scale, and, as Helen Wallace points out in Chapter 1 of this volume, security interdependencies have grown ever since. A response to these problems was offered at different levels: the United Nations and the Council of Europe drafted conventions and created special working groups to cope with the drugs problem;[1] and Trevi was created for the exchange of strategic information in the fight against terrorism. These initiatives were also necessitated because of the inadequacy of existing fora; Interpol was hampered by the varying willingness of national central bureaux to provide the requested criminal information, and by the disparity of political regimes, making it impossible to develop common strategies, against terrorism in particular.

Other initiatives emerged in the mean time, such as the Pompidou Group which specialized in the drugs problem,[2] the Police Working Group on Terrorism,[3] and the *Ad Hoc* Group on Immigration,[4] which became closely associated with the work of the Group of Coordinators. This group was created at the European Council summit in Rhodes in 1988 and drafted the 'Palma document', which was adopted by the European Council in Madrid in June 1989, and which contained a programme of compensatory measures for the completion of the single market and the realization of free movement. The planned compensatory measures were divided between five groups: the *Ad Hoc* Group on Immigration; Trevi; the Judicial Cooperation Group; the Mutual Assistance Group '92 (customs); and the European Commission. These reported to their own authorities, which were coordinated by a group of national coordinators, consisting of one representative per member state. In turn, this group reported to the European Council on a biannual basis (Van der Wel and Bruggeman 1993: 67).

All these initiatives somehow failed to match newly arising needs. Cooperation within the frameworks of the UN, Interpol, and to a certain extent also the Council of Europe exposed the troublesome, but necessary, connection between politics and law enforcement. The objective of combining intensified international criminal justice cooperation with a basic level of international political consensus became increasingly entangled with questions about the framework of the EC. The promotion of a 'European Judicial Space' by Giscard d'Estaing and Felipe González in the 1970s was an expression of this. Except for the need to build intensified law-enforcement cooperation on political consensus, there was a growing desire for cooperative mechanisms. These would be more closely associated with the practical and operational needs of law enforcement and tuned in to the specific security situation of the EC, which would arise from the abolition of internal border controls (Den Boer 1994a: 280).

In practical terms, the prospective abolition of border controls on 1 January 1993 provided the most concrete rationale for improving transnational collaboration for criminal justice in the EC framework (M. Anderson 1994: 10). Politicians, senior police officers, and civil servants argued that the removal of border controls would have negative repercussions for the internal security situation of the member states, and hence that international crime would rise without compensatory measures. Although few of these arguments were based on reliable statistics on the effectiveness of border controls for law-enforcement purposes, they tended to have great rhetorical appeal (Den Boer 1994b: 184–92). The proposals did not remain confined to the exchange of criminal intelligence and the policing of external borders, but eventually pertained to the free movement of persons in general (immigration, asylum, visa, identity cards), judicial cooperation, customs cooperation, cooperation against fraud, and intensified cooperation against western Europe's newest enemy: international organized crime (M. Anderson 1993).

The substitution of security threats (east European communism was replaced by fears about 'mass' immigration and international organized crime after the collapse of the bi-polar system) was a final—be it more indirect—rationale for intensified criminal justice cooperation (Den Boer 1994b: 182), specifically because it provided the member states with a common objective, and thus reinforced the organic feeling of a community. Apart from this organic effect, the transformation of the security situation brought with it two special effects. The first was that the former 'external security' issues—traditionally closely associated with foreign policy, military functions, and state security aspects—became a responsibility for national law-enforcement bodies, and therefore acquired the status of 'internal security' issues. The second effect was that as a result of their close association with the threat from outside the Community, such as immigration from the Maghreb and the Russian Mafia, security threats became more and more amalgamated (Anderson et al. forthcoming: ch. 5). This gave rise to the construction of a 'security continuum' (Bigo 1994; Bigo and Leveau 1992) and made the analysis of organized crime in terms of ethnic origin more popular. Criminal justice cooperation—and in particular police cooperation—was thus given a higher priority status as a result of the homogenization of security threats and as a result of the waning significance of common military action.

Trevi: preparing the ground for justice and home affairs cooperation

At the European Council meeting in Rome in December 1975, justice and home-affairs ministers decided to create a forum against international terrorism within the framework of European Political Cooperation. Interpol did not (yet) have terrorism on its agenda, and indigenous and Middle Eastern terrorism continued. In a resolution of 29 June 1976 Trevi's objectives were stated as being: cooperation in the fight against terrorism and exchange of information about the (terrorist) organization, and the equipment and training of police organizations, in particular tactics employed against terrorism.

Trevi, cryptically named after its first chairman, A. R. Fonteijn, and the Trevi fountain in Rome (where its first meeting was convened), was divided into working groups (Van der Wel and Bruggeman 1993: 43). The first Working Group (WG I) was created in 1977 and carried responsibility for anti-terrorism, information, and security. Its principal activities concerned the exchange of information about terrorist plans and actions, and the security aspects of air traffic, nuclear installations, and transport. Central contact bureaux were created in each EC member state for international information exchange about terrorism. The second Working Group (WG II) carried responsibility for police tactics, organization, and equipment. Its mandate was widened in 1985, and WG II then also covered technical and tactical cooperation issues in relation to public order, such as football vandalism (Van der Wel and Bruggeman 1993: 48, 49).

A new WG III was created in the mid-1980s to enhance cooperation in the area of serious organized international crime, such as drug trafficking, bank robbery, and arms trafficking. Its objective was to improve the infrastructure for criminal investigation, and to develop common repressive and preventive strategies (Van der Wel and Bruggeman 1993: 49–51).

In 1988 Trevi '1992'[5] was created to study the consequences of the abolition of internal border controls in the EC. The group reported to the Group of Coordinators (see above). The tasks of Trevi '1992' to some extent resembled the cooperative measures in the Schengen Agreements, including a compulsory mutual notification in the case of suspected cross-border criminal activity, the creation of a system of liaison officers in charge of the exchange of information about organized crime, the possibility of cross-border sur-veillance, and so on. Trevi '1992' was also responsible for overseeing activities in the *Ad Hoc* Group on Europol. The creation of the European Drugs Intelligence Unit, with a National Drugs Intelligence Unit, was agreed between the Trevi Ministers in June 1990 in Dublin, when Trevi issued one of its rare public documents, the Programme of Action. Trevi '1992' was dis-banded in 1992 during the UK Presidency upon completion of its tasks, and its remaining responsibilities were allocated to WG III. The *Ad Hoc* Group on Europol then had to report directly to the senior officials.

Finally, an *Ad Hoc* Group on Organized Crime was created in September 1992 to cope with the spread of the Mafia across the European continent and the infiltration of east European organized crime groups. Its first report appeared at a Trevi ministerial meeting on 6 and 7 May 1993 in Denmark. This contained an analysis of the structure and nature of the Mafia and other criminal organizations, and also offered an assessment of the organized-crime threat.

Before Trevi was integrated in Title VI, its organizational structure used to be tripartite. The first level was ministerial, the second that of senior officials, and the third that of the working groups. Which ministers represented a member state depended on the national constitution. Usually both the min-ister of justice and the minister of the interior were present at the biannual meetings. The senior officials, representing the chiefs of police at ministerial level, prepared the biannual ministerial conference. They were also responsi-ble for drafting recommendations and proposals for the working groups. The meetings of these groups, which represented the civil servants active within the relevant ministries, were held two to four times annually (Van der Wel and Bruggeman 1993: 44). Out of this grew a *wining and dining culture*. Many officials now active within the 'K4'-structure (see below, Fig. 15.1) reminisce about Trevi with warmth.

At the Hague summit in 1986 it was decided that the work of the Trevi working groups and senior officials would be coordinated by a permanent committee, the so-called *troika*, which consisted of the current, previous, and successor presidencies. In 1989 this was extended to a *piatnika*, a quinary committee which now also included the presidencies on either side of the *troika*. Despite these attempts to guarantee consistency in the long-term development of common policies, Trevi remained without a fixed secretariat and never became a genuine institution. Trevi was created under the auspices of the European Council and its chair changed with the Council presidency, but it was not controlled by the European Commission, Parliament, or Court of Justice, because of its intergovernmental status. Members of the European Parliament and civil-rights lobbies often criticized Trevi for its lack of transparency and external democratic control, especially given the weight of its decisions and the sensitivity of its topics (J. Benyon *et al.* 1993: 168; Van Outrive 1992*b*; Spencer 1990). This criticism was also voiced by the British House of Commons, but it modified the point by noting that the Trevi ministers were accountable to their national parliaments (House of Commons 1990: xxi). Police authorities criticized Trevi for its remoteness from the operational practice of policing: it was primarily a political forum run by senior officials and politicians (Van der Wel and Bruggeman 1993: 55–6). Except for the realization of the Europol initiative, it is difficult to assess which of Trevi's proposals were successfully implemented, because negotiations took place behind the scenes.

The Schengen Agreements

When the SEA was adopted in 1986, a political declaration was made to the effect that the promotion of the free movement of persons within the EC would have to coincide with cooperation between the member states with regard to the entry, movement, and residence of nationals of third countries, and with regard to the combating of terrorism, crime, drug trafficking, and illicit art and antiques trading. Another general declaration was appended to Articles 9 to 13 of the SEA, to the effect that nothing in their provisions would affect the right of the member states to take (national) law-enforcement measures or measures to control immigration from third countries. Member states thus maintained the option of introducing measures to improve their internal security, as long as these were compatible with EC legislation. This opened the way for intergovernmental cooperation in the criminal justice field.

Developments towards the realization of the free movement of people were slow and the target of 31 December 1992 looked unrealistic: it was doubtful whether the Twelve would be able to achieve a speedy and smooth consensus. Five of the EC member states—notably those that had experience with bilateral and multilateral agreements[6]—then decided to accelerate the negotiations (Belgium, France, Germany, Luxemburg, and the Netherlands). In June 1985 they signed an agreement in the Luxemburg town of Schengen, symbolically located on the border between Luxemburg, Germany, and France.

The core coalition of the five Schengen partners was not particularly appreciated by Italy, a founding signatory member of the Treaty of Rome. Italy had to wait for its accession until after the Schengen Implementing Agreement was signed in 1990. Spain, Portugal, and Greece acceded to the

Agreement subsequently, and Austria has long been keen to join. Denmark stated its interest in becoming a member in June 1994, anticipating the accession to the EU by Finland, Norway, and Sweden, with which it has a passport-free travel agreement (Nordic Union).[7] All of these are keen to join, provided that their relationship with Norway can be safeguarded.

The new Schengen signatories were not allowed to renegotiate the terms of the Agreements. The metaphor *Europe à deux vitesses* became popular to signify the two different speeds of internal security cooperation between the Schengen states and the non-Schengen states. The metaphor is, however, beginning to lose its accuracy because only seven of the nine Schengen members implemented the Agreement on 26 March 1995, some retained controls at borders or airports and because among the seven, smaller groups (Benelux) are negotiating separate but parallel forms of cross-border police cooperation.

Negotiations were undertaken by a central group consisting of senior officials from the different countries. Four steering groups elaborated relevant themes in further detail: police and security, movement of people, transport, and movement of goods. There was a separate steering group on the Schengen Information System (SIS), which included working parties on data-processing problems and the supplementary information service (SIRENE). Political decisions and mandates were agreed on at the meeting of ministers and secretaries of state. The European Commission had observer status in these meetings.

The 1985 Schengen Agreement contained short-term measures for the relaxation of border controls, such as the introduction of visual checks on EC citizens (implemented in June 1985), the green sticker behind the front window of passenger motor vehicles, and joint checks at the borders. The Schengen partners intended to coordinate the fight against illegal drug trafficking, serious international crime, and illegal immigration by improving cooperation between police and customs authorities. Long-term objectives were also agreed on, such as the intention to harmonize visa policies, firearms and ammunition laws, and rules on the hotel registration of travellers (Taschner 1990: 113). After four years of negotiations, Schengen had matured from an agreement primarily intended to relax the movement of goods, services, and persons to an agreement loaded with compensatory measures in the sphere of law enforcement. The Schengen Implementing Agreement included detailed measures on hot pursuit, cross-border surveillance, asylum, the status of refugees, illegal immigration, and the common computerized system for the exchange of personal data (SIS).

The Schengen Implementing Agreement has eight Titles, which include issues such as: definitions of concepts used in the Agreement; measures on the removal of controls at internal frontiers and movement of persons; law enforcement matters and compensatory measures in the sphere of police cooperation, drugs, and firearms (including 'hot pursuit', cross-border surveillance, extradition, mutual assistance in criminal justice matters, and the joint 'war' against drugs); the SIS; transport and movement of goods; data protection; the Executive Committee; and the final provisions, which contain the precondition that only EC/EU member states alone can accede.

The Implementing Agreement was supposed to be signed on 15 December 1989, but West German signature was postponed due to concern about the effects of the fall of the Berlin Wall on the movement of east Europeans across the border with East Germany.[8] Even before this significant historical event, the Schengen parties had failed to reach agreement at their meeting in Paris in June 1989, which particularly concerned issues such as extradition, the harmonization of drugs policies, visa policy, and police activities on foreign territory. Also the harmonization of asylum procedures, the lifting of bank secrecy in Luxemburg, and the location and creation of the central SIS were troublesome negotiating points.

Negotiations were resumed in March 1990, when the Schengen partners had agreed to ratify the Council of Europe Convention of 28 January 1991 for the Protection of Individuals with Regard to the Automatic Processing of Personal Data, to introduce national data-protection legislation, and to introduce complaint procedures, before the SIS could become operational. They also agreed that the rules on asylum would be explicitly linked with the 1951 Geneva Convention of the United Nations on the status of refugees (amended by the New York Protocol of 31 January 1967), and that Luxemburg's bank secrecy would have to change as a consequence of Luxemburg's signature of the Money Laundering Directive.[9] It was furthermore expected that West and East Germany would be unified, and it was agreed that the external border controls would apply at the borders between former East Germany and Poland and Czechoslovakia. This culminated in a readmission agreement between the Schengen states and Poland (signed on 29 March 1991) concerning the return of illegal aliens to Polish territory. The Agreement would come into effect in 1990 after the German unification, and extended the same rights and obligations to all Germans.

After the Implementing Agreement was signed on 19 June 1990, several deadlines for its implementation were postponed, due to technological difficulties and disagreement about the stringency of drugs and immigration policies. The rising costs resulting from the renovation of airports and the introduction of new forms of control also created an obstacle to its implementation. Previously Schengen had been considered as a more favourable policy-making arena than the EC because of the smaller number of partners.

This made it seem easier to achieve consensus about troublesome policy objectives. But Schengen lost much of its fresh flavour as negotiations about its implementation lasted nearly a decade. The general decline in enthusiasm—judged by the meagre results of the referenda on ratification of the TEU—for the single market affected the Schengen negotiations in a negative way. Second thoughts about Maastricht were projected on to Schengen. In particular the game of trust between the national governments of the member states about the sharing of responsibility for internal security issues nearly wrecked the basis of its common starting-points. Swings in the political composition of national governments (especially France) and the dramatic increase in the number of asylum-seekers furthermore meant that the political climate in which Schengen started off was much more favourable than the predominantly distrustful political climate during the time of its implementation.

Some national parliaments, the EP, lawyers, and civil-rights lobbies across Europe have criticized the Schengen Agreements for their democratic and legal deficit, which is mainly caused by their intergovernmental basis. Although the French Parliament was the first national parliament to ratify the Schengen Agreement on 3 June 1991, the Constitutional Council was challenged by M. Mazeaud, an MP, to judge whether the text of the Schengen Agreement was consistent with the Court's past jurisdiction. While Mazeaud argued that all constitutional principles of sovereignty were violated by the Schengen Agreement, the Court ruled on 25 July 1991 that it was not in conflict with the French Constitution.[10] Meanwhile, the Dutch Council of State recommended that the Dutch parliament should not ratify the Schengen Agreement because of discrepancies with international conventions, such as the Geneva Refugee Convention. The Council of State was of the opinion that the position of refugees would deteriorate as a consequence of the agreement. It further criticized the absence of a substantive asylum policy,[11] the lack of data-protection provisions, and the absence of a tribunal that could oversee the implementation of the provisions of the Schengen Agreements (J. Benyon et al. 1993: 138–9; Den Boer 1991: 10).

The creation of Schengen has therefore not been easy. The elements of the Schengen Agreement that overlap with Title VI have already been given a more solid existence within the TEU framework, although there, too, we witness the delay of several conventions due to political disagreement. More generally, the momentum of the Schengen Agreement may be outstripped by Title VI, as its implementation is deferred and as EU-legislation has been developed which eclipses some of Schengen's criminal justice issues (J. Benyon et al. 1993: 149–51; Den Boer 1991: 33). The EC Directives on Money Laundering and Firearms and Ammunition[12] supersede the relevant articles in the Schengen Implementing Convention (respectively Article 72 and Chapter 7, Articles 77–91).

Schengen is beginning to lose its experimental reputation: the question whether or not the Schengen Implementing Agreement can be integrated in the TEU becomes more pressing with time. There are a number of parallels between Schengen and Title VI. The most important is the precondition for accession to Schengen (Article 140). Only EU member states are entitled to accede, although candidate member states may have observer status in the Schengen negotiations. The EU member states that have not signed the Schengen Agreement—the United Kingdom, Ireland, and Denmark—are linked with the Agreement through their signature of the Dublin Asylum Convention,[13] which seeks to prevent an individual from submitting more than one request for asylum in EU member states.[14] The Asylum Convention is also integrated in Title VI (Article K.1(1)). Furthermore, the non-signatories have been involved indirectly in the drafting of the Schengen Implementing Convention, because of their involvement in the Working Programme of Action for the completion of the Internal Market ('Palma Document') (Schutte 1990: 10).

Article 134 of the Schengen Implementing Agreement provides another parallel with Title VI, as it demands that the Schengen provisions are applicable only if they are compatible with Community law. Article 142 of the Schengen Implementing Agreement determines that if a European-wide convention for the creation of an area without internal frontiers is concluded (the External Frontiers Convention; Article K.1(2) of Title VI), the Schengen Agreement will be subject to incorporation, replacement, or alteration.[15] If Schengen were to be integrated in the SEA, the European Court of Justice could assume jurisdiction. This is in line with the provision in Article K.3.2(c) of Title VI, which says that conventions drawn up by the Council 'may stipulate that the Court of Justice shall have jurisdiction to interpret their provisions and to rule on any disputes regarding their application . . .'. There is also a practical convergence between Schengen and Title VI, which is most visible in the creation of the SIS, which will probably be extended to the European Information System (EIS). Furthermore, the national SIRENE-units,[16] responsible for the verification and legal validation of requests for information and action, may combine with the National (Drugs) Units, which have been installed in connection with the creation of the Europol Drugs Unit.[17]

As we shall see in the next section, however, there is a rough distinction between Schengen and Title VI. Schengen provides rules for practical and operational cooperation between law-enforcement authorities, often consolidated through a network of bilateral agreements. Title VI does not have these provisions, although it does not exclude the possibility that two or more member states establish or develop closer cooperation (Article K.7 of Title VI). In general terms, this means that Schengen concentrates on the development of 'horizontal' contacts between law-enforcement authorities, while Title VI provides a more hierarchical or 'vertical' structure, especially through the establishment of Europol. There may as such be a problem with the integration of Schengen's cross-border executive police powers. In the event of integration, however, these may be arranged by means of an additional series of bilateral or multilateral agreements. Some commentators claim that the disadvantage of integrating Schengen into Title VI is its intergovernmental status. The difference between Schengen and Title VI is minimal in this respect, however, because the conventions drafted under Title VI (External Frontiers, European Information System, establishment of Europol, etc.) have also been developed outside the legal or democratic control of the EU. The only difference is that Title VI has been negotiated between twelve EU member states and not between the initial five Schengen partners (Van Outrive 1992a: 17). The analogy between the enlargement cycles of Schengen and the TEU is striking, in that justice and home affairs issues were not negotiated with the new EU Members: Austria, Finland, and Sweden just 'accepted' what had been established, including the Schengen Agreements.

Criminal justice cooperation under Title VI and the 'third pillar'

At the Maastricht meeting in December 1991, which laid the basis for the TEU, the European Council decided to complement the EC with two intergovernmental 'pillars', which concern common foreign and security policy (CFSP) and justice and home affairs cooperation (JHA).

As already noted in the introduction, Title VI is a compromised and heterogeneous construction. Some matters of 'common interest' are considered to be more intergovernmental and thus less suitable for transfer to the first pillar than others. These are judicial cooperation in criminal matters, customs cooperation and police cooperation for the purposes of preventing and combating terrorism, unlawful drug trafficking and other serious forms of international crime in connection with a European Police Office (Article K.1(7)–(9)).[18] The three matters all refer to some form of criminal justice cooperation, which is a novelty as they have always been closely associated with national sovereignty (Walker 1994). The common matters of interest that seem more eligible for transfer to the first pillar are: asylum policy; the crossing of external borders and external border controls; immigration policy and policy with regard to citizens from third countries; combating drug addiction; combating fraud; and judicial cooperation in civil matters (Article K.1(1)–(6)). Some of these matters have a footing in already signed or drafted conventions, such as the Dublin Asylum Convention (see above), the External Frontiers Convention, and the European Information System Convention.

The heterogeneous character of Title VI is due to the implicit distinction between 'transferable' and 'non-transferable' matters. Under Article K.3(2), for instance, the Commission does not have a right of initiative with regard to Article K.1(7) to (9), but the member states do. A further observation is that the link between EC matters and Title VI matters remains unexplained. The heterogeneity of the TEU is particularly reinforced by this separation of issues. This concerns visa policy (Article 100c and Article K.1(3-a); Hailbronner 1994); drug addiction (the relationship between Article 129(1), Article K.1(4), and the European Drugs Monitoring Centre); customs cooperation (the relationship between the EC generally or K.1(8); (M. Anderson, Den Boer, and Miller 1994: 117); and combating fraud (the relationship between Article K.1(5) and the UCLAF[19]). The inter-statutory integration with other parts of the TEU is therefore rather vague (Curtin 1993: 26), but so are aspects of institutional, functional, and policy integration (Müller-Graff 1994).

Activities under the JHA pillar have to be in accordance with the European Convention for the Protection of Human Rights and Fundamental Freedoms (4 November 1950) and the Geneva Refugee Convention (28 July 1951). This is provided for in Article K.2, which also epitomizes the principle of subsidiarity; the exercise of the responsibilities of member states in the maintenance of law and order and the safeguarding of internal security will not be affected. The balancing of national versus transnational action in the criminal justice field is very much a matter of interpretation. In concrete terms, the member states themselves decide whether or not a criminal justice matter should be dealt with as part of JHA. Criminal justice issues will not be entrusted to supranational bodies unless they concern criminal offences that require cooperation between at least two member states. The subsidiarity principle has also been exemplified in the draft Convention on the Establishment of Europol, namely that this organization should primarily target crimes that require a common rather than a national approach.[20] Moreover, this principle leaves open the possibility for member states to retain or reclaim control over law enforcement issues (see also Ch. 1).

The contents of Article K.3 have already been discussed to some extent. It gives the member states practical guidance on how to inform and consult each other and how to coordinate their action. The platform for this collaboration is the JHA Council, which can adopt joint provisions and joint action, and which can draw up conventions to be recommended to the member states for adoption. These conventions can, as discussed above, stipulate that the ECJ has jurisdiction. The member states tend to disagree about the extent to which the ECJ should be given jurisdiction (M. Anderson *et al.* forthcoming: ch. 6). The UK, in particular, had problems when the October 1994 draft Convention on the Establishment of Europol which contained an article concerning the arbitration of the ECJ in the event of differences of opinion between the member states or between the member states and Europol.[21] It preferred a wording in which the conflict would be discussed within the Council. The compromised character of Title VI is demonstrated quite clearly in Article K.3: the level of the lowest common denominator has been accepted, and thus the evolution of JHA proceeds at the speed of the slowest—read least federation-minded—member.

Most of the organizational details of Title VI are covered by Article K.4, which states that a Coordinating Committee (the 'K4' committee) consisting of senior officials is to be created. As explained above, this Committee reconstitutes the old Trevi in many ways, but also absorbs the *Ad Hoc* Group on Immigration and the Group of Coordinators. It is responsible for giving opinions to the JHA Council and contributes to the Council's discussions. The European Commission is fully associated with the work of the 'K4' Committee, which means it has observer status. The JHA Council is the central body within the Title VI structure, which is illustrated in the detailed structure of the 'K4' committee.[22] The meetings of the three steering groups are all held in Brussels and attended by civil servants of the relevant ministries. Although the new structure is more centralized and coordinated, when compared with the pre-Maastricht policy arena on justice and home affairs issues, it has also added more filters to the negotiating structure, making it more difficult to push issues up to the highest level of the JHA Council's agenda. (See Fig. 15.1.)

Article K.6 makes up slightly for the previous democratic deficit, which was one of Trevi's greatest shortcomings. It rules that the Presidency and the Commission shall regularly inform the EP of discussions in the areas of Title VI. Furthermore, the Presidency shall consult the EP on the principal aspects of activities and the EP may ask questions of the Council or may make recommendations to it. In the case of Europol, for example, the Parliament may direct questions to the Council which will then be answered by the Director of Europol by means of a statement. The EP shall not have a controlling task, nor will it be able to ask questions directly, thus leaving political control on

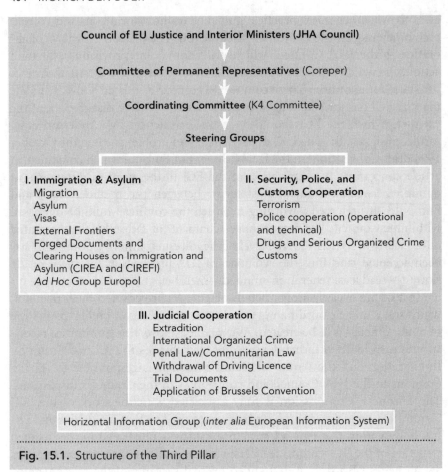

Fig. 15.1. Structure of the Third Pillar

justice and home affairs issues rather flimsy. The principal problem lies, how-ever, with the overall intergovernmental status of Title VI, which prevents Community institutions from playing a more substantial role.

The principal position of the Council comes to the fore in the so-called *passerelle* article of Title VI, namely Article K.9. Members of the Council can decide unanimously—on the initiative of the Commission or a Member State—that Article 100a should be applied to Article K.1(1)–(6). Criminal justice matters (Article K.1(7)–(9)) are not eligible for transfer. If the Council decides to change a 'matter of common interest' into a Community matter, the member states would have to adopt that 'in accordance with their respective constitutional requirements, hence a ratification-like procedure similar to that of the enactment of amendments to the Treaties establishing the communities or the Union (Article N(1) sub.3 TEU)' (Müller-Graff 1994:

508). The treaty amendment procedure, laid out in Article N of the TEU, can thus be bypassed, but the similarity is that it requires unanimity at Council level followed by unanimous approval of all member states in accordance with their constitutional requirements.

This may have the consequence that the future development of Title VI will be uneven. Even though certain matters of common interest have been given priority status and are most likely to be transferred to the first pillar, the Council may still decide that one matter of common interest is more advanced than the other. The transfer of justice and home affairs issues will therefore depend on whether or not common interest matters are considered to be controversial; intense negotiation will precede the establishment of consensus about these matters. The rule of unanimity determines again that the speed at which this occurs equals the speed of the most reluctant member state. The procedure of unanimous voting also applies to the amendment of the Convention on the Establishment of Europol. This is specifically relevant for the expansion of the list of criminal activities that it seeks to deal with and for the potential attachment of executive powers to the organization. Title VI and the matters of common interest listed under Article K.1 will be subject to review at the Intergovernmental Conference in 1996, when it will be decided whether this pillar should remain intergovernmental, or whether certain aspects of it should be transferred to the first pillar.

The future development of the 'third pillar'

Despite chronic reticence demonstrated by certain EU governments, justice and home affairs have undergone a significant development and a substantial body of policies and legislation has been coined, be it outside the official EU machinery. The argument in favour of intergovernmentalism as the appropriate method to develop cooperation is beginning to be outweighed by its disadvantages:

- First, there is confusion at the level where decisions are implemented. The similarity between agreements on asylum, for instance, make it difficult for civil servants and immigration-control officers to understand which of these agreements should be applied with priority, and how the new rules ought be interpreted for the (re-)design of airports and other entrances.

- A second disadvantage has been discussed above, namely the lack of coherence between the first pillar and the third pillar, making it unclear, for example, whether EU member states should adopt primarily a pre-ventive or a repressive approach to criminal activities (Den Boer 1994*a*: 283).

- A third disadvantage is that the intergovernmental status of Title VI affords insufficient protection against the profound impact of transfor-mations in justice and home affairs cooperation on EU citizens (M. Anderson, Den Boer, and Miller 1994: 118–20). There will be no supranational tribunal to which EU citizens can turn for judicial redress when their privacy has been infringed by Union-wide information sys-tems (M. Anderson *et al.* forthcoming: chs. 6 and 8). The EP will only be informed and consulted, but it will not have a legislative or supervisory role. This means that the onus of democratic and legal control will rest on the national parliaments and courts. In the absence of criminal jus-tice harmonization this may lead to uneven jurisdiction and therefore to an unequal protection of EU citizens.

This problem is aggravated by a number of other developments. As police organizations are gradually shifting their focus towards international orga-nized crime, and law-enforcement activities are beginning to take place on the regional, national, and international level, community policing may lose its priority status, which may have negative repercussions for the traditional links between the police and the public. The loosening of this linkage is accompanied by a change in policing methods: law-enforcement authorities consider conventional policing methods to be ineffective means for the fight against international organized crime, and thus have more frequent recourse to proactive policing methods which are not always legitimate. Whether or not these developments are a direct consequence of intensified efforts to coordinate police cooperation internationally, they are issues that ought to be contemplated in relation to the social legitimacy and public acceptance of international criminal justice cooperation (M. Anderson *et al.* forthcoming: ch. 5; Walker 1993*a*; 1993*b*).

As the EU has recently been enlarged to include an even richer pool of policing cultures, a struggle may be expected with regard to the permissive-ness of the mandate and minimal accountability standards: the divide between North and South over professionalism, repression versus preven-tion, accountability procedures, and so on is already significant (Zanders 1994: 14), but may even become stronger as Austria, Finland, and Sweden add their experience to the justice and home affairs forum.

All this takes place in the context of a still changing security situation. The enlargement of the EU has implications for the location and control of the external border: with the accession of Finland, Russia now borders the territory of the EU. Threat assessments will have to be drawn up to predict the movement and development of crime. The recent experiences with asylum have taught us that the number of asylum requests in EU member states is mainly a consequence of security imbalances in the European region. The same is probably true for transfrontier criminal activities, although the importance of global developments should not be underestimated. The spread of criminal activities is closely related to the availability of a favourable political, economic, social, legal, and geographical infrastructure. Although these arguments tend to suggest that criminal justice initiatives at the level of the EU are perhaps already too small in scale to cope with global crime, it is the EU framework that has the potential to render stability to coordinated law-enforcement strategies. The permanence, reliability, and consistency of justice and home affairs cooperation could benefit, however, from firmer integration in the TEU, thereby consolidating coherence between all relevant EU policy areas.

Notes

1. Selected examples are the 1988 UN Convention against Illicit Traffic in Drugs and Psychotropic Substances, the European Convention on Mutual Assistance in Criminal Matters (1959), the European Convention on Extradition (1957), the European Convention on the Transfer of Proceedings in Criminal Matters, the European Convention on the Suppression of Terrorism (1977), the European Convention on Insider Trading (1989) and the European Convention on Laundering, Search, Seizure, and Confiscation of the Proceeds from Crime (1990). (See Anderson *et al.* forthcoming: ch. 7).
2. This group was created in 1971 at the initiative of the French President Pompidou. It was officially called Groupe de coopération en matière de lutte contre l'abus et le traffic illicite des stupéfiants. In 1980 the Committee of Ministers of the Council of Europe decided that interested countries could sign a partial agreement. Since November 1980 the secretariat of the Pompidou Group has been under the auspices of the Council of Europe as part of the Partial Agreement Division under the Directorate of Economic and Social Affairs. In 1993 twenty-two countries, which included all EU member states, participated in the Pompidou Group (Van der Wel and Bruggeman 1993: 72, 73).
3. The PWGT is a forum of West European police and security chiefs, created in 1983. As the forum includes Finland, Norway, and Sweden, and as it has a predominantly operational orientation, it has not yet been integrated in the Title VI framework (Riley 1993: 43–5).
4. This group was created in London in October 1986. Its task was to coordinate asylum and refugee policies between EU member states.

5. For a comprehensive overview, see J. Benyon *et al.* (1993: 152–64).

6. Belgium, the Netherlands, and Luxemburg had already signed the Benelux Treaty on Extradition and Mutual Assistance in Criminal Matters (1962), and France and Germany were signatories to the Convention of Paris (1977) and the Saarbrücken Convention (1984), respectively on police cooperation in the German-French border area and on the easing of controls on individuals at the common border.

7. Norway voted against accession in a referendum held in November 1994, but is interested in joining Schengen.

8. Taschner (1990: 29) observes that this situation posed a problem for the partners of Germany rather than for Germany itself.

9. Council Directive 91/308 (EEC) of 10 June 1991 on the prevention of the use of the financial system for the purpose of money laundering.

10. *Le Monde*, 27 July 1991.

11. i.e. common substantive criteria employed to evaluate an asylum request.

12. Council Directive 91/308 (EEC) (see above, n. 9) and Council Directive 91/477 (EEC) of 18 June 1991 on control of the acquisition and possession of weapons.

13. Convention Determining the State Responsible for Examining Applications for Asylum Lodged in one of the member states of the European Communities, Dublin, 15 June 1990.

14. Chapter 7 (Articles 28–38) of the Schengen Implementation Agreement.

15. For a comparison of the visa policy in the External Frontiers Convention and the Schengen Convention, see Hailbronner (1994: 990).

16. Title IV of the Schengen Implementing Convention. The direct contact between national SIRENE offices requires a secondary electronic information system which transfers complementary dossiers (Discussion paper of the Interlabo GERN, Louvain, 18 Mar. 1994—mimeo; Circulaire No. SAEI 93-1-L2/27-09-93, Ministère de la Justice, Paris).

17. The Europol Drugs Unit (EDU) is the forerunner of the European Police Office (Europol). The former works within the limits of the Copenhagen Ministerial Agreement, and concentrates on organized crime in connection with drug trafficking and related money laundering. Europol's remit is much wider and is based on the Convention on the Establishment of Europol. Agreement on the final text, planned for the European Council at Cannes in June 1995, was delayed by a British objection to allowing jurisdiction over the Convention to the ECJ. It was decided to seek resolution of this issue at the European Council in June 1996.

18. It is to some extent surprising that customs cooperation has been included in the list of areas that cannot be transferred to the first pillar, particularly as the EU has developed exchange programmes for customs officers. On the other hand, however, customs cooperation in the context of justice and home affairs cooperation, i.e. in relation to criminal investigations, is a new matter within the EU.

19. Unité de la Coordination de la Lutte Anti-Fraude (Reinke 1992: 21–4).

20. The condition for action by Europol is that there are 'factual indications that an organized crime structure is involved and two or more member states are

affected by the forms of crime in question in such a way as, owing to the scale, significance and consequences of the offence, to require a common approach by the member states'. Source: 10324/94 EUROPOL 112 (restricted), 27 Oct. 1994 (Brussels).

21. See previous note, Europol Convention, Article 37.

22. Compiled from: M. King, 'Conceptualizing "Fortress Europe": a consideration of the processes of inclusion and exclusion', paper presented to the ECPR planning session on 'Police and Immigration: towards a Europe of Internal Security', Madrid, 17–22 Apr. 1994: 2 (mimeo); J. Peek, 'Police Cooperation' (mimeo) and L. Drüke, 'The Position of the UNHCR' (mimeo), contributions to Second Expert Meeting on Third Pillar of the Union Treaty, College of Europe, Bruges, 19/20 Sept. 1994.

Further Reading

On the development of the third pillar generally see Monar and Morgan (1994) and Bieber and Monar (1995). On the development of European police cooperation see Anderson and Den Boer (1994) and Anderson *et al.* (forthcoming). Müller-Graff (1994) provides a helpful account of the legal issues and see Hailbronner (1994) on the impact on third-country nationals. For Schengen see Den Boer (1991) and Van Outrive (1992*a*).

Anderson, M., and Den Boer, M. (1994) (eds.), *Policing Across National Boundaries* (London: Pinter).

—— —— Cullen, P., Gilmore, W. C., Raab, C., and Walker, N. (forthcoming), *Policing the European Union* (Oxford: Oxford University Press).

Bieber, R., and Monar, J. (1995) (eds.), *Justice and Home Affairs in the European Union—The development of the Third Pillar* (Brussels: European Interuniversity Press).

Den Boer, M. (1991), 'Schengen: Intergovernmental Scenario for European Police Cooperation', Working Paper No. V (Edinburgh: University of Edinburgh, Department of Politics).

Hailbronner, K. (1994), 'Visa Regulations and Third-Country Nationals in EC Law', *Common Market Law Review*, 31/5: 969–95.

Monar, J., and Morgan, R. (1994) (eds.), *The Third Pillar of the European Union—Cooperation in the fields of justice and home affairs* (Brussels: European Interuniversity Press).

Müller-Graff, P.-C. (1994), 'The Legal Bases of the Third Pillar and its Position in the Framework of the Union Treaty', *Common Market Law Review*, 31/3: 493–510.

Van Outrive, L. (1992*a*), 'The Entry into Force of the Schengen Agreements' (Brussels: European Parliament Committee on Civil Liberties and Internal Affairs).

CHAPTER SIXTEEN

COMMON FOREIGN AND SECURITY POLICY: A NEW POLICY OR JUST A NEW NAME?

Anthony Forster and William Wallace

European Political Cooperation (EPC) began in 1970 as an intergovernmental network among the foreign ministries of the member states, coordinating civilian diplomacy within the Atlantic security framework. By 1990 it had transformed the working patterns of national foreign ministries and embassies and developed a rudimentary secretariat, while the Commission had become gradually more involved. Extensive meetings and intensive communications had not, however, generated any effective common policies; preoccupation with procedures in themselves served as a substitute for policy. The end of the cold war transformed the context as the 1991 IGC proceeded. The compromise embedded in the Maastricht TEU registered a formal shift from cooperation to common foreign and security policy. But the underlying tension between sovereignty and common policy remains, despite the ambitious rhetoric.

Introduction

'A common foreign and security policy [CFSP] is hereby established.' The confident language of Article J of the Maastricht Treaty on European Union (TEU) appeared to end forty years of hesitation over the principle and practice of integrating national foreign and defence policies among west European states. Closer examination of the text of Article J, its sub-clauses, and protocols, however, reveals that ambiguities continue to cloak cooperation among Community governments in this field in spite of the transformation of the international context, marked by German unification and Soviet collapse.

Defence and diplomacy, like border control, policing, and citizenship—and currency—are after all part of that core of state sovereignty around which practitioners of functional integration had tiptoed throughout most of the intervening years. Transfer of effective authority (and budgetary allocation) over foreign policy and defence would create a European federation. Policy cooperation in this field has therefore operated under contradictory pressures. The rationality of maximizing influence through common action and of sharing scarce resources has been balanced by concern for the preservation of national sovereignty and of diverse national traditions and taboos.

Preoccupation with procedures has overshadowed attempts to develop common policies from the outset. Twenty-five years after the establishment of EPC in 1970 there remain unresolved dilemmas over the distribution of competences and the balance to be struck between the preservation of state sovereignty and the efficient pursuit of joint objectives. This is reflected in the weak institutionalization and marginal policy output of the process. It has substantially reshaped the working practices of EU foreign ministries and embassies, and entirely transformed the diaries of EU foreign ministers. But it has not moulded fifteen national foreign policies into one.

Atlantic alliance, civilian power

Foreign policy and defence had been coordinated among European Community (EC) member states since the time of the Schuman Plan within the broader framework of the North Atlantic Treaty, under American leadership. Successive American administrations had continued to guard the supremacy of Nato consultations against repeated French-led challenges, up to the early months of the intergovernmental conference which led to the TEU. Under intense American pressure to accept the rearmament of Germany, and in the face of an evident Soviet threat, the Six had attempted to tackle the question of common foreign and defence policy as early as 1951. Article 38 of the European Defence Treaty, signed in Paris in May 1952, committed the signatories to examine within six months the form of the new political superstructure needed to give direction and legitimacy to the future European Defence Community (EDC).

The resulting 'de Gasperi proposal' for a European Political Community included a new 'European Executive', accountable to a directly elected European Parliament. This direct attack upon the core issues of national sovereignty collapsed with the failure of the French Assembly to ratify the EDC Treaty in August 1954 (Fursdon 1980). An intergovernmental alternative, promoted by the British government, provided a looser framework for German rearmament under agreed conditions: transforming the 1948 Brussels Treaty of 'Western Union' into the Western European Union (WEU), with the addition of Germany and Italy to the original five signatories, while firmly subordinating the new organization's military functions to those of Nato.

President de Gaulle took foreign policy cooperation as the ground on which to make his twin challenge to American hegemony and to the supranational pretensions of the infant European Economic Community (EEC). On French initiative, the very first conference of heads of state and government and foreign ministers of the Six met in Paris in February 1961 'to discover suitable means of organizing closer political cooperation' as a basis for 'a progressively developing union'. There followed twelve months of argument primarily between the French and the other five on what were labelled the Fouchet proposals: about the shape of the proposed 'Political Committee' or 'Commission', about the status and site of its secretariat, about the association or exclusion of Britain from the discussions, about whether defence questions should be included, and about its future relationship with Nato and with the USA. Intrinsically linked to the British application for EEC membership and the Kennedy Adminstration's plans for a new 'Atlantic Partnership', the Fouchet proposals found little support except within the German government, were vigorously opposed by the Dutch, and collapsed amid general suspicion of Gaullist objectives.

De Gaulle then pursued an alternative—bilateral—means of harnessing German economic strength to French international ambitions, through the Franco-German Treaty of the Elysée, rapidly negotiated and signed in March 1963, weeks after the French President had blocked British negotiations to enter the EEC. But the promise of its defence clauses was blighted by the

Bundestag's addition of an Atlanticist preamble during the ratification debate, against Chancellor Adenauer's wishes, leaving those clauses unimplemented until their revival in 1982 by President Mitterrand and Chancellor Kohl (Grosser 1980).

The 'relaunch' of European integration which followed de Gaulle's departure, at the summit meeting in The Hague in December 1969, was a carefully crafted package deal. French acceptance of negotiations for British accession was balanced by insistence on consolidation of the system of agricultural finance within the Community budget, by commitment to economic and monetary union (EMU), and by agreement on a renewed effort at 'political cooperation' on foreign policy. The initial scepticism of other member governments, compounded by suspicions of French intentions to undermine transatlantic cooperation, gave way as foreign ministers discovered the utility of informal consultations, and as their diplomats learned to appreciate a framework for multilateral diplomacy which was entirely under their shared control.

Yet EPC in its early years more clearly served German international interests than French. It provided western multilateral support for *Ostpolitik* in a period when American policy-makers were preoccupied with Vietnam; and it provided a caucus within which to operate when the Federal Republic was admitted into the UN in 1973. The machinery of EPC first proved its worth in preparations for the Conference on Security and Cooperation in Europe (CSCE) in 1972–4. The Ten (later Nine—the four applicants had joined EPC ahead of accession, with Norway dropping out again after its referendum rejected membership) played a more active role in CSCE than the USA in shaping the agenda and in negotiating the complex package of the Helsinki Declaration (W. Wallace and Allen 1977; de Schoutheete 1986).

The evolution of EPC during the first twenty years of its operation can be seen as a cycle of hesitant steps to strengthen the framework, followed by periods of increasing frustration at the meagre results achieved, culminating in further reluctant reinforcement of the rules and procedures. Relations with the USA were a significant factor in this cycle. Secretary of State Kissinger provoked a debate on the links between European and Atlantic political cooperation in his 'Year of Europe' speech of April 1973. Divergent reactions to the Arab–Israeli War of October 1973 escalated the debate into a bitter Franco-American confrontation, with other west European governments caught in between. The dispute was resolved in the 'Ottawa Declaration' of June 1974, in the context of a Nato Summit, which set up an additional consultative mechanism between the rotating Presidency of EPC and the US State Department before and after each EPC ministerial meeting (W. Wallace 1983b).

European dismay at the drift of US policy towards confrontation with the USSR in 1979–81, and at the failure to manage a concerted response to the Soviet invasion of Afghanistan, led to renewed efforts to promote cooperation. The British Presidency of July–December 1981 sponsored the 'London Report', formalizing and extending the informal practices which had grown up during the previous five years, and tentatively expanding foreign policy consultations to cover security questions. The German and Italian foreign ministers followed this with the much more ambitious Genscher–Colombo Plan: calling for a 'European Act' which would 'reaffirm' member governments' commitment to a common foreign policy and to 'the coordination of security policy and the adoption of common European positions in this sphere' (Nuttall 1992: 186). But other governments—in particular the Danish, whose foreign minister took his turn in the rotating Presidency in the second half of 1982, the neutral Irish, and the unsocialized Greeks, who had joined the EC and the EPC procedures in 1981—retained strong reservations over sharing sovereignty in such a sensitive area, in which the views of the larger member governments were likely to prevail. The 1986 Single European Act (SEA) thus provided for only limited reinforcement of foreign policy consultations among member governments (de Schoutheete 1986).

Extension of such consultations among west European governments into security policy and defence raised even more acute anxieties in Washington, and therefore also in those European capitals strongly committed to the Atlantic alliance. West European integration had developed as a civilian process under American protection and sponsorship; the directness of the Gaullist challenge to American security leadership had only sharpened sensitivities elsewhere. The British–German initiative to establish the Eurogroup in 1970 had been firmly rooted within the Nato framework, intended to demonstrate to members of the US Congress the willingness of their European allies to shoulder a substantial share of the common defence.

It took accumulated dissatisfaction with the quality of American leadership at the end of the 1970s to lessen inhibitions a little. In 1980–1 the German, Italian, and British foreign ministers repeatedly called for the coordination of security policy within the EPC framework, though carefully circumscribing their definition to exclude the 'hard' security dimension of defence policy. French concern about apparent pacifist tendencies in German policy, and German concern about cut-backs in conventional spending in French defence, led to a relaunching of a Franco-German defence dialogue in 1982, extended through a trilateral meeting with the British into a revival of six-monthly WEU ministerial meetings (of foreign and defence ministers) in 1983 (W. Wallace 1984). WEU membership expanded to nine with the accession of Spain and Portugal in 1987, following their accession to the EC, after negotiation among the original seven members of a Declaration underlining the political and military obligations of membership. But warnings from Washington continued to accompany every gesture towards closer European cooperation, with the German, Dutch, and British governments in particular anxious to reassure the Atlantic hegemon of their prior loyalty to the Atlantic alliance (Menon *et al.* 1992).

EPC after twenty years: real progress, or shadow-boxing?

By the end of the 1980s the procedures of EPC had evolved into an extensive network, drawing in some thousands of diplomats in the foreign ministries of the member states, in their embassies outside the EC and in missions to international organizations. Its working practices still reflected the determinedly intergovernmental character of its origins. A hierarchy of committees and working groups had grown up beneath the meetings of foreign ministers, bringing together national officials across a wide range of geographical and functional responsibilities. Memories of the Fouchet Plan proposals to set up a Political Secretariat in Paris had led to initial resistance from other governments to any such institutionalization, relying on the efforts of the foreign ministry holding the Presidency to convene meetings. The pressures that this imposed on the smaller member states, and the discontinuities created—not least in representations to third countries and international organizations—by six-monthly changes in the chair led first to the development of the 'rolling troika' (of the presidency in office, its immediate predecessor and successor) and then to the slow emergence of a secretariat out of the group of seconded officials from previous and succeeding presidencies who moved from foreign ministry to foreign ministry. The SEA, which formally brought together the separate EC and EPC hierarchies under the 'single' umbrella of the European Council, settled the EPC Secretariat in Brussels, nevertheless maintaining the Gaullist heritage of intergovernmental separation from the EC by leaving it independent of the Council Secretariat.

Looking back over twenty years, the transformation of diplomatic working practices was evident (see Table 16.1). The structures of foreign ministries themselves had converged: each now with its Political Director to represent its minister on the Political Committee, as their colleagues (and sometimes rivals) the Permanent Representatives did on Coreper, each with a Political Correspondent to monitor working groups and to manage the flow of paper. Direct communication among foreign ministries had been established in 1973 through a secure communications link, the Coreu network, managed from the Dutch foreign ministry. Traffic around this network grew from an initial two to three thousand telegrams a year to some 9,000 in 1989, short-circuiting the intermediary role of national embassies within the EC. Desk officers in foreign ministries now dealt directly with their opposite numbers, meeting them regularly in their appropriate working group, in touch by telephone and Coreu every day if necessary. Cooperation among embassies in third countries grew in the 1980s to regular joint reports, of particular value to smaller members without representation in the countries concerned. As the habits and assumptions of a generation of national diplomats were thus reshaped, so also joint training courses were initiated, the sharing of embassy facilities in some third countries was undertaken, and even exchanges of personnel between foreign ministries accepted (Nuttall 1994).[1]

The Commission, which had at the outset been rigorously excluded by Gaullist doctrine and French insistence from participation, had crept slowly into working group after working group, from its first invitation to join the discussions on those economic aspects of the CSCE negotiations which fell within Community competence in April 1971. The launching on French initiative of the Euro-Arab Dialogue in December 1973 also involved direct linkage between political and economic relations, necessitating a limited Commission involvement. The change of French Presidency in 1981 weakened the ideological resistance from Paris, bringing into the Quai d'Orsay as foreign minister Claude Cheysson, who was himself a former Commissioner. The small EPC office within the Secretariat-General grew after the SEA into a full Directorate, with its own 'Political Director' (Nuttall 1994).

For its defenders, in London and Paris, EPC in 1989 represented a working model of intergovernmental cooperation without formal integration. The model had indeed been extended to the equally sovereignty-sensitive area of justice and home affairs, as noted in Chapter 15. Foreign ministers, and foreign ministries, now spent much of their working life within this multilateral context, moving from EC Councils of Ministers to EPC ministerial meetings to WEU, each with their subordinate committee structures, meeting with each other more often than they met with their colleagues in national Cabinets.[2]

EPC's critics, outside foreign ministries, questioned whether the extensive scale of these activities was justified by the modesty of their output. Operating under the rule of consensus, without any commitment to common action, without military or financial resources except what participating states might volunteer, the structure resembled a diplomatic game, producing declarations and conclusions without visible effect. Confidential procedures, which were sparsely reported to national parliaments and virtually unnoticed by national publics, provided much satisfying work for officials

Table 16.1. From EPC to CFSP: a history of major developments

Event/Date	Aims	Procedures	Instruments
The Hague Summit December 1969	launch of political cooperation	preparation of a report on political cooperation	
Luxembourg Report October 1970	• to give shape and will to the Union • to exercise Europe's growing responsibilities • to match the political with economic policies • gradual action in areas of common agreement	• regular exchange of information • coordination of positions • foreign ministers meetings (2 per year) • political directors meetings (4 per year) • meetings in national capitals	common positions
Copenhagen Report July 1973	• to act in the world as a distinct entity • to seek common solutions • the aim of consultation is now common policies	• increased meetings of foreign ministers • presidency role elaborated • correspondents group confirmed • working groups formalized • COREU established	political dialogue
London Report October 1981	• goal of EPC is now joint action • coordination of political aspects of security	• strengthened presidency role • troika secretariat confirmed • national secondment to presidency • stronger commitment to consult • full association of Commission	sanctions, trade, aid

Solemn Declaration on European Union June 1983	• greater coherence and even closer coordination between EC and EPC • to consider economic aspects of security	• European Council issues general guidelines for EC and EPC • each presidency of the European Council presents a report to the EP	• economic and political instruments
Single European Act July 1987	• to transform relations as a whole into a European Union • to be achieved by consistency and solidarity • to reduce further the differences between the instruments of EPC and EC • treaty review by 1993	• legal treaty basis for EPC • common actions by information and consultation • establishment of EPC secretariat • decisions by consensus but governments must refrain from blocking consensus	
Treaty on European Union November 1993	• a common foreign and security policy is established • CFSP is a Union and not a Community responsibility • instruments of EC and CFSP fully combined but institutional distinction maintained through 'pillared' structure • treaty review by 1996 • subcontracts defence to WEU	• merging of General Affairs Council and EPC • merging of EPC secretariat into Council Secretariat • DG A and Commissioner created for External Political Affairs • Coreper directly involved • majority voting is permissible • all meetings held in Brussels • synchronization of procedures with WEU	• common action not just common positions • joint actions on all issues except defence • request WEU action on defence issues

without engaging wider political or public opinion. It thus failed to promote any substantial convergence of national attitudes. There was little evidence that EPC had exerted any direct influence on Arab–Israeli relations, for example, or on events in sub-Saharan Africa or the Persian/Arabian Gulf. These were procedures without policy, activity without output, while American arms and American diplomacy still determined the course of western interests throughout the regions to Europe's immediate south.

Political union and European revolution: 1990–1991

The Intergovernmental Conference (IGC) planned for 1990–1 was initially intended to focus on monetary union and its institutional consequences, not directly on political union defined in terms of foreign and defence policy. It was the revolutions in central and eastern Europe in the course of 1989, the unanticipated shift of post-Communist opinion within East Germany towards reunification, and the rapid moves towards German unification which followed in 1990, which forced foreign and security policy on to the IGC agenda. German unification unnerved the political class in almost all other member states, raising the spectre of a new 'German problem' as well as more fundamental questions about the future internal equilibrium of the EC.[3] One of the underlying purposes of west European integration from the Schuman Plan on had after all been to constrain the sovereignty of a reconstructed Germany (Soetendorp 1994: 103).

Washington, the established leader of the western alliance, was the first to respond. Secretary of State James Baker, in his Brussels speech of 12 December 1989 (Baker 1989), proposed a redefined transatlantic bargain to reflect the end of western Europe's security dependence, with North American and European pillars and an agenda extended to cover the full range of politico-military, economic, and environmental issues. West European governments resisted the idea of incorporating this redefined relationship in a formal new treaty; the Transatlantic Declaration, which was signed in the autumn of 1990, more modestly formalized and extended the network of contacts between the EC, the EPC Presidency, and the US administration (J. Peterson 1994).

Both British and French governments initially expressed strong reservations about the prospect of German unification. In March 1990, as political developments across central Europe continued to move faster than west European governments wished, the Belgian government proposed a second IGC on political union to consider strengthening the EC, to run in parallel with the IGC on EMU already agreed in December 1989. Overcoming French reservations, Mitterrand and Kohl issued a joint letter the following month in which they endorsed the Belgian initiative and called for the second IGC to formulate a common foreign and security policy as a central feature of the European Union (Laursen and van Hoonacker 1992). This gained support amongst other member states as the pace of German unification quickened, with other governments recognizing the logic of strengthening the EC as the most effective institutional framework for containing the potential regional hegemony of a united Germany. Thus, against London's preference, the majority view at the Strasburg European Council in July 1990 acknowledged a formal link between German unification, political union, and EMU.[4]

It soon became clear, however, that there were significant differences over the character of that formal link. Chancellor Kohl and the German political élite saw their acceptance of monetary union as part of a package which had to include substantial strengthening of the political institutions of the Community, moving towards a political union which would include the development of common foreign and security policies within an integrated (and democratically accountable) Community framework.

This intention to end the legacy of Gaullism, the separation of EPC from EC, was shared by the Dutch and the Belgian governments. The French and British governments, however, resisted the transfer of authority over foreign policy from a confidential intergovernmental framework to the Community proper, with the accretion of power to the Commission (and indirectly to the European Parliament) which this would imply, and with pressure to move from consensual decisions towards majority voting and joint action. This position was supported at the outset by the Danes and the Portuguese (Cloos et al. 1994). On security policy and defence there was a different divide between Atlanticists (Britain, the Netherlands), resisting any substantial transfer from the Nato framework, into which the WEU was closely integrated, and Europeanists (France), with the German government in the middle (Gnesotto 1990; Tiersky 1992). Negotiations over preferred policy outputs were thus entangled with constitutional questions throughout the IGC.

As important a divide, though one less willingly admitted by many delegations, lay between states with the capacity and the domestic support for active foreign policies, and those for which foreign policy (and even more defence) was surrounded with political inhibitions. Here France and Britain lay at one end of the spectrum and Germany—the government most determinedly pushing for a common foreign and security policy—at the other. With the Iran–Iraq war dragging on through the 1980s, British and French ships had contributed naval patrols in the Red Sea and the Gulf, partly in response to

American pressure; Dutch, Italian, and Belgian ships had later joined them, operating under a WEU umbrella. Germany sent a frigate no further than the Mediterranean, to replace vessels detached by other countries from the Nato Standing Force. On Iraq's invasion of Kuwait in August 1990 Britain responded to American calls for military support by sending an armoured division; France (to the embarrassment of its military and political leaders, who wished to demonstrate a comparable commitment) could only assemble and despatch from its depleted conventional forces a smaller contingent. Germany contributed (substantially) only to the financial costs of the military operation. This reflected historical and constitutional inhibitions about the projection of military power beyond its borders, firmly entrenched in public opinion and the opposition parties within the Bundestag.

Any attempt to negotiate on bringing security and defence closer to the Community framework immediately raised the central issue of future relations with the USA under transformed strategic circumstances. European members of Nato were not willing to take decisions about defence integration without first establishing how these might affect the relationship between the alliance and the EC (Forster 1994: 58). The USA was thus an active external player throughout the IGC, across the whole common foreign and security policy dossier. Successful agreement on the conclusions of the Alliance Strategic Review, launched in April 1990 and running in parallel with the EC deliberations, was a precondition for successful agreement among the Twelve (Menon *et al.* 1992). Bilateral Franco-American conversations were an important element in the reconciliation of different views.

Negotiations thus proceeded in 1990–1 in three different fora, Nato, WEU and EC–IGC, with overlapping but non-identical memberships; the same foreign ministers and foreign ministry officials might on occasion adopt inconsistent positions across these fora.[5] Since these were politico-military negotiations, focusing on security and defence policy rather than on defence *per se*, the political sections of foreign ministries led in all three negotiating fora, with defence ministries in second place in Nato and WEU discussions and absent from the EC–IGC.

If rational negotiators had been able to focus on the issues at stake undistracted by extraneous developments, the Maastricht package on CFSP might conceivably have been tied up more neatly. But external developments intruded from beginning to end, to preoccupy ministers with more immediate crises and with their longer-term implications. As the Iraq/Kuwait crisis evolved several of the former socialist states of central and eastern Europe volunteered non-combat forces; their aim, self-evidently, was to demonstrate their commitment to early membership of the EC, WEU, and Nato, raising the unwelcome prospect of rapid and extensive enlargement and of an east-

ward extension of western Europe's security commitments. US forces transferred from Germany to the Gulf returned in 1991 directly to the USA, accelerating a run-down of American troops in Europe from 350,000 in 1989 to a
target of 100,000 by 1994. Within western European states, as in the USA, the
promise of a 'peace dividend' pushed governments towards uncoordinated
cuts in defence spending, forcing defence ministries towards closer regional
integration as a means of both reducing costs and saving commitments from
their finance ministers.

Foreign ministers were also preoccupied by the fraught atmosphere of
US–EC negotiations (and intra-EC differences) in the final stages of the
Uruguay Round (see Ch. 12). Tensions in the Baltic states in January 1991
between Soviet forces and nationalists overhung the initial stages of IGC
negotiations on the CFSP dossier. When the Yugoslav crisis broke in June
1991, many of the most sensitive issues remained unresolved. Ministers
assembled to discuss the principles of future common policy, but found
themselves disagreeing over immediate and pressing problems. Their attempt
to establish a cease-fire in Croatia quickly moved beyond the traditional
instruments of EPC, to the deployment of EC peace monitors and then reluctantly to the dispatch of peacekeeping forces.[6] The Moscow *putsch* of
19 August and the progressive disintegration of the Soviet Union from then
until the declarations of independence in its constituent states in December
1991 accompanied the final stages of the IGC negotiations.

A busy Community agenda for political leaders and national officials, in
which immediate problems displace long-term strategy, is so routine as to be
unremarkable. The scale of the overload in foreign policy in the turbulent circumstances of 1991, was however, exceptional by any standard. Foreign ministers and their representatives were caught up in negotiating the terms under
which they might act together, while under acute external pressures to take
common action in a rapidly changing international environment. This
increased the incentives to find a basis for agreement within the time-scale
agreed: to settle the policy framework so that they could turn their attention
more fully to policy content.[7]

The IGC negotiations

In early discussions a number of metaphors were introduced to describe
the appropriate structure to bring together EPC and the Community: a tree
with many branches, a Greek temple with many pillars. The Luxemburg
Presidency circulated a questionnaire in January to other governments,

drawing on their responses in drafting a 'non-paper', circulated in April. This sketched out a 'pillar' model as the basis for a compromise with French and British (and Danish and Portuguese) opposition to formal integration. It attracted strong criticism from the Commission, as guardian both of the treaties and of its own institutional interests. Those in favour of an integrated framework talked of the need for a 'global' approach to foreign policy, from EC external relations to common defence. The French and British argued, against this, that it was illusory to imagine that an effective foreign policy, which included the 'hard' issues of security and defence, could be built upon the weak legitimacy of the Community; it could rest only on the commitment of governments representing states. The formal paper produced by the Luxemburg Presidency in June 1991 incorporated the comments and criticisms offered by other players over the previous eight weeks, and formed the basis for the final Maastricht agreement (Cloos *et al.* 1994; Buchan 1993).

The Dutch Presidency, which took office in July, was more *communautaire*; its ministerial negotiator, Piet Dankert, in particular opposed British and French emphasis on a pillared structure. Exploiting the absence of a clear consensus (and of his own foreign minister), Dankert produced a new paper in September, proposing an integrated structure which would bring policy-making and implementation of foreign and security policy within the Treaty of Rome institutions. After nine months of negotiations, with only ten weeks to go before their intended conclusion, the radical nature of these proposals aroused general dismay, and apart from Belgium, no other government was sufficiently committed to the integrity of the Community institutions. It was quickly agreed, to the Dutch Presidency's embarrassment, to return to the Luxemburg 'pillared' text (Pryce 1994; Wester 1992).

Three other issues were caught up in the negotiations over institutional structure: the question of consensus versus majority voting as a decision rule; the role and status of the Commission and of the EPC secretariat; and the desirability of moving beyond declaratory diplomacy and voluntary implementation to a commitment to follow up decisions with 'joint actions'. Behind all these lay different assumptions about whether the pillar structure was intended as a long-term alternative to the 'Community method' or as a temporary expedient, from which *passerelles* should be provided for the progressive transfer of functions and power to the Community proper.

Negotiations on the appropriate link between EC and WEU proceeded in parallel, focusing on similar issues of institutional coherence, decision rules, and authority over implementation. The SEA had brought EPC within the 'single' framework of the European Council, though not within the Rome Treaty framework; WEU had so far remained entirely autonomous. The opening question was therefore whether this framework of foreign and defence ministries should be directly subordinated to the European Council, thus also shifting defence ministers' orientation from Nato to the framework of the Twelve. On this the French and Dutch governments exchanged positions, with the Dutch (and the British) visualizing WEU as a permanent 'bridge' linking the EU and Nato, and the French (supported by the Belgian, Italian, and Spanish delegations) preferring the image of a 'ferry' which would gradually transfer defence functions from Nato to the Union. The French, supported ambivalently by the Germans, wanted WEU to acquire an operational capability with the right to operate within, as well as outside, the Nato area; the British, supported by the Dutch, wanted it to remain primarily an institutional forum for discussion among European defence and foreign ministers.

France's long-standing resistance to Nato's integrated military framework (from which President de Gaulle had withdrawn in 1966) raised particular problems; extensive military cooperation among other EU members had developed since the 1960s within the Nato framework (W. Wallace 1989). Franco-British competition for leadership and command in European defence, as allied structures were adjusted to fit post-cold war needs, persisted. The Franco-German commitment was to maintaining the initiative of their defence dialogue and of the Franco-German brigade to which it had led (raising the question of whether such joint forces were committed to Nato). The two pressures pushed the French government into modifying its position (Richardson 1994: 150). The underlying division was over whether a WEU integrated into the EU would develop into an alternative defence organization to Nato, or whether the object was to construct a more effective European pillar of the Atlantic alliance (Delors 1991).

A WEU ministerial meeting at Vianden preceded the Luxemburg European Council at the end of June 1991 (where the British prime minister also attempted to raise the issue of a potential conflict in Yugoslavia, a subject about which only his Greek colleague appeared actively concerned). Britain and Italy developed from consultations that had begun at Vianden a joint proposal, launched on 4 October, one day before an informal 'Gymnich' foreign ministers' meeting. This envisaged WEU as the European pillar of Nato, with a WEU Rapid Reaction Force based on 'double-hatted' Nato and national contributions.[8] Determined not to lose the initiative, the French (to Dutch presidential fury) issued an invitation to other governments to meet in Paris to discuss the further integration of European defence. They followed this ten days later with a Franco-German proposal (rapidly prepared by their foreign ministries without consultation with their defence colleagues), proposing an organic link between WEU and EU. Its ultimate aim was to assimilate WEU into the Union, transforming the Franco-German Brigade into the Eurocorps as the basis for an integrated European military structure (Menon *et al.* 1992; Forster 1994).[9]

But it was the Nato Rome summit of 7–8 November which built the basis for a compromise, after some sharp exchanges between French and American leaders, from President Bush down. The new Nato 'Strategic Concept' which it agreed approved the development of European multinational forces, but also reaffirmed the primacy of Nato as the forum for defence cooperation. Under intense pressure to reconcile opposing British and French positions in the four remaining weeks, with divergent signals still coming from the foreign and defence ministries in Bonn, heads of government arrived in Maastricht to find square brackets and alternative drafts scattered throughout the CFSP text.

The Maastricht compromise

The deliberate ambiguities of the final text thus reflect distracted political engagement and inattention to detail as well as continuing symbolic and substantive differences. Heads of government at Maastricht were occupied for much of their time with other politically sensitive chapters, leaving to foreign ministers and political directors the task of negotiating mutually acceptable language. The outcome satisfied all participants, partly because it represented real progress towards a common view, and partly because it left many major questions open. Policy initiative, representation, and implementation were explicitly reserved to the rotating Council presidency, 'assisted if need be by the previous and next member states to hold the Presidency'. The commitment to intergovernmental pillars was counterbalanced by assurances that the Commission would be 'fully associated' and 'the views of the European Parliament . . . duly taken into consideration', though the European Court of Justice is entirely excluded (Articles J.5, J.9, J.7, TEU; Fink-Hoojer 1994).

The WEU is described as 'an integral part of the development of the Union', but the Council is entitled only to 'request' (not instruct) it to carry through the defence implications of EU decisions (Article J.4.2). Complex language allows for, but does not compel, joint actions in pursuit of agreed common aims, and suggests that the member states will move on from formulation of a common security policy to 'the eventual framing of a common defence policy, which might in time lead to a common defence' (Articles J.3, J.4.1). An attached declaration created a WEU military operational planning 'cell', and provided that its civilian secretariat would move from London to Brussels. A long-standing dispute between the British and the French over

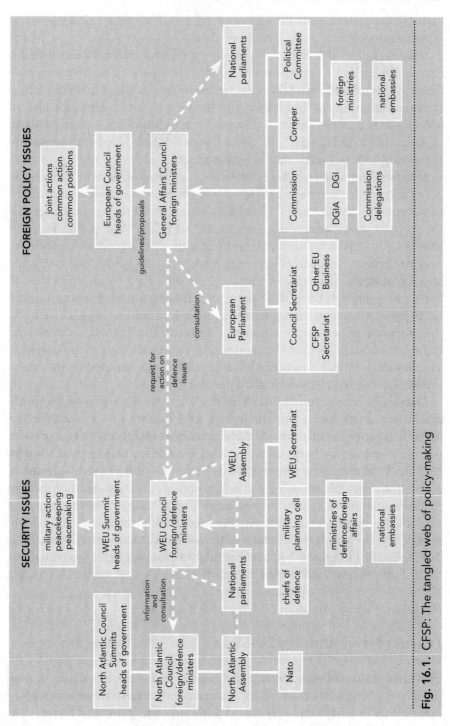

Fig. 16.1. CFSP: The tangled web of policy-making

further enlargement of WEU was resolved, in effect, by the Greek govern-ment's declaration in early December that it would veto the entire Treaty unless it was allowed to join. This forced negotiators to offer associated status also to Turkey and to other European Nato members. British resistance to funding expenditure under this Title (primarily to cover its expanding secre-tariat and meetings in Brussels) out of the Community budget, for fear of allowing the European Parliament to gain leverage over CFSP, was recognized in Article J.11, which left it to the Council to decide later whether or not a sep-arate budget for this pillar should be introduced.

The most remarkable aspect of the whole CFSP negotiations in 1990–1 was how successfully they were contained within the network of foreign min-istries established through EPC, and how little attention was paid to them by the press, by politicians outside government, or by the wider public. Defence ministries in Paris and Bonn were at times entirely excluded from consulta-tion on Franco-German policy initiatives, a factor which explained the absence of detail in successive French proposals. Nationalist politicians in Paris and London trained their sights on other institutional issues and on social policy, the major stumbling-block of the summit, leaving unchallenged this slippage of sovereign autonomy towards constrained cooperation. Indirectly, debates over political and military commitments to the develop-ing conflict in former Yugoslavia, in different capitals, touched on the issues raised; but they were not confronted openly.

One major innovation of Title V of the TEU was the provision for 'joint action in matters covered by the foreign and security policy', allowing for decision by qualified majority once 'the principle of joint action' on a specific issue had been agreed (Article J.3.1). A year-long debate on what issues might be considered suitable for joint actions, however, failed to reach agreement on an initial list: the so-called Asolo list, named (like Gymnich, Schengen, Ioannina, and other items of Eurojargon) after the site of a meeting, in this case the informal foreign ministers' meeting convened by the Italian Presidency. Up against the deadline of the Maastricht European Council, the Twelve decided to leave the issues that might be considered for joint actions to the Lisbon European Council in June 1992.

Learning by doing: 1992–1994

The same pressures of external events which had complicated the task of negotiating the Maastricht package pushed member governments to imple-ment much of what had been agreed without waiting for ratification. The most rapid progress was made in the defence field, as both Nato and WEU responded to the intractable problems posed by the Yugoslav conflict and by divergent American and European views of the appropriate response. The excellent personal relations between Nato's German Secretary-General, Manfred Wörner, and his Dutch WEU counterpart, Willem van Eekelen, moderated the potential rivalry between the two organizations; the two spelt out in a series of speeches how they thought their respective roles should develop. The British and French defence ministers modified their respective governments' positions on the links between Nato joint forces and the Eurocorps.

The June 1992 WEU ministerial meeting, in Bonn, registered in its 'Petersberg Declaration' the useful further progress made in the previous six months (WEU 1992). The WEU Council and Secretariat, with its planning cell of forty military officers, were moving to Brussels. The Franco-German Eurocorps, which was now to be joined by double-hatted Spanish and Belgian contingents, was in future to be 'assigned' to WEU alongside other double-hatted 'European' forces in Nato 'made available' for possible WEU use (Foster 1992). The plethora of European defence agencies was also reduced and a West European Armaments Agency created from IEPG and Eurogroup. But the question of a common procurement policy, which logistic considerations and interoperability suggested was essential to make multi-national units effective, was left open.

The underlying tension between Atlantic and Community frameworks for politico-military integration nevertheless remained, made more acute by confusion both in Washington and in most European capitals about the appropriate response to instability in eastern Europe and the Mediterranean and about appropriate divisions of responsibility among Nato and EU members. The evolution of the Yugoslav crisis was a painful learning process for governments on both sides of the Atlantic. West Europeans had instinctively looked to the USA to provide leadership, while the US Administration had firmly signalled that this was a regional issue for which the west Europeans should take responsibility. Despite the permissive conclusions of the Rome Nato summit, both the US State Department and the Pentagon pressed for an enlarged role for Nato in east–west security. They regarded WEU's creation of a consultative forum with the foreign and defence ministers of eight central and east European states as a competitor to Nato's recently created Advisory Council for Cooperation (NACC), which also included Russia and other CIS states.[10]

Confusion and competition were most evident in responses to Yugoslavia. WEU lacked the command and control needed to mount the complex intervention needed in Bosnia; the UN Headquarters there was therefore constituted out of Nato personnel, with French but without American or German officers. Franco-American institutional rivalry led to the despatch of two naval forces to the Adriatic, one under Nato and one under WEU, each commanded by Italian admirals. Aerial surveillance was provided by Nato's multinational-crewed AWACs, provoking an agonized debate within Germany over whether the German air force personnel among their crews could be permitted to overfly Bosnia. The German foreign minister, who was declaring his deep commitment to common foreign policy in the final stages of the IGC, was at the same time threatening to act unilaterally in recognizing Croatia, in response to domestic pressure.

Americans were sharply critical of hesitant and incoherent west European policies in former Yugoslavia; but it was not evident that any clearer approach could have been hammered out within Nato. Manœuvring between three (sometimes four) local protagonists, with American and Russian sensitivities, Islamic pressures, and UN politics in the background, made consistency hard to maintain. The French and the British provided the largest single contingents on the ground; the Spanish and Dutch also contributed substantial contingents. Five of the other eleven member states had troops in Bosnia or Croatia in early 1995, including Danes with Swedes, Norwegians, and Finns in a joint Nordic battalion; ten out of fifteen had ships maintaining the naval blockade in the Adriatic. French attitudes both to Nato and to Britain shifted further under the experience of cooperation with British forces in the field and closer appreciation of the utility of Nato military assets, now partly under French command (La Serre and Wallace 1995). An active, though confidential, Franco-British defence dialogue was under way by the end of 1993, and was publicly announced in November 1994 by the two foreign ministers with the setting up of a joint air wing. The EC/EPC/CFSP machinery was actively employed in the Yugoslav crisis, sponsoring initiatives, nominating mediators, and supervising the imposition of economic sanctions. Its ineffectiveness in resolving the conflict reflected both the inherent weaknesses built in to its mechanisms and the peculiar intractability of this post-socialist ethnic conflict (van Eekelen 1993; Gnesotto 1994).

On the three central issues of European political cooperation—Atlantic relations, European security, and the Middle East—the transition to CFSP appeared still to leave the EU on the margins. President Clinton's first year in office was marked by transatlantic strains over a range of economic and political issues, but it remained for Washington again to propose a reformulation of Atlantic politico-military cooperation at the Brussels Nato summit of January 1994. The central role played by the USA (with diplomatic support from Norway) in the Arab–Israeli dialogue, on an issue that had preoccupied the EC for over twenty years, underlined western Europe's secondary role.

Management of relations with the former socialist states of central and eastern Europe remained disjointed (see Ch. 14). Hard bargaining over the protection of entrenched agricultural and industrial interests within the EU undermined ambitious rhetoric about long-term transformation. It held member governments back from offering any timetable for future membership. As a group the member governments of the EU and their common institutions were providing over half of Western public and private financial flows to eastern Europe; but relations with Russia and Ukraine and approaches to Nato enlargement towards Poland continued to be driven primarily by American policy-makers.

The institutions and the personnel involved in CFSP continued to expand, in Brussels and elsewhere. The EPC Secretariat now became an autonomous unit within the Council Secretariat. The argument continued over a separate versus a common budget and served as the battleground between a British government which wished the pillar structure to endure and others who saw it as a transitional measure.

For the European Commission, external relations and foreign policy had now become one of its most important fields of operation. The rapid expansion of Community activities in central and eastern Europe and in the former Soviet Union necessitated a rapid build-up of Commission representations in those states. With over one hundred missions, it had a more dense network than many member states, all under tight financial constraint to reduce diplomatic costs. The Commission's prestigious foreign policy portfolios were fiercely fought over by ambitious commissioners. In the allocation for the new Commission in January 1992 the Commission President created a new Directorate General out of the EPC directorate within the Secretariat General: DG IA External Political Relations, alongside DG I for External Economic Relations. DG I reported to Leon Brittan, Commission Vice-President and former British Secretary of State for Trade and Industry; DG IA was allocated to Hans van den Broek, former Dutch foreign minister. This went against the logic of the Commission's earlier arguments for an integrated approach to economic and political issues, institutionalizing competition among Commissioners and officials for control of dossiers.

Conclusion: the unstable compromise?

'We now have four years to demonstrate that the intergovernmental model can work', the British foreign secretary declared in February 1992, looking ahead to the review already conceded in the 1996 IGC (Hurd 1992). 'The structure was never intended to work', a senior Commission official remarked more sceptically two years later. The compromise struck at Maastricht between the integrationist rhetoric of governments with passive foreign policies and the insistence of the EU's two self-conscious 'powers' on retaining their freedom of action in foreign policy and defence was inherently unstable.

Uncertainty over the extent of continuing US involvement in west European security and foreign policy made for greater instability, compounded by the determination of American policy-makers to assert a continuing leadership role in spite of radically reduced financial and military commitments. The absence of any serious engagement of public opinion, in any member state, left the half-commitments made in Title V without the domestic foundations needed for the successful conduct of common foreign policy. It was characteristic of the gap between appearance and reality that Franco-German military cooperation was vested in public symbolism, culminating in the 14 July 1995 parade when German troops (in the Eurocorps) marched down the Champs Elysée. The Franco-British defence dialogue in contrast continued largely in secret for fear of exciting nationalist opposition. France and Germany were nevertheless almost incapable of jointly deploying troops outside the two countries concerned, while the French and British were building on common experience in the field.

The limited procedural innovations of the TEU marked a further stage in the formalization of the informal consultations with which EPC began, and in the erosion of the Gaullist divide between political foreign policy and economic external relations. The Commission was now present in all discussions, with the right to table proposals alongside others (Nuttall 1994). Secretariats and Council meetings had now come together. Majority voting was at least on the agenda, with further EU enlargement to fifteen in 1995 helping to concentrate ministerial minds. The WEU had been clearly recognized as a part of the broader European Union, with its closer integration (ahead of the fifty-year review clause in the 1948 Brussels Treaty) into the EU a major agenda item for the 1996 IGC.

There was, however, a difference of time-scale between this slow process of adjustment to the pooling of sovereignty and the pressure of external developments. Mediterranean security problems and the tensions and conflicts involved in eastern Europe's transition from authoritarian socialism could not be put off until after the next set of national elections, nor managed discreetly without national parliaments or publics being drawn in. The transformation of western Europe's international environment since the 1991 IGC was first proposed had been far more radical than the modest improvements in policy-making capabilities made in the TEU.

Until 1989 American political leadership and the Atlantic security framework had shielded west European governments from hard choices in foreign and security policy. They had been able to avoid explicitly addressing the issue of leadership within western Europe. Franco-British rivalry continued to shape the agenda for some time after German reunification, which returned a German state, still inhibited about active foreign policy, to the position of regional hegemon which its predecessor had occupied after 1871. The balance of interests and perceptions between large and small states, between northern and Mediterranean, between those which still conceived of an active national role in foreign policy and those which had accepted a position of limited international influence, all remained undefined.

Though each of the fifteen member states has formally accepted a commitment to CFSP, none has spelt out the meaning or implications of such a commitment. In the absence of an immediate and evident external threat it is unlikely that they would carry the support of their parliaments or publics if they were actively to move towards one. Pusillanimous in presenting unwelcome choices to their electors, preoccupied with immediate domestic and foreign policy problems, their governments therefore see little alternative to moving forward from one ambiguous compromise to another. The twenty-five years since the launch of EPC have moved west European foreign policy-making from independent diplomacy to information-sharing, limited common analysis, extensive interpenetration of governments, and the beginnings of integrated west European defence. The policy networks are in place, constituting a powerful interest in maintaining the momentum. The symbol of sovereignty, however, hampers movement; and the ambivalence of political leaders and publics about the desirability of the declared objective slows it further. Short of acute external crisis to mobilize and transform public sentiment, the 1996 IGC will start from the same contradictions.

Notes

1. British exchanges with the Auswärtiges Amt had by 1995 included German attachments to the Foreign Secretary's private office and the FCO Planning Staff. In May 1995 three German diplomats, one French, and one Danish were in posts in the FCO; two British diplomats were working in the Quai d'Orsay, two in the Auswärtiges Amt—and one in the US State Department. A larger number of German trainees came for shorter attachments to the FCO each summer. Co-location and sharing of embassy facilities between the British and Germans in some of the new states of the former Soviet Union had led in addition to informal exchanges and attachments 'in the field'. The Franco-German proposal to set up a fully joint embassy in Ulan Bator, in 1989–90, however, had to be

referred to the Conseil d'État, which declared it incompatible with the French constitution for French diplomats to serve under German ambassadors: the sovereignty block at its most immovable.

2. Douglas Hurd began a public lecture in Paris in 1990 by remarking that this was the tenth time that he had met his French opposite number in two weeks, over five different countries.

3. For a flavour of the debate see Howorth (1990: 126–30), and the comments made by the British Trade and Industry Minister Nicholas Ridley (Lawson 1990).

4. Some commentators argue that a back-room deal took place between France and Germany linking EMU with the 'two plus four' talks (Valence 1990). For a similar argument see Tsakaloyannis (1991).

5. Representatives of some EC governments remarked in particular that the British foreign secretary, Douglas Hurd, struck a much more Atlanticist line in Nato meetings and a more Europeanist line in EC Councils.

6. At one ministerial meeting in September 1991 the German foreign minister, Hans Dietrich Genscher, is said to have made a passionate speech insisting that troops be sent to Yugoslavia; to which his British counterpart is reported to have replied 'You mean *you* want to send *our* troops'.

7. As *Agence Europe* noted on 11 Oct. 1991, there was 'widespread acceptance [that] no agreement would have been worse than an imperfect one'. For the imposed costs of bargaining on participants see Stein (1993).

8. The practice of informal meetings of foreign ministers, usually over a weekend in a pleasant country house, without a detailed agenda and without officials sitting in, was initiated by Hans Dietrich Genscher during the German Presidency of 1974, with an invitation to his colleagues to join him at the former archbishop's palace of Gymnich. Regularized under successive presidencies, it brought foreign ministers together for extensive discussions in pleasant surroundings every six months, acting to socialize new foreign ministers into the EPC 'club', with its many long-serving members. In 1991 Genscher himself was (by some distance) its longest-serving member.

9. Only the Spanish and German governments accepted the French invitation; it was to other governments a serious breach of the procedural conventions of the IGC.

10. This left some confusion as to the configuration in which the WEU Council should meet, in effect leaving it as neither the defence arm of the Union nor a European pillar of Nato. WEU could meet at ten (the full WEU members); at fifteen (the ten plus the two EU observers and the three European Nato associates); at twenty-three (plus the eight central and east European associates). Numbers of associates and observers grew to a total of twenty-seven for full meetings with the division of Czechoslovakia and the EU accession of Sweden, Finland, and Austria.

Further Reading

For overviews of EPC see Nuttall (1992) and de Schoutheete (1986). For the development of both institutions and policy in the 1980s see Pijpers *et al.* (1988). On the Maastricht negotiations see de Schoutheete (1993). On the emergence of CFSP see Nuttall and Edwards (1994) and Soetendorp (1994).

Nuttall, S. J. (1992), *European Political Co-operation* (Oxford: Clarendon Press).

—— and Edwards, G. (1994), 'Common Foreign and Security Policy', in A. Duff, J. Pinder, and R. Pryce (eds.), *Maastricht and Beyond* (London; Routledge, 1994), 84–103.

Pijpers, A., Regelsberger, E., and Wessels, W. (1988) (eds.), *European Political Cooperation in the 1980s: A Common Foreign Policy for Western Europe?* (Dordrecht: Nijhoff).

Schoutheete, P. de (1986), *La Coopération politique européenne*, 2nd edn. (Brussels: Labor).

—— (1993), 'Reflexions sur le Traité de Maastricht', *Annales de Droit de Louvain*, 1: 73–90.

Soetendorp, B. (1990), 'The Evolution of the EC/EU as a Single Foreign Policy Actor', in W. Carlsnaes and S. Smith (eds.), *European Foreign Policy* (London: Sage, 1990), 103–19.

PART THREE: CONCLUSIONS

CHAPTER SEVENTEEN

GOVERNMENT WITHOUT STATEHOOD: THE UNSTABLE EQUILIBRIUM

William Wallace

The EU after forty years is in some ways a remarkably stable system, although facing some immense challenges. Its multi-level processes, well displayed in the handling of regulatory and distributive issues, are best understood as a loose federation. EU policy-making in the 1980s and 1990s has, however, had to grapple increasingly openly with redistributive and constitutional issues, as the policy agenda has touched on issues central to national sovereignty. Successive enlargements and the transformation of the European order since 1989 threaten to destabilize a system built to handle second-order issues among a small group of states within a stable geopolitical context. Globalizing trends have weakened the links between national identities and state capacities. The attempt to build an intermediate level of European governance has not gained sufficient popular legitimacy to support all the demands now placed upon it.

Continuity and discontinuity

Forecasting the progress of integration is . . . a risky business: for that progress is the net outcome of opposing forces. Haas's 'spillover' effect no doubt exists. . . . But counterforces have also been at work and have in fact been aroused by the very success of integration, so that the net outcome or balance is continually in doubt (Hirschman 1981: 270).[1]

This volume has set out to illustrate both the complexity and the diversity of European policy-making in the late 1980s and early 1990s. The cases chosen have ranged from high politics to low politics: from issues like defence, currency, and border controls, which touch directly on national sovereignty, to issues like the regulation of competition or trade in bananas, which in principle are matters of day-to-day politics and administration. Several of the cases reflect the remarkable expansion of the European Union (EU) agenda since the last edition of this volume was sent to the press in 1982. Environmental concerns were already disrupting national politics in many member states, but were only beginning to impinge on the Community level. Cooperation among police and intelligence services, coastguards and border-guards, and ministries of interior and justice, was still contained within discreet intergovernmental committees. Security and defence were still taboo topics within the then European Community (EC), and were only indirectly covered within European Political Cooperation (EPC).

The cases presented here do not support the assertion that the processes of European integration stagnated throughout the 1970s, to revive again in the mid-1980s. As the case-studies make plain, the coordination of national policies and development of common European policies have emerged over extended periods from successive partial bargains, compromises, incremental drift, and responses to external demands over extended periods. Sometimes policy development has resulted from deliberate political initiative, sometimes from the unintended consequences of previous decisions working their way through multilateral policy-making within a complex institutional framework. There were no 'dark ages' for European integration between the glad confident morning of the EEC Commission's first surge towards a common market, blotted out by the shadow of President de Gaulle, and the renaissance represented—for some—by the SEA and the 1992 programme (Hoffmann 1990).

We share Hirschman's perception of integration as a process which emerges out of the tension between opposing forces. The functional logics of economic and technological transformation have vied with the political logics of statehood, sovereignty, national identity, and political accountability (Milward 1992; W. Wallace 1994b). The pendulum has swung repeatedly, as

argued in Chapter 1, between the search for common policies and the assertion of national priorities. The inadequacy of national action has been
accepted, even though the legitimacy of European decision-making has been
repeatedly questioned.

Restatements of functional logic have seen in this a dynamic process, in
which 'stagnation is a regular stage in the integration process', each forward
step towards common policies provoking resistance and countermeasures at
the national level (Dorbey 1995: 281). These in their turn give rise to
demands for further common policies to resolve the tensions created by competitive national policies. The case-studies provide extensive evidence of such
underlying tensions within the EU and its member states, but also of wider
pressures and counter-pressures from external developments, from technological and economic innovation, and from social and cultural change. These
undermine the foundations of existing (national and European) policies and
push policy-makers towards new partial bargains (W. Wallace 1990c,
Keohane and Hoffmann 1990).

It remains, however, open to question whether these in-built tensions can
be depicted in a historicist fashion, as a dialectical process following a wind-
ing path from national sovereignty to European federation. 'Si la
Communauté Européenne réussissait à survivre et même à progresser, c'était
au prix d'une ambiguité persistente sur son objectif ultime' (Giscard
d'Estaing 1995). West European integration is marked by underlying ambiguities, not only about the relationship between the institutions created at the
European level and the constitutional and political structures of its member
states but also about its geographical extent and future extension, about its
economic and political priorities, and about the compatibility of regional
integration with Atlantic and global integration (W. Wallace 1994a).

The image presented in Chapter 1 was of a pendulum, held in place by the
institutional structures of the EU, swinging between the magnetic fields of
national sovereignty and of common action, the one with its comforting
autonomy and identity and the other offering the benefits of integration, but
at a price. The image presented by the neofunctionalist theorists of the 1960s
was of a continuum, in which common policies in limited areas attracted economic interests and spilled over into other areas, accumulating functions and
attracting new supporters until the focus for European government had
clearly moved beyond the nation state. The image presented by those who
have returned to a functional analysis in recent years is of a broken continuum, the path from national autonomy to European integration punctuated
by the renegotiation of partial bargains or by efforts by national governments
to resist the functional logic of further integration (Keohane and Hoffmann
1990; Dorbey 1995). Those who have focused on long-term trends, on the

modernization of European economies and societies, and on the growth of cross-border interactions also imply a continuum. The 'great transformation', the industrial, technological, and socio-economic revolution which swept across Europe from west to east from the early 19th century on, created the context within which modern centralized nation states could develop, and generated the resources which they required. Such long-term trends now, however, operate to weaken the autonomy and centrality of the nation state, pushing reluctant governments to cooperate within a wider framework (Polanyi 1944; Puchala 1971; W. Wallace 1990a, 1990c, 1994b).

The reader should, however, consider the appropriateness, in the mid-1990s, of an alternative image of discontinuity. The crucial test of neofunctionalist incrementalism comes when spillover reaches issue areas which are central to statehood and national identity. At that point incrementalism is no longer enough; what is required for further development of common policies is a transformation of institutional structures and patterns of loyalty. If the political community, promoted by common policies and increased interactions, will not bear such weight, then further development is blocked; popular resistance or resentment may even unravel common policies established earlier. The 'supranational compromise' struck by national negotiators forty years ago had transferred substantial powers to the new European institutions, but in strictly limited fields, with the most difficult and divisive issues of national sovereignty and the future role of national institutions deferred until a later stage (Lindberg and Scheingold 1970).

The institutions were designed to stress administration and regulation, to minimize the visibility of the political choices at stake, and to operate on the basis of a permissive popular consensus rather than of active or informed participation (W. Wallace and Smith 1995). The case-studies make clear that it has proved harder and harder to maintain this compromise, as the issue areas brought within the European institutions have widened, and as the political and symbolic challenges to the autonomy of the nation state have become more and more difficult to disguise. Member governments have been driven to resolve the contradictions by successive package deals and treaty adjustments, repeatedly attempting to strike new compromises between sovereignty and integration and between effective policy-making and political accountability.

A second dimension of discontinuity derives from geopolitical change. West European integration, as several earlier chapters have argued, was firmly embedded in the stabilizing framework of the Western alliance. Its eastern limits were set by the Iron Curtain, its potential enlargement limited by the reluctance of democratic but neutral European states to join such an evidently 'western' institution. Successive enlargements in the 1970s and 1980s had necessitated some adjustments in decision rules and policy packages, but had not required fundamental changes. The transformation of the international context since 1989 is of a different quality from the marginal changes of the previous thirty years. It raises the question of whether the EU is capable of radical change, in terms of internal and external policies, of membership, and of institutional structure; or, if not, whether this loose structure of government without statehood will buckle under the strains.

Most radical of all in transforming the context within which west European integration and its patterns of policy-making have developed has been the geopolitical revolution. The EC had its origins in the creation of a multilateral framework within which West Germany was to be allowed to rebuild its economy, to recover its sovereignty, and to reconstruct its armed forces. It was established under direct American sponsorship and pressure, within the wider framework of US security protection and political and economic hegemony (Duchêne 1994). It was, in many ways, itself a product of the cold war: a framework established to provide a political and economic anchor for fragile democracies emerging from occupation and authoritarian rule. It served to reinforce German and Italian democratic stability in the 1950s, as it served for Greece, Spain, and Portugal in the 1980s.

The end of the cold war and the unification of Germany altered the internal balance of the EC, placed the US–European relationship in question, and threw central and eastern Europe with its acute political, economic, and security needs on to the west European agenda. Perhaps the greatest achievement of European integration over the past decade has been to provide a framework within which the unification of Germany could be accepted by its partners and neighbours. East German accession to the EU came automatically with German unification, imposing severe strains on the German economy (and therefore on the other economies of western Europe which revolved around Germany), but not causing a rupture in west European politics (Spence 1991). Germany's greater weight within the EC, and its lessened dependence on its western allies for security, were important background factors in the approaches of most other governments to the 1991 IGC. For France in particular, Germany's greater freedom of manœuvre made it all the more important to construct a political union within which Europe's potential hegemon could safely be contained.

Over the twelve years since the second edition of this volume—and the six years since the Berlin Wall came down—the context within which EU policies are made and managed has thus changed markedly. The policy agenda has widened. New issues have been pushed into the EU framework, sometimes by one or more member governments or by particular political leaders, sometimes on Commission initiative, sometimes by advocacy coalitions, sometimes by the need to formulate positions in international negotiations. Few issues have been pushed off the agenda. The inertia of established policies and of established policy communities has been compounded by the lack of consensus among member governments about the appropriate hierarchy of policy responsibilities to which the principle of subsidiarity should lead. Interaction among governments through the extended EU substructure of committees and working groups has intensified, reaching a plateau beyond which ministers and officials would lose their ability to balance multilateral engagements against continuing national tasks (Wessels 1990; 1992). Interaction among interest groups has also intensified, as EC deregulation and reregulation have entered new fields (Mazey and Richardson 1993b).

There have been two treaty revisions of the EC's constitutional framework, in the SEA and the TEU. The EU budget has grown slowly but steadily, as Chapter 3 showed, to reach some 1.28 per cent of EU GDP (2.5 per cent of EU public expenditure) in 1994. The institutions themselves have grown both more diverse and more complex, with two intergovernmental pillars alongside the Community framework, with intricate (and arcane) patterns of conciliation on different issues between the Council of Ministers and an increasingly influential European Parliament. There have been two rounds of enlargement, first to the south and then to the north, with the shadow of impending enlargement to the east (as Ch. 14 shows) hanging over the evolution of EU policy and over preparations for the 1996 Intergovernmental Conference (IGC), the third process of treaty revision in a decade.

The long-term implications of this transformation of the EU's internal balance, size, and external context were not directly addressed in the Maastricht IGC. The inertia of established assumptions and institutions carried member governments through to Maastricht without confronting the awkward choices posed. The shadow of the future nevertheless hung over many of the dossiers under negotiation, and in particular drove negotiations on foreign policy and defence. The 1996 IGC will find it difficult to avoid a more radical approach to policy and institutional reform in a transformed European environment. What conclusions can be drawn from the evidence of these fourteen case-studies, most of them stretching across the 1989–92 divide, about the stability or instability of EU policy-making processes and the institutions in which they are embedded? How far has the analysis offered in the final chapter of *Policy-making in the European Community* been outdated by internal and external developments over the subsequent twelve years (W. Wallace 1983a)?

Multi-level governance: the federal analogy

The conclusions of the previous edition, that the EC 'incontestably represents a new level of government', are reinforced by the case-studies in the present volume (W. Wallace 1983*a*: 406). A substantial academic literature has developed in the interim on multi-level governance (Marks 1993; Scharpf 1994*a*; Bulmer 1994*a*; J. Peterson 1995a). Drawing on a range of detailed studies, the common themes of this literature are: that institutions shape preferences and outcomes; that the complexity and contested character of policy-making on multiple levels makes for dispersed and disjointed decisions, and for incomplete implementation; and that national governments (as Chs. 3, 4, 7, and 8 in particular confirm) have struggled both to use the EC level to serve their own national objectives and to maintain control over inputs and outcomes, with varying success in different areas.[2]

A similar wealth of studies has supported the appropriateness of the federal analogy as a starting-point for understanding the institutions, politics and policy-making of the EU. The first generation of American students of west European integration started from the assumption—and the hope—that the institutions designed under American sponsorship in the years after World War II would lead to the creation of a United States of Europe on the American model. A few of those responsible for the development of institutionalized integration in its early years shared the same objective. Walter Hallstein, first president of the EEC Commission, who had studied the American constitution as a prisoner of war in the USA and had returned after the war to teach constitutional law at Georgetown University, entitled his study of the EEC *Der unvollendete Bundesstaat* (Hallstein 1969; Loth *et al.* 1995).

Judged in comparison with the post-New Deal American federal model, the institutional structure and political system of the EU clearly falls short of that highly developed federal model. Comparison with the 'Philadelphian system'—the looser federal structure under which the United States were governed from 1787 to 1860—provides a different perspective. The pre-Civil War USA, Daniel Deudney argues, 'had a government but was not a state': it was a union of states (a *Staatenbund*) rather than a integrated federal state (a *Bundesstaat*) (Deudney 1995: 207; Forsyth 1981). The domestic politics of its constituent states were diverse, its policing and most of its military forces remained under state control. Its internal balance, furthermore, was repeatedly disrupted by the process of enlargement and expansion (Poole and Rosenthal 1993).

Comparison with the Canadian model also offers insights into a multi-level system with diverse member provinces, within which politics revolves around the constitutional rules of the federation, and around provincial autonomy and identity, as much as around distributive and redistributive policies or patterns of regulation (Leslie 1991; Sbragia 1992a: 282–3; W. Wallace 1994a: 32–3). The German model offers further insights, with its intensive interpenetration—even fusion—of political and administrative élites from different levels in policy formulation and implementation, building *lourdeur*, or inertia, or joint decision-traps into the system (Scharpf 1988; Wessels 1992). The considerable contribution which the German constitutional and administrative tradition made to the development of the EC system was less immediately visible than the French; Monnet, who initiated the Community model, is better remembered than Hallstein, who organized the EEC Commission.[3]

Patterns of policy-making

The previous edition drew on Theodore Lowi's fourfold classification of policy-making patterns within the American federal system. It distinguishes between different clusters of EC policy-making: *constituent*, in which the rules and priorities of the system are themselves under negotiation, leading to a pattern of high-level bargaining among shifting political coalitions; *redistributive*, in which system-wide organizations and political coalitions clash over structural costs and benefits; *distributive*, in which sectoral organizations build policy networks with sponsoring bureaucracies and with sections of the political legislature and executive; and *regulatory*, characterized by disaggregated decisions, legal or quasi-legal processes, and specialized interests (Lowi 1972). The growing literature on EC policy-making since then has largely confirmed this approach, while offering additional insights into the characteristic styles of each of these patterns (J. Peterson 1995a).

Constituent issues—renegotiation of the treaty framework, redefinition of the overall balance of EC priorities or core goals—are negotiated among governments, with each government attempting to maintain a coherent approach (Moravcsik 1991). Redistributive issues also pitch governments against each other, attracting and holding the attention of their national publics as they bargain over broad gains and losses (H. Wallace 1983a). Distributive issues are the stuff of functional politics, of sectoral interests cooperating with national and European administrators, as classically represented by the common agricultural policy in the 1970s and early 1980s (Pearce 1983). Regulatory issues exhibit expert technical committees and epistemic communities, with a strong legal bias and often an active role for the ECJ (Dashwood 1983).

The categorization of issues among patterns of policy-making only partly follows rational principles. Questions of identity, or of ideology, or of taste, intervene, bringing in passion to clash with rational interests. Negotiations on monetary union, as Tsoukalis emphasizes in Chapter 12, raise both constituent and redistributive issues, bringing in heads of government to overrule sectional interests and to weigh up the broad national and European costs and benefits. They also arouse passions over national symbols—the Deutschmark as the symbol of German prosperity, the pound as a familiar aspect of British identity—which reinforce the sensitivity of monetary politics to national governments. As this book went to press the debate on monetary union was distracted by demands from within Germany to change the name agreed for the common currency: a nominalist debate in some ways reminiscent of medieval theology, but reflecting public attachment to the symbolism of the mark and widespread association within the German élite of the term 'ecu' with French inflationary tendencies.[4]

The passion with which Germans approached the issue of trade in bananas, reflected in newspaper coverage and in political attention in the Bundestag, raised the issue from the quiet backwaters of regulatory politics to the post-bags of ministers. It was rooted in the symbolism of a fruit unobtainable in Germany during and after the war, thereafter associated with the recovery of prosperity. When the Berlin Wall was demolished East Germans flocked to buy quantities of the bananas which socialism had denied them from the open markets of the west. Cooperation in combating crime, drugs, terrorism, and illegal immigration might in contrast be regarded as high politics, touching on the fundamental security and territorial integrity of the state. Yet the secretive professionalism which Monica den Boer describes in Chapter 15 has carefully contained most cooperation in this field within technical and official committees for much of the past twenty years. It broke out into higher politics when symbolic issues were engaged in some member states. Identity cards carry connotations of wartime occupation in the Netherlands, and of authoritarian government (and racial discrimination) in Britain, attracting widespread press, public, and political attention; agreements on cross-border arrest and the exchange of files on suspected criminals have passed almost unobserved.

Differences of taste arouse particular passions, transforming regulatory issues into questions of the redistribution—or redefinition—of values (Mint and Simeon 1982). Regulation of the common market in food and drink has proved a peculiarly sensitive area, over which technical considerations of hygiene and purity repeatedly clash with national idiosyncrasies and traditions, easily elevated into symbols of national identity: German beer versus Belgian or British; unpasteurized cheeses; acceptability of local varieties of apples. Hunting and animal welfare carry similar undertones of values and identity. The British government's strategy of repatriating policy competences from the EU institutions, after the Maastricht TEU, was swept aside by the popular (and media) support generated by animal-welfare campaigners who were determined to stop British calves being reared in Dutch veal crates for the French market. Prevailing British and French values were pitched against each other over French proposals to loosen the Wild Birds Directive, with British bird-lovers determined to enforce restrictions on French hunters.[5]

The distribution of issues among different patterns of policy-making is thus neither necessarily predictable nor stable. Policy entrepreneurship may succeed in redefining and repackaging issues to raise their political salience and saleability. Chapter 5 illustrates the successful repackaging of internal market regulation and deregulation into the 1992 programme, taking advantage of shifts in the focus of national policies to capture public imagination. On energy policy, as noted in Chapter 10, the deregulation of national utilities in the 1980s provided an opening for the Commission to redefine and to make politically more compelling an issue which had been central to west European integration in its early years (in the ECSC and Euratom) but had thereafter sunk to the technical level of official committees and Commission proposals for incremental regulatory change. Agricultural policy, carefully insulated from political challenge by the institutional separation of sectoral policy-making from General Affairs Councils and Coreper, was pushed up to heads-of-government level after 1990 both by the external pressures of the Uruguay Round and by the sharp rise in expenditure from the EC budget.

Fisheries policy, primarily distributive in the 1970s, became unavoidably redistributive when enlargement negotiations with fishing countries—such as Spain—threatened to alter the balance of advantage (Shackleton 1983). This developed into an increasingly bitter redistributive struggle in the early 1990s as declining stocks forced cut-backs in quotas. Parliaments and public opinion had become inflamed in Spain, Britain, France, and Ireland in 1994–5, and member governments fell out publicly over negotiations with Canada on reductions in Atlantic fishing. Even budgetary policy was contained during the 1970s (after the conclusion of the package deal on British entry) within a distributive framework which the majority of member governments, and the Commission, defended against British efforts to reopen the question of its redistributive effects. Resistance weakened from 1980 on in the face of a determined British position articulated by a vigorous head of government, until other governments partly conceded British demands in the redistributive Fontainebleau package of 1984. This set a precedent (as Ch. 3 demonstrates) which the Spanish successfully followed on entry, transforming the budget into an instrument of redistribution from rich north to developing south. David Allen argues in Chapter 8 that the increase in resources provided for the structural funds led national governments to attempt to claw back control from the alliance which had developed between the Commission and regional and local authorities. With larger sums to bargain over, national governments struggled to define the terms for negotiation (cohesion, solidarity) to suit their preferred outcome; the characteristic patterns of redistributive politics thus came into play at Council of Ministers meetings and at European Councils.

The structure of Community policy-making was designed from the outset to disaggregate issues wherever possible, to disguise broader political issues, to push decisions down from ministerial confrontation to official *engrenage* within the hierarchy of committees which formulated proposals for ministers to approve and the parallel hierarchy of committees which cooperated in their implementation (Wessels 1990 and 1992). The rhetoric of technocracy and rational administration reinforced this tendency to de-politicize issues. The members of the Council of Ministers, to whom contentious issues are referred, are actors at the same time within their own national political systems, coming to Community negotiations often preoccupied with domestic dossiers and short of time.

The common tendency in day-to-day national (and international) policy-making for the management of policy to be contained within sectoral policy communities has thus been repeated within the Community framework. Ring-fenced by the rules of compulsory budgetary expenditure, much agricultural policy was handled throughout the 1970s and early 1980s as distributive politics, despite British efforts to highlight the perversity of these distributive flows. In this one sector, as Elmar Rieger argues in Chapter 4, welfare functions and costs were successfully transferred from member states to the Community level. Some would argue that this has also been the case with EU expenditure on research and development (J. Peterson 1995a). The small size of the Community budget has, however, severely limited the possibilities for distributive politics, leading to a marked dependence on regulation as a preferred basis for common policy. This preference is reinforced by the shift of emphasis within member-state political systems from direct intervention to regulatory policies during the 1980s (see Ch. 5; Majone 1994b). The legalistic processes of regulatory policy provide a compartmentalized framework within which the expert and the interested may bargain, insulated for extended periods from broader political controversy.

'The stupendous growth of EU regulation' in recent years, which has placed regulatory policy 'at the core of EU policy making', partly reflects a recurrent coalition between the Commission as policy entrepreneur and industrial interests preferring to move regulation to the EU level (Majone 1994b: 98).

Basically, the interests have recognized the reality of a federal Europe. Whatever terms we may use to describe the EC—whether it be federal, neo-federal, would-be state, or whatever—the reality for interests is that much of the regulatory framework under which their own policy areas are governed is now affected by EC intervention. (Mazey and Richardson 1993b: 255)

But it partly reflects, as well, the preference of governments for an approach to joint policy which has minimal implications for public expenditure, within a European framework in which implementation and enforcement are relatively flexible. 'The serious implementation gap that exists in the European Community may make it easier for the member states, and their representatives in the Council, to accept Commission proposals which they have no serious intention of applying' (Majone 1994b: 86).

Stability and instability

A dispersed framework for shared government, without the extended apparatus of a state to provide information as the basis for intelligent policy formulation, with limited and indirect sources of revenue, and with only extremely weak authority to ensure implementation and enforce compliance, represents the weakest form of confederation. The obscurity of EU procedures, intensified by the SEA and the TEU, has in many ways suited the member states well, serving to maintain their role as intermediaries between the public and the EU institutions. The institutional and jurisdictional tensions which are built in to the structure, however, carry costs. The magnetic fields between which European governance oscillates are also charged by unresolved conflicts: over underlying objectives; over authority and legitimacy; over distinctive core national values; over the appropriate balances to be maintained between Germany and its neighbours; between rich states and poor; between large states and small.

A stable environment for policy allows stable policy communities to develop. But, as the case-studies have made clear, the environment for policy has become increasingly unstable during the past decade. The EU has served as a framework within which changing assumptions about the role of the state in the economy and in social policy have been tested, both between different member governments and between governments and domestic actors within different states, and between public and private actors across state boundaries. The shift of paradigm for economic policy from the Social-Democratic/Christian-Democratic/Keynesian consensus of the 1960s has not been uniform across the EU. Divergent assumptions about the boundaries between the public and the private sectors, and about the role of the state in providing welfare and employment, have therefore unsettled all the more policy-making on economic and social issues.

Changes in the external environment have also unbalanced the Community *acquis*, altered the priorities of national governments and of industrial and other interests, and first weakened and then destroyed the geopolitical framework within which the Community institutions were established. The EU now faces a succession of constituent and redistributive issues for which its institutions are poorly adapted: monetary union; common defence policy; enlargement to bring in a significant group of poorer countries; agricultural reform; environmentalist challenges to accepted ideas about economic growth and open markets; and the redefinition of citizenship.

This volume offers much evidence of the difficulties of coping with such broad issues within this loose political system—part framework for a continuous process of multilateral bargaining among state governments, part apex of a pattern of multi-level governance which has significantly weakened the centrality of the state. It remains, as we described it in the conclusions to the second edition, an incomplete political system: a 'quasi-state', without the coherent articulation of interests and political preferences characteristic of a well-developed polity (Streeck and Schmitter 1991: 159).[6]

Politics and policy-making in the European Community force one to rethink the role of government. At the Community level, governments bargain with other governments—and, significantly, with the Commission—rather than with business or labor. The role of political parties is problematic at best. . . . The political system they are gradually constructing, therefore, will be an original one. The insistence on the representation of territorially defined governments, in fact, will lead to a reconceptualization of federalism, representation, and the functions of government. (Sbragia 1992*a*: 290–1)

The question for policy-makers (as well as for students of policy-making) is whether such a pattern of governance can produce policy outcomes that are sufficiently innovatory to maintain their own stability within such an unstable environment. The development of west European integration was itself a response to the inadequacy of the nation-state model, seeking to strike a balance between the maintenance of the national framework for legitimacy and accountability and the construction of a wider framework for shared security, prosperity, and (it was hoped) political community. There is room for doubt as to whether that balance, poised as it has been for the past forty years between the European and the national level, will prove sufficient to satisfy either domestic publics or external *demandeurs* in the transformed European and global environment of the 1990s.

The declining centrality of the nation state

In such an intensive and diverse pattern of political and administrative interactions, it has been possible for member governments to hold to the principles of sovereignty and unity in their relations with each other only with respect to constituent decisions on changes of structure or policy framework. This was apparent to representatives of member governments from the outset. Paul-Henri Spaak, as foreign minister in the Belgian government-in-exile in the winter of 1941, had spelt out to his colleagues the limitations to any post-war re-establishment of state autonomy: 'What we shall have to combine is a certain reawakening of nationalism and an indispensable internationalism' (quoted in Milward 1992: 320).

Negotiations on the treaties themselves and on their revision have brought heads of government together, bargaining on the basis of perceived national (and party or coalition) interests to reach a multilateral compromise among governments. But for the regular business of government, now more and more caught up within a multilateral framework of rules and negotiations, the lowering of state boundaries has severely weakened the ability of national governments to define and pursue coherent 'national interests' in their relations with other states. Transborder economic and political networks have intensified, expert communities, cooperation between police forces and local and regional authorities in adjacent countries, even transgovernmental coalitions, have developed. It is possible for national governments to hold the gate between domestic and international politics only for a shrinking number of policy areas. More and more, public servants and private actors have developed their own paths across national boundaries; goods, services, money, and people flow across such boundaries less and less affected by state action or control. Governments thus have to decide which issues they choose to define as key to the preservation of sovereignty, autonomy, or national idiosyncrasy, conscious of the political costs of defining too many issues in the symbolic terminology of high politics. Different governments, with different traditions of statehood and different myths of national identity, choose different issues, further complicating the management of Europe's multilateral and multi-level government.

The reader should bear in mind that the processes of European integration were not imposed upon self-confident states, secure within their boundaries and in control of their economies. American international-relations scholars, who approach the EU as a particularly intensive example of an international regime, bring with them assumptions about state capacity drawn from the exceptional experience of their own superpower state. The ECSC was designed five years after a war in which all the states of continental Europe had collapsed, to be reconstituted after occupation under American (or Soviet) tutelage. The implications of the bargain struck were seen to be different within still-occupied Germany from within weakly imperial France, different again within politically divided Italy from within the consociational Netherlands. But for all the original signatories there was a trade-off between the acceptance of some limitations on sovereignty and the provision (through the pursuit of shared prosperity and security) of the economic and fiscal means with which to rebuild the damaged structures of the democratic industrial and welfare state (Milward 1992). The reconstruction of the (west) European state around a Social-Democratic/Christian-Democratic model, with rising public expenditure providing higher standards of welfare, and with steady economic growth providing higher standards of living, went side by side with the construction of (west) European institutions to manage the interdependence which was seen as the necessary condition for continuing growth (W. Wallace 1994b).

'Arguably the most significant change in Europe, justifying appelations such as "transformation" and "metamorphosis",' since the establishment of the Communities on the basis of this limited bargain, 'concerns the evolving relationship between the Community and its member states' (Weiler 1991: 2406). The character of this transformation is disputed, among governments as well as among academic experts. Weiler emphasizes the immense shift in the balance of legal and jurisdictional authority between state and EU over the past forty years, giving rise to problems of legitimacy, accountability, and compliance, and provoking both popular and governmental demands for the pendulum to swing back to national jurisdiction in many fields. Scharpf (1994a) emphasizes the gap between the loss of state control at the national level and the limited policy-making capacities of the EU, leaving markets and external actors freer from political constraint. Wessels (1992), more optimistically, sees national governments operating collectively at the EU level to achieve objectives which they can no longer attain within the confines of the single state.

The editors of this volume argue that the centrality of the European state has been progressively reduced both by the transformational effects of technological, economic, and social change and by the reluctant response of national governments in transferring some previously core functions of national sovereignty to the European level (see above, Ch. 1; W. Wallace 1990*a*, 1994*b*). The political sensitivities and contradictions involved in multilateralizing monetary policy, foreign policy, border controls, policing, citizenship, and defence have been noted in earlier chapters of this volume. The difficulties which governments face in agreeing on common policies in these areas reflect the agony which the transfer of such central elements of sovereignty involves for players whose legitimation rests upon their claim to embody the authority of the nation incorporated in the state.

The perceived challenge posed by further formal integration to British national identity, and to the unwritten constitution in which that identity was seen to be embedded, was threatening to tear its Conservative government apart as this volume went to press. The problems posed for Danish identity and statehood were vigorously argued in their two post-Maastricht referenda. The Belgian state, brought together by resistance to Dutch domination and the threat of incorporation into France, has progressively weakened as external threats have disappeared and as more and more of the functions which national government used to fulfil have been transferred either to the EU or to its mutually mistrustful regions. The only remaining policies clearly under the control of the national government, apart from Belgian representation within the EU itself, are the social security system and the armed forces. The Swiss state, similarly dependent historically on the perception of external threat (and on the symbolism of its citizen army) to hold together its diverse linguistic communities, has found itself divided: between political and business élites which recognize the advantages of EU membership and the costs of continued exclusion, French-speaking cantons which see little to fear in closer association with their neighbours, and German-speaking cantons which cling to the independence their ancestors won from the Habsburgs, the Hohenzollerns, and their successors. Norwegian anxiety over the threat which formal integration posed to identity and statehood led to its electorate's rejection of EU membership in November 1994, twenty-two years after the first refusal to ratify its government's decision to join.

But even in the core countries of the EU, the TEU raised questions which both touched the heart of national identity and in effect required their incorporation into the national constitution. In France the Decision of the *Conseil Constitutionnel* of 9 April 1992 focused in particular on its implications for guarantees of the rights and liberties of citizens, and on the implications of the concept of EU citizenship. Both questions struck at 'the nature of French national identity in a post-1992 Europe' (Ladrech 1994: 69). In Germany a number of challenges were made to the Federal Constitutional Court, on the transfer of powers from one level of government to another and on the inadequacy of democratic accountability over those powers transferred. Article 23 of the Basic Law was amended to strengthen the position of the Länder in EU decision-making, which will arguably 'make it increasingly difficult for Germany to act effectively at European level' (Ress 1994: 47; Bundesverfassungsgericht 1994).[7]

The optimistic language of the ECSC Treaty preamble looked forward to the development of 'a broader and deeper community among peoples long divided by bloody conflicts'. Four decades since then have dimmed memories of bloody conflicts. The majority of the population of western Europe in 1990 had grown up within the stable order of the Atlantic system. They had taken for granted the benefits of economic integration for which the EC and the wider OEEC/OECD had provided a framework of rules and institutions. A loose sense of political community had evolved, in which the possibility of renewed conflict among west European states was entirely excluded. But the style and content of the policy-making described above have not been such as to capture public imagination and transform popular self-identity: rather the reverse. 'The European idea', Hans Dietrich Genscher remarked in an article to mark the twenty-fifth anniversary of the Treaty of Rome,

is no longer a vibrant concept in the minds of our peoples. It has perhaps become degraded to the level of a somewhat annoying established notion. . . . I'm afraid that a much darker image of Europe has established itself in the minds of our citizens, characterized by farm production surpluses, budgetary problems and financial book-keeping'. (Genscher 1982)

The institutional design of the EC was intended to avoid confronting the awkward issues of legitimacy, accountability, and identity. Promoters of the supranational 'Community model' hoped that popular loyalty and identity would follow economic interest, as affluence undermined old ideologies. Practical governmental negotiators hoped to contain institutional integration within a limited field of technocratic administration, bounded by intergovernmental approval, with legitimacy and accountability continuing to flow through the representative processes of the nation state (W. Wallace and Smith 1995). Neither hope has been fulfilled. The hierarchy of committees through which European policy-making operates engages officials and interested groups, but looks an impenetrably bureaucratic process to the public outside. The directly elected European Parliament has gained only limited influence and visibility, and is still only a minor player in most of the policy sectors we have described. The legitimacy and authority of the state has been weakened, and narrowed, by the evident transfer of authoritative decision-making to multilateral negotiations within the European institutions. But the legitimacy and authority of those European institutions has not commensurately strengthened, leaving a void within which popular mistrust and discontent may collect.

This diffusion and weakening of legitimacy is not, however, a phenomenon of European integration alone. It represents one aspect of the crisis of the modern state, to which the construction of formal European integration was intended to respond.[8]

The growth of the postwar state, with its complex, mixed polity, inter-institutional networks, and hybrid organizations, has not been founded on nor followed by any coherent reassessment of the legitimate constitutional principles of authority, power and accountability. Legitimacy claims have become increasingly instrumental'. (Olsen 1995: 18)

Claims to legitimacy which rest upon the provision of public and private goods—of welfare, employment, rising prosperity—and of efficient and rational administration are much less firmly founded than those which rest also on the symbolic bases of shared history or myth, or of common ethnic or citizen identity. Member governments have so far resisted allowing the EU to develop its own political symbols. It is part of the underlying contradiction of their approaches to monetary union and common defence that neither objective is likely to be sustainable without substantial symbolic support to strengthen the popular sense of shared political community. The development of the concept of European citizenship, institutionalized in the TEU, was a response by political leaders from several member states to this perceived need, attempting through such measures as common passports, common border controls, and (limited) common voting rights to associate national publics with this broader community. Chapter 15 explored some of the difficulties and contradictions involved in negotiating on such nationally sensitive issues. Here is a constituent issue, in which fundamental political values are at stake with which the structures and policy-making processes of the EU are ill equipped to deal (Garcia 1993; A. D. Smith 1992).

Europeanization and globalization

For those who assume that the nation state is the natural and permanent framework for political life, this weakening of both its capability and its symbolic solidarity is a threat to political order and a charge to be laid against the EU. For those for whom the bringing together of nation and state, of economic management and security provision within defined territorial boundaries, was always a contingent (even artificial) achievement, the problems which west European states—and the European institutions—face in adjusting to external transformation and to technological, economic, and social change are part of a wider 'crisis' of the nation-state model (Dunn 1994; Müller and Wright 1994; *Daedalus* 1995). Neither the inadequacy of the state nor the phenomenon of globalization are specific to Europe, as Helen Wallace argues in Chapter 1. The processes of modernization which provided the foundation for the rise of the centralized European nation state in the late nineteenth century now make for globalization, sweeping across the boundaries of the densely populated and geographically concentrated states of western Europe.

The pendulum which has swung between Europeanization and the maintenance of national autonomy acted as an internal balance within the west European system established under American leadership and protection in the decade after 1945. There is, however (as we noted earlier), a continuum of modernization, partly impelled by developments within western Europe but powerfully affected by forces outside. Many of the innovations and changes in policy examined in this volume were responses to modernization: from the remarkable and continuing growth in agricultural productivity to shifting patterns of work, welfare, and demographic balance; from the explosion in border-crossing generated by affluence, automobiles, and aircraft to the explosion in trade in services generated by electronic communications. What has been specific to Europeanization has been the geographical concentration and—at least until 1989—clear outer boundaries of the region, and the much tighter framework of rules and institutions which have attempted to channel, and thus moderate, these forces for modernization (W. Wallace 1994a).[9]

One paradox of west European integration has been this juxtaposition of economic and social modernization, a process which has transformed west European societies and economies since the 1950s (leaving behind those central European societies and economies which found themselves forced into a socialist system), with a stable regional political and security order. The stability of that regional order depended upon the equilibrium between opposing forces along several dimensions. Externally the balance was maintained by Soviet hostility and American hegemony, by the division of Europe's potential hegemonic power between western and eastern alliances, and by delicate adjustments of relative weights in US relations with western Europe. Internally it was maintained by Germany's continuing willingness to act as the EC's main funder (*der Zahlmeister Europas*) in a system in which (until the 1990s) almost all other states were net beneficiaries, gaining both cash and power from German contributions. Political and geopolitical inhibitions held German governments back from exerting their full weight in EC policy-making. The dynamism of the core European economy, increasingly centred around the German economy, drew a wider circle of countries into its beneficial influence. Successive French governments remained determined to maintain a close Franco-German relationship, in spite of continuing disagreements between Paris and Bonn over the pace and direction of European integration. The delicate balance between sovereignty and integration was somehow adjusted through successive conferences of heads of government, revising from time to time the treaties.

That regional equilibrium has now been overturned. Member governments within the EU attempted between 1990 and 1992 to adjust their policies and priorities to this revolution without reopening the underlying bargains on which the EU's institutions and redistributive arrangements rested. Several of the chapters above examine the strains which this transformation exerted on the Community's *acquis*, and the stumbling efforts of EU institutions and governments to come to terms with its implications. Preparations for the next IGC planned for 1996 were overshadowed by awareness among governments and political leaders of the inadequacy of their response so far, and of the massive scale of policy and institutional adjustment required.

For the present members of the European Union, eastward expansion constitutes both a challenge and a test not only in terms of the material contribution they are able and willing to make but also in terms of their moral and spiritual self-conception. The Union's response will show whether it is able and willing to become the main pillar of a continental order. (CDU/CSU 1994)

The high political issues of European order were managed, throughout the development of west European integration, within the Atlantic framework under American leadership. The EC institutions were designed to manage low politics: deregulation and re-regulation, modest distribution and marginal redistribution. As tasks have expanded, so these institutions have provided a framework strong enough for the negotiation and implementation of a succession of partial bargains on increasingly broad policy packages. The institutional and policy dilemmas facing the EU, its member governments, and its national publics in the mid-1990s were, however, fundamental. Governments and EU officials were now facing issues of high politics which their publics were reluctant to recognize, working together through an institutional structure unsuited to the tasks in hand but which their publics were reluctant to strengthen or reform. It remains to be seen whether they can maintain their increasingly precarious balance.

Notes

1. Hirschman adds (269) that 'Haas and his co-workers need not really have become so despondent over their inability to predict correctly the meandering movement and dynamics of integration. Economists make wrong forecasts all the time and have learned how to thrive on them.'

2. Most of those who have written of European policy-making in terms of multi-level governance have avoided using the language of federalism. Since, however, the sharing of functions among different levels of government within a framework of institutions and rules is the essence of federal government, the difference between explicit and implicit federal analogies is unimportant—at least to the academic.

3. Monnet was by nature an anti-bureaucrat, preferring to work with small groups outside administrative systems (Duchêne 1994). Hallstein had been first a law professor and then a *beamte* in the German Chancellery and Foreign Office (Löth *et al.* 1995).

4. German political leaders and officials in 1994–5 were fond of remarking that the British were insisting on anglicizing their European currency by including as their national symbol the head of the House of Saxe-Coburg-Gotha (the Queen). But German attitudes to a single currency, among the élite and among the wider public, were also deeply affected by symbolic factors. The similarity of 'ecu' and 'Kuh', the German word for cow, was also said to conjure up images of inflation-prone neighbours milking German monetary reserves.

5. In the early 1990s the Royal Society for the Protection of Birds was (in terms of membership) the second largest voluntary organization in Britain, after the National Trust: an influential organization with a predominantly middle-class membership and a substantial full-time staff, several of whom were regularly in Brussels to lobby on this issue. In 1994–5 demonstrations at ports and airports against the shipment of live calves were effective enough to push the British government into arguing for the strengthening of animal-welfare measures at EU level. At the same time it maintained the position that it had taken over the social protocol in the TEU that human welfare was more appropriately a matter for national regulation.

6. The conclusions to the second edition of this volume noted how many of its case-studies had stressed the need to move on from distributive to redistributive politics. It noted the institutional and political tendencies built into the EC system 'for postponing awkward issues or for pushing the costs of decisions on to third countries'. 'Redistribution demands a greater degree of underlying consensus about the structure and explicit objectives of the policy-making system. . . . Redistribution raises underlying questions about political values' (W. Wallace 1983*a*: 421, 419).

7. Note also the unguarded comment of Douglas Hurd, the British Foreign Secretary, before the full storm of Conservative protest over the Maastricht TEU

broke, that it was now 'right to think of the Treaties as part of the Constitution' (Hurd 1992).

8. The USA was suffering a not dissimilar crisis of legitimacy in the early 1990s, in which substantial groups of citizens had come to reject the authority of the federal government, and to regard federal officials almost as agents of a foreign power. American students of European integration should note that the US federal government appears as impenetrably bureaucratic and deadlocked to many of their fellow citizens as the EU does to its citizens.

9. ' "Europeanization" refers to processes which make "Europe" a more significant political community and thereby European boundaries more relevant politically' (Olsen 1995: 21).

REFERENCES

Adams, W. J. (1992) (ed.), *Singular Europe* (Ann Arbor, Mich.: University of Michigan Press).

Addison, J. T., and Siebert, W. S. (1993), 'The EC Social Charter: The Nature of the Beast', *National Westminster Bank Quarterly Review*, Feb., 13–28.

Aguilar, S. (1993), 'Corporatist and Statist Designs in Environmental Policy: The Contrasting Roles of Germany and Spain in the European Community Scenario', *Environmental Politics*, 2/2: 223–47.

Allen, D. (1983), 'Managing the Common Market: The Community's Competition Policy', in H. Wallace *et al.* (1983): 209–36.

—— (1993), 'Dividing the Spoils for 1994', *European Brief*, 1/2: 38–9.

Altenstetter, C. (1992), 'Health Policy Regimes and the Single European Market', *Journal of Health Politics, Policy and Law*, 17/4: 813–46.

Alter, K. J., and Meunier-Aitsahalia, S. (1994), 'Judicial Politics in the European Community: European Integration and the Pathbreaking *Cassis de Dijon* Decision', *Comparative Political Studies*, 26/4: 535–61.

Anania, G., Carter, C. A., and McCalla, A. F. (1994*a*), 'Agricultural Policy Changes, GATT Negotiations, and the US.–E.C. Agricultural Trade Conflicts', in Anania *et al.* (1994*b*), 1–39.

—— —— —— (1994*b*) (eds.), *Agricultural Trade Conflicts and GATT: New Dimensions in U.S.–European Agricultural Trade Relations* (Boulder, Col.: Westview Press).

Andersen, S. S., and Eliassen, K. A. (1993) (eds.), *Making Policy in Europe: The Europeification of National Policy-Making* (London: Sage).

Anderson, K., and Hayami, Y. (1986), *The Political Economy of Agricultural Protection: East Asia in International Perspective* (Sydney: Allen and Unwin).

Anderson, M. (1993), 'Control of Organized Crime in the European Community', Working Paper No. IX (Edinburgh: University of Edinburgh, Department of Politics).

—— (1994), 'The Agenda for Police Cooperation', in Anderson and Den Boer (1994), 3–21.

—— and Den Boer, M. (1992) (eds.), *European Police Cooperation: Proceedings* (Edinburgh: University of Edinburgh, Department of Politics).

—— —— (1994) (eds.), *Policing Across National Boundaries* (London: Pinter).

—— —— and Miller, G. (1994), 'European Citizenship and Cooperation in Justice and Home Affairs', in Duff *et al.* (1994), 104–22.

—— —— Cullen, P., Gilmore, W. C., Raab, C., and Walker, N. (forthcoming), *Policing the European Union* (Oxford: Oxford University Press).

Andriessen, F. (1991), 'Towards a Community of Twenty-Four?', speech to the 69th Assembly of Eurochambers, Brussels, 19 Apr.

Arkleton Trust (1992), *Farm Household Adjustment in Western Europe, 1987–1991: Final Report on the Research Programme on Farm Structures and Pluriactivity* (Luxemburg: Office for Official Publications of the European Communities).

Arp, H. A. (1993), 'Technical Regulation and Politics: The Interplay between Economic Interests and Environmental Policy Goals in EC Car Emission Legislation', in Liefferink *et al.* (1993*b*).

Atkinson, M. M., and Coleman, W. D. (1989), *The State, Business and Industrial Change in Canada* (Toronto: University of Toronto Press).

Attwood, E. A. (1963), 'The Origins of State Support for British Agriculture', *Manchester School of Economics and Social Studies*, No. 31 (Manchester), 129–48.

Averyt, W. F., Jr. (1977), *Agropolitics in the European Community: Interest Groups and the Common Agricultural Policy* (London: Praeger).

Axelrod, R., and Keohane, R. O. (1988), 'Achieving Cooperation Under Anarchy: Strategies and Institutions', *World Politics*, 38/1: 226–54.

Baker, J. (1989), 'A New Europe, a New Atlanticism: Architecture for a New Era', speech to the Berlin Press Club, 12 Dec., *Europe Documents*, No. 1588, 15 Dec.

Baker, S., Milton, K., and Yearly, S. (1994) (eds.), *Protecting the Periphery: Environmental Policy in Peripheral Regions of the European Union* (Essex: Frank Cass).

Balassa, B. (1962), *The Theory of Economic Integration* (London: Allen and Unwin).

—— (1975), *European Economic Integration* (Amsterdam: North-Holland).

Baldwin, R. (1994), *Towards an Integrated Europe* (London: CEPR).

—— et al. (1992), *Monitoring European Integration*, iii: *Is Bigger Better? The Economics of EC Enlargement* (London: CEPR).

Bassompierre, G. de (1988), *Changing the Guard in Brussels: An Insider's View of the EC Presidency* (New York: Praeger).

Bayliss, B. T., *et al.* (1994): Committee of Enquiry, *Road Freight Transport in the Single European Market* (Brussels: European Commission, Directorate General for Transport).

Benelux (1990), 'Future EC Regulation on the Import of Bananas', Memorandum of the governments of the Benelux countries, BEB (90) 5, 27 Sept.

Bennett, G. (1992), *Dilemmas: Coping with Environmental Problems* (London: Earthscan).

Benyon, F. (1992), 'Les "Accords européens" avec la Hongrie, la Pologne et la Tchécoslovaquie', *Revue du Marché Unique Européen*, 2: 25–50.

Benyon, J., Turnbull, L., Willis, A., Woodward, R., and Beck, A. (1993), *Police Cooperation in Europe: An Investigation* (Leicester: University of Leicester Press).

Berenz, C. (1994), 'Hat die betriebliche Altersversorgung zukünftig noch eine Chance?', *Neue Zeitschrift für Arbeitsrecht*, 11/9: 385–90 and 11/10: 433–8.

Beyme, K. von, and Schmidt, M. G. (1985) (eds.), *Policy and Politics in the Federal Republic of Germany* (New York, NY: St. Martin's Press).

Bieback, K.-J. (1991), 'Harmonization of Social Policy in the European Community', *Les Cahiers de Droit*, 32/4: 913–35.

—— (1993), 'Marktfreiheit in der EG und nationale Sozialpolitik vor und nach Maastricht', *Europarecht*, 28/2: 150–72.

Bieber, R., and Monar, J. (1995) (eds.), *Justice and Home Affairs in the European Union—The development of the Third Pillar* (Brussels: European Interuniversity Press).

Bigo, D. (1994), 'The European Internal Security Field: Stakes and Rivalries in a Newly Developing Area of Police Intervention', in Anderson and Den Boer (1994).

—— and Leveau, R. (1992), 'L'Europe de la sécurité intérieure', end of study report for the Institut des Hautes Études de la Sécurité Intérieure, Paris.

Black, R. A., Jnr. (1977), '*Plus ça change, plus c'est la même chose*: Nine Governments in Search of a Common Energy Policy', in H. Wallace *et al.* (1977): 165–96.

Blackman, D. (1994), 'Aid to the Democratically Elected Parliaments of Central and Eastern Europe: The European Parliament's Programme', paper presented to the Workshop of Parliamentary Scholars and Parliamentarians, Berlin, 19–20 Aug.

Boehmer-Christiansen, S., and Skea, J. (1992), *Acid Politics: Environmental and Energy Policies in Britain and Germany* (London: Belhaven).

—— and Weidner, H. (1992), 'Catalyst versus Lean Burn: A Comparative Analysis of Environmental Policy in the Federal Republic of Germany and Great Britain with Reference to Exhaust Emission Policy for Passenger Cars, 1970–1990', FS II 92-304 (Berlin: Wissenschaftszentrum Berlin für Sozialforschung).

Boltho, A. (1989), 'European and United States Regional Differentials: A Note', *Oxford Review of Economic Policy*, 5/2: 105–15.

Bonvicini, G., *et al.* (1991), *The Community and the Emerging European Democracies* (London: Royal Institute of International Affairs).

Boons, F. (1992), 'Product-Oriented Environmental Policy and Networks: Ecological Aspects of Economic Internationalization', *Environmental Politics*, 1/4: 84–105.

Borrell, B., and Yank, M. (1992), *EC Bananarama 1992: The Sequel*, Working Paper 958, International Economics Department (Washington DC: The World Bank).

Bowler, I. R. (1985), *Agriculture under the Common Agricultural Policy: A Geography* (Manchester: Manchester University Press).

Bresserts, H., Huitema, D., and Kuks, S. M. M. (1994), 'Policy Networks in Dutch Water Policy, *Environmental Politics*, 3/4: 24–51.

Brinkhorst, L. J. (1991), 'Subsidiarity and European Environmental Policy', in *Subsidiarity: The Challenge of Change* (Maastricht: European Institute of Public Administration).

Brouwer, H. J., *et al.* (1995), *Do We Need a New Budget Deal?* (Brussels: The Philip Morris Institute).

Buchan, D. (1993), *Europe: The Strange Superpower* (Aldershot: Dartmouth).

Bueno de Mesquita, B., and Stokman, F. N. (1994) (eds.), *European Community Decision-Making* (New Haven: Yale University Press).

Buigues, P., and Sheehy, J. (1994), 'European Integration and the Internal Market Programme', paper delivered to the ESRC/COST A7 Conference, University of Exeter, 8–11 Sept.

Buigues, P., Ilzkovitz, F., and Lebrun, J.-F. (1990), 'The Impact of the Internal Market by Industrial Sector: The Challenge for the Member States', *European Economy: Social Europe*, Special Edition, 3–7.

Buksti, J. A. (1983), 'Bread-and-Butter Agreement and High Politics Disagreement: Some Reflections on the Contextual Impact on Agricultural Interests in EC Policy-Making', *Scandinavian Political Studies*, 6/4: 261–80.

Bulmer, S. (1984), 'Domestic Politics and EC Policy-Making', *Journal of Common Market Studies*, 21/4: 349–63.

—— (1994a), 'The Governance of the European Union: A New Institutionalist Approach', *Journal of Public Policy*, 13/4: 351–80.

—— (1994b), 'Institutions and Policy Change in the European Communities: The Case of Merger Control', *Public Administration*, 72/3: 423–44.

—— and Scott, A. (1994) (eds.), *Economic and Political Integration in Europe: International Dynamics and Global Context* (Oxford: Blackwell).

Bundesverfassungsgericht (1994), 'The Maastricht Decision', in Winckelmann (1994), 751–99.

Burley, A.-M., and Mattli, W. (1993), 'Europe Before the Court: A Political Theory of Legal Integration', *International Organization*, 47/1: 41–76.

Cafruny, A. W., and Rosenthal, G. G. (1993) (eds.), *The State of the European Community*, ii: *The Maastricht Debates and Beyond* (Boulder, Col.: Lynne Riener).

Cameron, D. R. (1992), 'The 1992 Initiative: Causes and Consequences', in Sbragia (1992b), 23–74.

Caporaso, J. A., and Keeler, J. T. S. (1993), 'The European Community and Regional Integration Theory', paper presented at the Third Biennial International Conference of the European Community Studies Association, Washington, DC, 27–9 May.

Cappelletti, M., Secombe, M., and Weiler, J. H. H. (1986) (eds.), *Integration through Law: Europe and the American Federal Experience* (Berlin: Walter de Gruyter).

Carlsnaes, W., and Smith, S. (1990) (eds.), *European Foreign Policy* (London: Sage).

Cawson, A. (1992), 'Interests, Groups and Public Policy-Making: The Case of the European Consumer Electronics Industry', in Greenwood *et al.* (1992), 99–118.

CDU/CSU-Fraktion des Deutschen Bundestages (1994): Christian Democratic Union/Christian Social Union, 'Reflections on European Policy', Bonn, 1 Sept.

Cecchini, P., with Catinat, M., and Jacquemin, A. (1988), *The European Challenge 1992: The Benefits of a Single Market* (Aldershot: Wildwood House).

CEPR (1992): Centre for Economic Policy Research, 'The Association Process: Making it Work: Central Europe and the EC', CEPR Occasional Paper No. 11 (London: CEPR).

Cini, M. (1994), 'Policing the Internal Market: The Regulation of Competition in the European Commission', Ph.D. thesis (Bristol).

Cloos, J., Reinesch, G., Vignes, D., and Weyland, J. (1994), *Le Traité de Maastricht: Genèse, analyse, commentaires* (Brussels: Bruylant).

Cochrane, W. W. (1965), *The City Man's Guide to the Farm Problem* (Minneapolis: University of Minnesota Press).

Cockfield, Lord (1994), *The European Union: Creating the Single Market* (London: Wiley Chancery Law).

Collins, D. (1975), *The European Communities: The Social Policy of the First Phase*, 2 vols. (London: Martin Robertson).

Commission (1985), 'Completing the Internal Market: White Paper from the Commission to the European Council', COM (85) 310 final.

—— (1987a), 'Making a Success of the Single Act', *EC Bulletin*, Supplement, 1/87.

—— (1987b), 'Report by the Commission to the Parliament on the Financing of the Community Budget', COM (87) 101 final.

—— (1988a), *Research on the 'Cost of Non-Europe': Basic Findings*, 16 vols. (Luxemburg: Office for Official Publications of the European Communities).

—— (1988b), *Seventeenth Report on Competition Policy* (Luxemburg: Office for Official Publications of the European Communities).

—— (1988/9), 'The List of Obstacles to be Overcome to Complete the Single Energy Market: An Initial Commission Analysis', supplement to *EEC Energy Policy and the Single Market of 1993* (Brussels: Prometheus).

—— (1989a), *Community Public Finance: The European Budget after the 1988 Reform* (Luxemburg: Office for Official Publications of the European Communities).

—— (1989b), 'Draft Directive on Electricity Transportation', COM (89) 336 final, 29 Sept.

—— (1989c), 'Draft Directive on Natural Gas Transportation', COM (89) 334 final, 6 Sept.

—— (1989d), 'Transparency of Consumer Energy Prices', COM (89) 123 final, 7 Mar.

—— (1990a), 'Association Agreements with the Countries of Central and Eastern Europe: A General Outline', COM (90) 398 final, 27 Aug.

—— (1990b) 'Commission Communication to the Council on the Uruguay Round Negotiations—Progress and Prospects', Feb.

—— (1990c), 'Commission Opinion of 21 October 1990 on the Proposal for Amendment of the Treaty Establishing the EEC with a View to Political Union', COM (90) 600 final, 23 Oct.

—— (1990d), 'The Development of the Community's Relations with the Countries of Central and Eastern Europe', SEC (90) 194 final, 1 Feb.

—— (1990e), 'One Market, One Money', *European Economy*, 44, Oct.

—— (1990f), *Twenty-Fifth General Report on the Activities of the European Communities, 1990* (Luxemburg: Office for Official Publications of the European Communities).

—— (1992a), 'Europe and the Challenge of Enlargement', *Bulletin of the EC*, suppl. 3/92.

—— (1992b), 'The European Community and the Uruguay Round', Nov.

—— (1992c), 'The Operation of the Community's Internal Market After 1992: Follow-up to the Sutherland Report', SEC (92) 2277 final, 2 Dec.

—— (1992d), 'Towards a Closer Association with the Countries of Central and Eastern Europe', SEC (92) 2301 final, 2 Dec.

Commission (1993a), 'Commission Report to the European Council on the Adaptation of Community Legislation to the Subsidiarity Principle', COM (93) 545 final.

—— (1993b), *The Community Budget: The Facts and Figures*, (Luxemburg: Office for Official Publications of the European Communities).

Commission (1993c), 'Conducting the Uruguay Round', Dec.

—— (1993d), 'Growth, Competitiveness, Employment: The Challenges and Ways Forward into the 21st Century', COM (93) 700 final.

—— (1993e), 'Stable Money—Sound Finances', *European Economy*, 53.

—— (1993f), 'Towards a Closer Association with the Countries of Central and Eastern Europe', SEC (93) 648 final, 18 May.

—— (1993g) *Twenty-Second Report on Competition Policy* (Luxemburg: Office for Official Publications of the European Communities).

—— (1994a), 'The Europe Agreements and Beyond: A Strategy to Prepare the Countries of Central and Eastern Europe for Accession', COM (94) 320 final, 13 July.

—— (1994b), *The European Union's Cohesion Fund* (Luxemburg: Office for Official Publications of the European Communities).

—— (1994c), 'Follow-up to Commission Communication on "The Europe Agreements and Beyond: A Strategy to Prepare the Countries of Central and Eastern Europe for Accession" ', COM (94) 361 final, 27 July.

—— (1994d), *Twenty-Third Report on Competition Policy* (Luxemburg: Office for Official Publications of the European Communities).

—— (1994e), 'The Uruguay Round Background Brief', updated, May.

—— (1995a), 'Commission's Work Programme for 1995', COM (95) 26 final.

—— (1995b), 'Preparation of the Associated Countries of Central and Eastern Europe for Integration into the Internal Market of the Union', COM (95) 163 final, 3 May.

—— (1995c) 'Report on the Operation of the Treaty on European Union', Brussels, 10 May.

—— (1995d) 'Transposition du Livre Blanc: Le seuil de 90% est atteint', *DG XV News*, 1, Mar.

Conférence de la charte européene de l'énergie (1994), *Traité de la charte de l'énergie*, 14 Dec.

Cot, J.-P. (1989), 'The Fine Art of Community Budgeting Procedure', *Contemporary European Affairs*, 1: 227–39.

Council of the European Union (1995), 'Report on the Functioning of the Treaty on European Union,' SN 1821/95, Brussels, 14 Mar.

Coutu, D., Hladik, K., Meen, D., and Turcq, D. (1993), 'Views of the Business Community on Post-1992 Integration in Europe', in Jacquemin and Wright (1993b), 47–80.

Cowles, M. Green (1994), 'The Politics of Big Business in the European Community: Setting the Agenda for a New Europe', Ph.D. thesis (American University, Washington, DC).

Cox, A. (1992), 'Implementing 1992 Public Procurement Policy: Public and Private Obstacles to the Creation of the Single Market', *Public Procurement Law Review*, 2: 139–54.

Curtin, D. (1993), 'The Constitutional Structure of the Union: A Europe of Bits and Pieces', *Common Market Law Review*, 30/1: 17–69.

Daedalus (1995), 'What Future for the State?', proceedings of the American Academy of Arts and Sciences, 124/2.

Dahrendorf, R. (1973), *Plädoyer für die Europäische Union* (Munich: Piper).

Dam, K. W. (1967), 'The European Common Market in Agriculture', *Columbia Law Review*, 67/2: 209–65.

—— (1970), *The GATT: Law and International Economic Organization* (Chicago and London: University of Chicago Press).

Dashwood, A. (1977), 'Hastening Slowly: The Communities' Path towards Harmonization', in H. Wallace *et al.* (1977), 273–99.

—— (1983), 'Hastening Slowly: The Communities' Path towards Harmonization', in H. Wallace *et al.* (1983), 177–208.

Davidow, J. (1977), 'EEC Fact-Finding Procedures in Competition Cases: An American Critique', *Common Market Law Review*, 14/2: 175–89.

De Grauwe, P. (1992), *The Economics of Monetary Integration* (Oxford: Oxford University Press).

—— (1994), 'Towards European Monetary Union without the EMS', *Economic Policy*, 18 (Apr.), 147–85.

—— and Papademos, L. (1990) (eds.), *The European Monetary System in the 1990's* (London: Longman).

Dehousse, R. (1994) (ed.), *Europe after Maastricht: An Ever Closer Union?* (Munich: C. H. Beck).

Delors, J. (1991), 'European Integration and Security', *Survival*, 23/2: 99–109.

—— *et al.* (1989): Committee for the Study of Economic and Monetary Union, *Report on Economic and Monetary Union in the European Community* (Luxemburg: Office for Official Publications of the European Communities).

Den Boer, M. (1991), 'Schengen: Intergovernmental Scenario for European Police Cooperation', Working Paper No. V (Edinburgh: University of Edinburgh, Department of Politics).

—— (1994a), 'Europe and the Art of International Police Cooperation: Free Fall or Measured Scenario?', in O'Keeffe and Twomey (1994), 279–91.

—— (1994b), 'The Quest for European Policing: Rhetoric and Justification in a Disorderly Debate', in Anderson and Den Boer (1994), 174–96.

—— and Walker, N. (1993), 'European Policing after 1992', *Journal of Common Market Studies*, 31/1: 3–28.

Deudney, D. H. (1995), 'The Philadelphian System: Sovereignty, Arms Control, and the Balance of Power in the American States-Union, 1787–1861', *International Organization*, 29/2: 191–228.

Devos, S., Woolcock, S., Bressant, A., and Raby, G. (1993) 'Study on Regional Integration', paper to the OECD Trade Directorate's Trade Committee, TD/TC (93) 15.

Devuyst, Y. (1995), 'The European Community and the Conclusion of the Uruguay Round', in C. Rhodes and Mazey (1995).

Dinan, D. (1994), *Ever Closer Union* (London: Macmillan).

Dorbey, C. (1995), 'Dialectical Functionalism: Stagnation as a Booster of European Integration', *International Organization*, 49/2: 253–84.

DTI (1991): Department of Trade and Industry, *Competition Policy: How it Works* (London: HMSO).

Duchêne, F. (1994), *Jean Monnet: First Statesman of Interdependence* (New York: Norton).

—— Szczepanik, E., and Legg, W. (1985), *New Limits on European Agriculture: Politics and the Common Agricultural Policy* (London: Croom Helm).

Duff, A., Pinder, J., and Price, R. (1994) (eds.), *Maastricht and Beyond: Building the European Union* (London: Routledge).

Dunn, J. (1994) (ed.), Contemporary Crisis of the Nation State?', *Political Studies*, 42, special issue.

Dunnet, D. (1991), 'The European Bank for Reconstruction and Development: A Legal Survey', *Common Market Law Review*, 28/3: 571–97.

Dyson, K. H. F. (1992) (ed.), *The Politics of German Regulation* (Aldershot: Dartmouth).

Edwards, G., and Spence, D. (1994) (eds.), *The European Commission* (London: Longman).

Eekelen, W. van (1993), 'WEU Prepares the Way for New Missions', *NATO Review*, 5 (Oct.), 19–23.

Ehlermann, C. D. (1995), 'Reflections on a European Cartel Office', *Common Market Law Review*, 32: 471–86.

Eichenberg, R. C., and Dalton, R. J. (1993), 'Europeans and the European Community: The Dynamics of Public Support for European Integration', *International Organization*, 47/4: 507–34.

Eichener, V. (1993), 'Social Dumping or Innovative Regulation? Processes and Outcomes of European Decision-Making in the Sector of Health and Safety at Work Harmonization', EUI Working Papers in Political and Social Sciences (SPS) No. 92/28 (Florence, European University Institute).

—— and Voelzkow, H. (1994) (eds.), *Europäische Integration und verbandliche Interessenvermittlung* (Marburg: Metropolis Verlag).

Eichengreen, B. (1992), 'Should the Maastricht Treaty be Saved?', Princeton Studies in International Finance, No. 74 (Princeton: Princeton University Economics Department).

—— and Frieden, J. (1994) (eds.), *The Political Economy of European Monetary Unification* (Boulder, Col.: Westview Press).

Eichenhofer, E. (1992) (ed.), *Die Zukunft des koordinierenden Europäischen Sozialrechts* (Cologne: Carl Heymanns).

Emerson, M., Aujean, M., Catinat, M., Goybet, P., and Jacquemin, A. (1988), *The Economics of 1992: The EC Commission's Assessment of the Economic Effects of Completing the Internal Market* (Oxford: Oxford University Press).

ENDS Report (1994), 234, July.

Environment Watch: Western Europe, 4 Feb. 1994.

Eurobarometer (1994), 41 (July).

Euro-East, 23 Apr. 1993.

European Parliament (1984), *Draft Treaty Establishing the European Union* (Luxemburg: European Parliament).

—— (1993), 'The Powers of the European Parliament in the European Union', Directorate General for Research, Political Series Working Paper E-1 (Luxemburg: European Parliament).

Evans, P. B., Jacobson, H. K., and Putnam, R. D. (1993) (eds.), *Double-Edged Diplomacy: International Bargaining and Domestic Politics* (Berkeley: University of California Press).

Faini, R., and Portes, R. (1995), (eds.), *European Union Trade with Eastern Europe: Adjustment and Opportunities* (London: CEPR).

Falkner, G. (1993), 'Die Sozialpolitik im Maastrichter Vertragsgebäude der Europäischen Gemeinschaft', *SWS-Rundschau*, 33/1: 23–43.

—— (1994*a*), 'Die Sozialpolitik der EG: Rechtsgrundlagen und Entwicklung von Rom bis Maastricht', in Haller and Schachner-Blazizek (1994), 221–46.

—— (1994*b*), *Supranationalität trotz Einstimmigkeit: Entscheidungsmuster der EU am Beispiel Sozialpolitik* (Bonn: Europa Union Verlag).

Ferge, Z., and Kolberg, J. E. (1992) (eds.), *Social Policy in a Changing Europe* (Boulder, Col.: Westview Press).

Fink-Hoojer, F. (1994), 'The Common Foreign and Security Policy of the European Union', *European Journal of International Law*, 5/2: 173–98.

Fisher, R., and Ury, W. (1982), *Getting to Yes: Negotiating Agreement without Giving In* (London: Hutchinson).

Fitzpatrick, J., *et al.* (1993), *Trade Policy and the EC Banana Market* (Dublin: Economic Consultants).

Flemming, J., and Rollo, J. M. C. (1992) (eds.), *Trade, Payments, and Adjustment in Central and Eastern Europe* (London: Royal Institute of International Affairs and European Bank for Reconstruction and Development).

Flora, P. (1993), 'The National Welfare States and European Integration', in Moreno (1993), 11–22.

Forster, A. (1994), 'The EC and the WEU', in Moens and Anstis (1994), 135–58.

—— (1995), 'Empowerment and Constraint: Britain and the Negotiation of the Treaty on European Union', D.Phil. thesis (Oxford).

Forsyth, M. (1981), *Unions of States* (Leicester: Leicester University Press).

Foster, E. (1992), 'The Franco-German Corps: A "Theological" Debate?', *RUSI Journal*, 137/4: 63–7.

Franklin, M., Marsh, M., and McLaren, L. (1994), 'Uncorking the Bottle: Popular Opposition to European Unification in the Wake of Maastricht', *Journal of Common Market Studies*, 32/4: 455–73.

Frazer, T. (1994), 'The New Structural Funds, State Aids and Interventions in the Single Market', *European Law Review*, 20/1: 3–19.

Freeman, C., Sharp, M., and Walker, W. (1991) (eds.), *Technology and the Future of Europe* (London: Pinter).

Fuchs, M. (1994) (ed.), *Nomos Kommentar zum Europäischen Sozialrecht* (Baden-Baden: Nomos).

Fursdon, E. (1980), *The European Defence Community: A History* (London: Macmillan).

Garcia, S. (1993) (ed.), *European Identity and the Search for Legitimacy* (London: Pinter).

Garrett, G. (1992), 'International Cooperation and Institutional Choice: The European Community's Internal Market', *International Organization*, 46/2: 533–60.

GATT (1993): General Agreement on Tariffs and Trade, *The European Communities: 1993*, 2 vols. (Geneva: GATT).

—— (1994), *The Results of the Uruguay Round of Multilateral Trade Negotiations: The Legal Texts* (Geneva: GATT).

Gautron, J.-C. (1991) (ed.), *Les Relations Communauté Européenne—Europe de l'Est* (Paris: Economica).

Genscher, H. D. (1982), 'Redefining the European Idea', *Europäische Zeitung*, March (London: German Embassy Press Office).

George, S. (1991), *Politics and Policy in the EC* (Oxford: Oxford University Press).

Gerber, D. (1994), 'The Transformation of European Community Competition Law?', *Harvard International Law Review*, 35/1: 97–147.

Geroski, P., and Jacquemin, A. (1985), 'Industrial Change, Barriers to Mobility, and European Industrial Policy', in Jacquemin and Sapir (1991), 298–333.

Giscard d'Estaing, V. (1995), 'Europe: Les raisons de l'echec', *Le Figaro*, 10 Jan.

Gnesotto, N. (1990), 'Defense européenne: pourquoi pas les Douze?', *Politique Etrangère*, 55/4: 881–3.

—— (1994), 'Lessons of Yugoslavia', Chaillot Paper No. 14 (Paris: WEU Institute on Security Studies).

Goldstein, J., and Keohane, R. O. (1993) (eds.), *Ideas and Foreign Policy: Beliefs, Institutions and Political Change* (Ithaca, NY: Cornell University Press).

Golt, S. (1982), *Trade Issues in the mid-1980s* (London: North Atlantic Committee).

Golub, J. (1994), 'British Integration into the EEC: A Case Study in European Environmental Policy', D.Phil. thesis (Oxford).

González Sánchez, E. (1992), *Manual del Negociador en la Comunidad Europea* (Madrid: Oficina de Información Diplomática).

Goodin, R. E., and Klingemann, H.-D. (1996) (eds.), *New Handbook of Political Science* (Oxford: Oxford University Press).

Goodman, J. (1992), *Monetary Sovereignty: The Politics of Central Banking in Western Europe* (Ithaca, NY: Cornell University Press).

Grant, C. (1994), *Delors: Inside the House that Jacques Built* (London: Nicholas Brealey).

Grant, W. (1993), 'Transnational Companies and Environmental Policymaking: The Trend of Globalization,' in Liefferink *et al.* (1993b), 59–74.

Greenwood, J., Grote, J. R., and Ronit, K. (1992) (eds.), *Organized Interests and the European Community* (London: Sage).

Gros, D., and Thygesen, N. (1992), *European Monetary Integration: From the European Monetary System towards Monetary Union* (London: Longman).

Grosser, A. (1980), *The Western Alliance: European-American Relations since 1945* (London: Macmillan).

Haas, E. B (1958), *The Uniting of Europe: Political, Social, and Economic Forces, 1950–1957* (Stanford, Cal.: Stanford University Press).

—— (1990), *When Knowledge is Power* (Berkeley: University of California Press).

Haas, P. M. (1992), 'Introduction: Epistemic Communities and International Policy Coordination', *International Organization*, 46/1: 1–35.

—— (1993), 'Protecting the Baltic and North Seas', in Haas *et al.* (1993), 133–82.

—— Keohane, R. O., and Levy, M. A. (1993) (eds.), *Institutions for the Earth: Sources of Effective International Environmental Protection* (Cambridge, Mass.: MIT Press).

Hagen, K. (1992), 'The Social Dimension: A Quest for a European Welfare State', in Ferge and Kolberg (1992), 281–303.

Haigh, N. (1992*a*), 'The European Community and International Environmental Policy', in Hurrell and Kingsbury (1992), 228–49.

—— (1992*b*) *Manual of Environmental Policy: The EC and Britain* (Harlow: Longman).

Hailbronner, K. (1994), 'Visa Regulations and Third-Country Nationals in EC Law', *Common Market Law Review*, 31/5: 969–95.

Hall, P. (1986), *Governing the Economy: The Politics of State Intervention in Britain and France* (New York: Oxford University Press).

Haller, M., and Schachner-Blazizek, P. (1994) (eds.), *Europa—Wohin? Wirtschafliche Integration, soziale Gerechtigkeit und Demokratie* (Graz: Leykam).

Hallstein, W. (1962), *United Europe: Challenge and Opportunity* (Cambridge, Mass.: Harvard University Press).

—— (1969), *Der unvollendete Bundesstaat* (Düsseldorf: Econ Verlag).

Hancher, L., and Moran, M. (1989), 'Introduction: Regulation and Deregulation', *European Journal of Political Research*, 17: 129–36.

Hardy-Bass, L. (1994), 'The U.S.–E.C. Confrontation in the GATT from a U.S. Perspective: What Did We Learn?', in Anania *et al.* (1994*b*), 236–45.

Hawkes, L. (1992), 'The EC Merger Control Regulation: Not an Industrial Policy Instrument: The De Havilland Decision', *European Competition Law Review*, 13/1: 34–48.

Hayes, J. P. (1993), *Making Trade Policy in the European Community* (London: St Martin's Press).

Hayes-Renshaw, F., and Wallace, H. (1996), *The Council of Ministers of the European Union* (London: Macmillan).

Heidhues, T., Josling, T. E., Ritson, C., and Tangermann, S. (1979), *Common Prices and Europe's Farm Policy*, Thames Essay No. 14 (London: Trade Policy Research Centre).

Held, D. (1991*a*), 'Democracy, the Nation-State and the Global System', in Held (1991*b*), 197–235.

—— (1991*b*) (ed.), *Political Theory Today* (Stanford, Cal.: Stanford University Press).

Herbst, L., Bührer, W., and Sowade, H. (1990), *Vom Marshallplan zur EWG: Die Eingliederung der Bundesrepublik Deutschland in die westliche Welt* (Munich: Oldenbourg).

Héritier, A. (1993) (ed.), *Policy-Analyse: Kritik und Neuorientierung* (Opladen: Westdeutscher Verlag).

Héritier, A. (1994) ' "Leaders" and "Laggards" in European Policy-Making: Clean-Air Policy Changes in Britain and Germany', in Waarden and Unger (1994).

—— Knill, C., and Mingers, S. (1995), *The Changing State in Europe: Changes in Regulatory Policy* (Berlin: Walter de Gruyter).

Hesse, J. (1994) (ed.), *European Yearbook of Public Administration and Comparative Government, 1994* (Oxford: Oxford University Press).

Hibbs, D. A., and Madsen, H. J. (1981), 'Public Reactions to the Growth of Taxation and Government Expenditure', *World Politics*, 33/3: 413–35.

Hildebrand, P. M. (1992), 'The European Community's Environmental Policy, 1957 to "1992": From Incidental Measures to an International Regime?', *Environmental Politics*, 1/4: 13–44.

Hindley, B. (1992), 'Exports from Eastern and Central Europe and Contingent Protection', in Flemming and Rollo (1992), 144–53.

Hirschman, A. O. (1981), *Essays in Trespassing: Economics to Politics and Beyond* (Cambridge: Cambridge University Press).

Hix, S. (1994), 'The Study of the European Community: The Challenge to European Politics', *West European Politics*, 17/1: 1–29.

Hoffmann, S. (1966), 'Obstinate or Obsolete: The Fate of the Nation State and the Case of Western Europe', *Daedalus* (Summer): 862–915.

—— (1990), 'A New World and its Troubles', *Foreign Affairs*, 69/4: 115–22.

Hogan, M. J. (1987), *The Marshall Plan: America, Britain, and the Reconstruction of Western Europe, 1947–52* (Cambridge: Cambridge University Press).

Hooghe, L. (1995) (ed.), *The European Union and Subnational Mobilization* (Oxford: Clarendon Press).

—— and Keating, M. (1994), 'The Politics of European Union Regional Policy', *Journal of European Public Policy*, 1/3: 367–93.

House of Commons (1990), *Practical Police Cooperation in the European Community*, Home Affairs Committee, Seventh Report, vol. i (London: HMSO).

House of Lords (1985), *External Competence of the European Communities*, 16th Report (London: HMSO).

—— (1993), *Enforcement of Community Competition Rules*, Session 1993–94, HL Paper 7 (London: HMSO).

Howorth, J. (1990), 'France since the Fall of the Berlin Wall: Defence and Diplomacy', *The World Today*, 46/7: 126–30.

Hucke, J. (1985), 'Environmental Policy: The Development of a New Policy Area', in Beyme and Schmidt (1985), 156–75.

Hurd, D. (1992), speech to the Cambridge University Conservative Association, 7 Feb. (London: Foreign and Commonwealth Office Verbatim Service).

Hurrell, A., and Kingsbury, B. (1992) (eds.), *The International Politics of the Environment: Actors, Interests, and Institutions* (Oxford: Clarendon Press).

Hurwitz, L., and Lequesne, C. (1991) (eds.), *The State of the European Community: Politics, Institutions and Debates in the Transition Years* (Boulder, Col.: Lynne Riener).

Inotai, A. (1994), 'Die Beziehungen zwischen der EU und den assoziierten Staaten Mittel- und Osteuropas', *Europäische Rundschau*, 22/3: 19–35.

Ireland, P. (1995), 'Fragmented Social Policy: Migration', in Leibfried and Pierson (1995*b*), 231–66.

Jachtenfuchs, M. (1993), *Ideen und Interessen: Weltbilder als Kategorien der politischen Analyse* (Mannheim: Mannheim Centre for Social Research).

—— (1995), 'Theoretical Perspectives on European Governance', *European Law Journal*, 1/2: 115–33.

—— and Huber, M. (1993), 'Institutional Learning in the European Community: The Response to the Greenhouse Effect', in Liefferink *et al.* (1993*b*).

—— and Kochler-Koch, B. (1995) (eds.), *Europäische Integration* (Leverkusen: Leske and Budrich).

Jacobs, F. (1989), 'Anti-Dumping Procedures with Regard to Imports from Eastern Europe', in Maresceau (1989*b*), 291–308.

Jacquemin, A., and Sapir, A. (1991) (eds.), *The European Internal Market: Trade and Competition* (Oxford: Oxford University Press).

—— and Wright, D. (1993*a*), 'Corporate Strategies and European Challenges Post-1992', *Journal of Common Market Studies*, 31/4: 525–37.

—— —— (1993*b*) (eds.), *The European Challenges Post-1992: Shaping Factors, Shaping Actors* (Aldershot: Edward Elgar).

Jobert, B. (1994) (ed.), *Le Tournant néolibéral en Europe* (Paris: L'Harmattan).

Joerges, C. (1994), 'European Economic Law: The Nation-State and the Maastricht Treaty', in Dehousse (1994), 29–62.

Johnson, C., and Collignon, S. (1994) (eds.), *The Monetary Economics of Europe* (London: Pinter).

Joilet, R. (1981), 'Cartelization, Dirigism and Crisis in the European Community', *The World Economy*, 3, pt. 1.

Jones, B., and Keating, M. (1994), (eds.) *The European Union and the Regions* (Oxford: Clarendon Press).

Jordan, F. (1990), 'Policy Communities: Realism versus "New" Institutionalist Ambiguity', *Political Studies*, 38/3: 470–84.

Judge, D. (1992), ' "Predestined to Save the Earth": The Environment Committee of the European Parliament', *Environmental Politics*, 1/4: 186–212.

—— and Earnshaw, D. (1994), 'Weak European Parliament Influence? A Study of the Environment Committee of the European Parliament', *Government and Opposition*, 29/2: 262–76.

Kaelble, H. (1990), *A Social History of Western Europe, 1880–1980* (Dublin: Gill and Macmillan).

Kamieniecki, S. (1993) (ed.), *Environmental Politics in the International Arena: Movements, Parties, Organizations and Policy* (Albany, NY: SUNY Press).

Kapteyn, P. (1991), ' "Civilization under Negotiation": National Civilizations and European Integration: The Treaty of Schengen', *Archives Européenes de Sociologie*, 32/2: 363–80.

Kapteyn, P. (1995), *The Stateless Market: The European Dilemma of Integration and Civilization* (London: Routledge).

Katzenstein, P. J. (1987), *Policy and Politics in West Germany: The Growth of a Semi-Sovereign State* (Philadelphia, Pa.: Temple University Press).

Keating, M. (1992), 'The Rise of the Continental Meso: Regions in the European Community', in L. J. Sharpe (ed.), *The Rise of Meso Government in Europe* (London: Sage).

Kenis, P. (1991), 'Social Europe in the 1990s: Beyond an Adjunct to Achieving a Common Market?', *Futures*, 23/7: 724–38.

Kent, P. (1992), *European Community Law* (London: Longmans).

Keohane, R. O. (1986), 'Reciprocity in International Relations', *International Organization*, 40/1: 1–28.

—— and Hoffmann, S. (1990), 'Conclusions: Community Politics and Institutional Change', in W. Wallace (1990*b*), 276–300.

Keynes, J. M. (1933), 'National Self-Sufficiency' (1982), in *The Collected Writings of John Maynard Keynes*, xxi (Cambridge: Cambridge University Press), 233–46.

Kim, C. (1992), 'Cats and Mice: The Politics of Setting EC Car Emission Standards', CEPS Working Document No. 64 (Brussels: Centre for European Policy Studies).

—— (no date), 'The Making of Maastricht: EC Environmental Policy', unpublished paper.

Kleinman, M., and Piachaud, D. (1992*a*), 'Britain and European Social Policy', *Policy Studies*, 13/3: 13–25.

—— —— (1992*b*), 'European Social Policy: Conceptions and Choices', *Journal of European Social Policy*, 3/1: 1–19.

Kohler-Koch, B. (1994), 'The Evolution of Organised Interests in the EU', Paper presented at the IPSA Congress, Berlin, Aug.

—— and Woyke, W. (1995) (eds.) *Europäische Union* (Munich: C. H. Beck).

Kolankiewicz, G. (1994), 'Consensus and Competition in the Eastern Enlargement of the European Union', *International Affairs*, 70/3: 477–95.

Kolinsky, E. (1994), 'The German Greens: Neither Left nor Right but Backwards?', *Environmental Politics*, 3/1: 164–68.

Koppen, I. (1993), 'The Role of the European Court of Justice', in Liefferink *et al.* (1993*b*), 126–49.

Kramer, H. (1993), 'The European Community's Response to the "New Eastern Europe"', *Journal of Common Market Studies*, 31/2: 213–44.

Kramer, L. (1992), *Focus on European Environmental Law* (London: Sweet and Maxwell).

Krugman, P. (1990), 'Policy Problems of a Monetary Union', in De Grauwe and Papademos (1990), 48–64.

Krupp, H.-J. (1995), 'Die Rahmenbedingungen für die Sozialpolitik auf dem Weg zur Europäischen Wirtschafts- und Währungsunion', in Riesche and Schmähl (1995).

Küsters, H. J. (1982), *Die Gründung der Europäischen Wirtschaftsgemeinschaft* (Baden-Baden: Nomos).

Ladrech, R. (1994), 'Europeanization of Domestic Politics and Institutions: The Case of France', *Journal of Common Market Studies*, 32/1: 69–88.

Laffan, B. (1996), 'The Politics of Identity and Political Order in Europe, in M. Smith (ed.), 'The European Union and a Changing Order', special issue, *Journal of Common Market Studies*, 34/1.

Lamassoure, A. (1995), 'Press Conference given by M. Alain Lamassoure, Minister Delegate for European Affairs', *Statements*, SAC/95/48 (London: French Embassy).

Landabru, E. (1994), 'How to Make Ill-placed Regions Fit to Compete', *European Affairs*, 4/2: 97–9.

Lange, P. (1992), 'The Politics of the Social Dimension', in Sbragia (1992*b*), 225–56.

—— (1993), 'The Maastricht Social Protocol: Why Did They Do It?', *Politics and Society*, 21/1: 5–36.

Langhammer, R. (1992), 'Die Assozierungsabkommen mit der CSFR, Polen und Ungarn: Wegweisend oder abweisend?', Discussion Paper No. 182 (Kiel: Kiel Institute for World Economics).

La Serre, F. de (1994), 'A la recherche d'une Ostpolitik', in La Serre *et al.* (1994), 11–41.

—— and Wallace, H. (1995), *Franco-British Relations after the Cold War* (London: Franco-British Council).

—— Lequesne, C., and Rupnik, J. (1994) (eds.), *L'Union Européenne: Ouverture à l'Est?* (Paris, Presses Universitaires de France).

La Spina, A., and Sciortino, G. (1993), 'Common Agenda, Southern Rules: European Integration and Environmental Change in the Mediterranean States', in Liefferink *et al.* (1993*b*), 217–36.

Laursen, F., and Hoonacker, S. van (1992) (eds.), *The Intergovernmental Conference on Political Union* (Maastricht: European Institute of Public Administration).

Lawson, D. (1990), 'Saying the Unsayable about the Germans', *The Spectator*, 14 July, 9.

Legg, W. (1993/4), 'Direct Payments for Farmers?', *The OECD Observer*, 185, Dec./Jan., 26–8.

Leibfried, S. (1994), 'The Social Dimension of the European Union: En Route to Positively Joint Sovereignty?', *Journal of European Social Policy*, 4/4: 239–62.

—— and Pierson, P. (1992), 'Prospects for Social Europe', *Politics and Society*, 20/3: 333–66.

—— —— (1995*a*), 'Semi-Sovereign Welfare States: Social Policy in a Multi-Tiered Europe', in Leibfried and Pierson (1995*b*), 43–77.

—— —— (1995*b*) (eds.), *European Social Policy: Between Fragmentation and Integration* (Washington, DC: Brookings Institution).

Lenz, C. O. (1994) (ed.) with the collaboration of Birk, R., *et al.*, *EG-Handbuch, Recht im Binnenmarkt*, 2nd edn. (Herne/Berlin: Verlag Neue Wirtschafts-Briefe).

Lepsius, M. R. (1991), 'Nationalstaat oder Nationalitätenstaat als Modell für die Weiterentwicklung der Europäischen Gemeinschaft', in Wildenmann (1991), 19–40.

Lequesne, C. (1991), 'Les Accords de commerce et de coopération Communauté européenne—pays d'Europe de l'Est', in Gautron (1991), 357–71.

Leslie, P. M. (1991), *The European Community: A Political Model for Canada?* (Ottawa: Minister of Supply and Services).

Liefferink, J. D., Lowe, P. D., and Mol, A. J. P. (1993*a*), 'The Environment and the European Community', in Liefferink *et al.* (1993*b*), 1–13.

—— —— —— (1993*b*) (eds.), *European Integration and Environmental Policy* (London: Belhaven).

Lindberg, L. N. (1963), *The Political Dynamics of European Economic Integration* (Stanford, Cal.: Stanford University Press).

—— and Scheingold, S. A. (1970), *Europe's Would-Be Polity* (Englewood Cliffs, NJ: Prentice Hall).

—— —— (1971) (eds.), *Regional Integration: Theory and Research* (Cambridge, Mass.: Harvard University Press).

Lodge, J. (1993) (ed.), *The European Community and the Challenge of the Future*, 2nd edn. (London: Pinter).

Loth, W., Wallace, W., and Wessels, W. (1995) (eds.), *Walter Hallstein: Der vergessene Europäer?* (Bonn: Europa Union Verlag).

Lowi, T. (1972), 'Four Systems of Policy, Politics and Choice', *Public Administration Review*, 32/4: 298–310.

Ludlow, P. (1982), *The Making of the European Monetary System* (London: Butterworths).

—— (1989), 'Beyond 1992: Europe and its Western Partners', CEPS Working Paper No. 38 (Brussels: Centre for European Policy Studies).

—— (1995), *L'Équilibre européen: Études rassemblées et publiées en hommage à Niels Ersbøll* (Brussels: CEPS).

Maasacher, M. H. J. M. van, and Arentsen, M. J. (1990), 'Environmental Policy', in Wolters and Coffey (1990), 70–8.

McAleavey, P. (1994), 'The Political Logic of the European Community Structural Funds Budget: Lobbying Efforts by Declining Industrial Regions', EUI Working paper RSC No. 942 (Florence: European University Institute).

MacDougall, D., *et al.* (1977), 'Report of the Study Group on the Role of Public Finance in European Integration', Economic and Financial Series, A 13, Apr. (Brussels: Commission of the EC).

McGowan, F. (1993), *The Struggle for Power in Europe: Competition and Regulation in the Electricity Industry* (London: Royal Institute of International Affairs).

McLaughlin, A., Jordan, G., and Maloney, W. A. (1993), 'Corporate Lobbying in the European Community', *Journal of Common Market Studies*, 31/2: 191–212.

Majone, G. (1989), *Evidence, Argument and Persuasion in the Policy Process* (New Haven: Yale University Press).

—— (1991), 'Cross-National Sources of Regulatory Policymaking in Europe and the United States', *Journal of Public Policy*, 2/1: 79–106.

—— (1992), 'Regulatory Federalism in the European Community', *Environment and Planning C: Government and Policy*, 10/3: 299–316.

—— (1993), 'The European Community Between Social Policy and Social Regulation', *Journal of Common Market Studies*, 31/2: 153–70.

——(1994*a*), 'Independence vs. Accountability? Non-Majoritarian Institutions and Democratic Government in Europe', in Hesse (1994).

—— (1994*b*), 'The Rise of the Regulatory State in Europe', *West European Politics*, 17/3: 77–101.

—— (1996) 'Public Policy: Ideas, Interests and Institutions', in Goodin and Klingemann (1996).

Maresceau, M. (1989*a*), 'A General Survey of the Current Legal Framework of Trade Relations Between the European Community and Eastern Europe', in Maresceau (1989*b*), 3–20.

—— (1989*b*) (ed.), *The Political and Legal Framework of Trade Relations Between the European Community and Eastern Europe* (Dordrecht: Nijhoff).

—— (1994) (ed.), *The European Commercial Policy after 1992: The Legal Dimension* (Dordrecht: Nijhoff).

Marjolin, R. (1989), *Architect of European Unity: Memoirs 1911–1986* (London: Weidenfeld and Nicolson).

Marks, G. (1992), 'Structural Policy in the European Community', in Sbragia (1992*b*): 191–224.

—— (1993), 'Structural Policy and Multilevel Governance in the EC', in Cafruny and Rosenthal (1993), 39–410.

Marshall, T. H. (1963), 'Citizenship and Social Classes', in id. (ed.), *Sociology at the Crossroads, and Other Essays* (London: Heinemann), 67–127.

Marston, G. (1973), 'The Continental Can Case', *Journal of World Trade Law*, 7/4: 476–82.

Matlary, J. H. (1993), 'The Development of Environmental Policy-Making in Hungary: The Role of the EC', CICERO Working Paper No. 2 (Blinden: CICERO).

—— (forthcoming), *Energy Policy in the European Union* (London: Macmillan).

Matthews, D., and Mayes, D. G. (1995), 'The Role of Soft Law in the Evolution of Rules for a Single European Market: The Case of Retailing', paper delivered to the Fourth Biennial International Conference of the European Community Studies Association, Charleston, SC, 11–14 May.

May, Bernhard (1994), *Die Uruguay-Runde: Verhandlungsmarathon verhindert trilateralen Handelskrieg* (Bonn: DGAP).

Maydell, B., Baron von (1991), 'Einführung in die Schlußdiskussion', in Schulte and Zacher (1991), 229–36.

Mazey, S., and Richardson, J. (1992), 'Environmental Groups and the EC: Challenges and Opportunities', *Environmental Politics*, 1/4: 109–28.

—— —— (1993*a*), 'Introduction: Transference of Power, Decision Rules, and Rules of the Game', in Mazey and Richardson (1993*b*), 3–26.

—— —— (1993*b*) (eds.), *Lobbying in the European Community* (Oxford: Oxford University Press).

—— —— (1994), 'Policy Coordination in Brussels: Environmental and Regional Policy,' in S. Baker *et al.* (1994), 22–44.

Menon, A., Forster, A., and Wallace, W. (1992), 'A Common European Defence?', *Survival*, 34/3: 98–118.

Mény, Y., Muller, P., and Quermonne, J.-L. (1995) (eds.), *Les Politiques publiques en Europe* (Paris: L'Harmattan).

Messerlin, P. (1993), 'The EC and Central Europe: The Missed Rendez-Vous of 1992?', *Economics of Transition*, 1/1: 89–109.

Mestmäcker, E.-J. (1972), 'Concentration and Competition in the EEC', *Journal of World Trade Law*, 6/6: 615–47 and 7/1: 36–63.

Metcalfe, L. (1992), 'After 1992: Can the Commission Manage Europe?' *Australian Journal of Public Administration*, 51/1: 117–30.

Miall, H. (1994) (ed.), *Redefining Europe: New Patterns of Conflict and Cooperation* (London: Pinter).

Michalski, A., and Wallace, H. (1992), *The European Community: The Challenge of Enlargement* (London: Royal Institute of International Affairs).

Millstone, E. (1991), 'Consumer Protection Policies in the EC: The Quality of Food', in Freeman *et al.* (1991), 330–43.

Milward, A. S. (1981), 'Tariffs as Constitutions', in Strange and Tooze (1981), 57–66.

—— (1984), *The Reconstruction of Western Europe, 1945–51* (London: Methuen).

—— (1992), *The European Rescue of the Nation-State* (Berkeley: University of California Press).

—— Lynch, F. M. B., Romero, F., and Srensen, V. (1993) (eds.), *The Frontier of National Sovereignty: History and Theory, 1945–1992* (London: Routledge).

Mint, J., and Simeon, R. (1982), *Conflict of Taste and Conflict of Claim in Federal Countries* (Toronto: IIR).

Moens, A., and Anstis, C. (1994) (eds.), *Disconcerted Europe: The Search for a New Security Architecture* (Oxford: Westview).

Monar, J. and Morgan, R. (1994) (eds.), *The Third Pillar of the European Union— Cooperation in the fields of justice and home affairs* (Brusels: European Interuniversity Press).

Le Monde, 13 Mar. 1994.

Moravcsik, A. (1991), 'Negotiating the Single European Act: National Interests and Conventional Statecraft in the European Community', *International Organization*, 45/1: 19–56.

—— (1993), 'Preferences and Power in the European Community: A Liberal Inter-governmentalist Approach', *Journal of Common Market Studies*, 31/4: 473–524.

—— (1994), 'Why the European Community Strengthens the State: Domestic Politics and International Cooperation', unpublished paper.

Moreno, L. (1993) (ed.), *Social Exchange and Welfare Development* (Madrid: Consejo Superior de Investigaciones Científicas).

Morgan, R., and Bray, C. (1984) (eds.), *Partners and Rivals in Western Europe: Britain, France and Germany* (Aldershot: Gower).

Mortensen, J. (1990), 'Federalism vs. Coordination: Macroeconomic Policy in the European Community', CEPS Paper No. 47 (Brussels: Centre for European Policy Studies).

Mosley, H. G. (1990), 'The Social Dimension of European Integration', *International Labour Review*, 129/2: 147–64.

Müller, W. C., and Wright, V. (1994) (eds.), *The State in Western Europe: Retreat or Redefinition?* (Ilford: Frank Cass).

Müller-Graff, P.-C. (1994), 'The Legal Bases of the Third Pillar and its Position in the Framework of the Union Treaty', *Common Market Law Review*, 31/3: 493–510.

Murphy, A. (1990), *The European Community and the International Trading System*, 2 vols. (Brussels: Centre for European Policy Studies).

Myrdal, G. (1957), 'Economic Nationalism and Internationalism', *Australian Outlook*, 11/4: 3–50.

Noetzold, J. (1995), 'European Union and Eastern Central Europe: Expectations and Uncertainties', *Aussenpolitik*, 46/1: 14–23.

North, D. C. (1990), *Institutions, Institutional Change and Economic Performance* (Cambridge: Cambridge University Press).

Neville Brown, L. (1981), 'Agrimonetary Byzantinism and Prospective Overruling', *Common Market Law Review*, 18/4: 509–19.

Neville Brown, L., and Kennedy, T. (1994), *The Court of Justice of the European Communities*, 7th edn. (London: Sweet and Maxwell).

Nuttall, S. J. (1992), *European Political Co-operation* (Oxford: Clarendon Press).

—— (1994), 'The Commission and Foreign Policy-Making', in Edwards and Spence (1994), 287–303.

—— and Edwards, G. (1994), 'Common Foreign and Security Policy', in Duff *et al.* (1994), 84–103.

OECD (1993*a*): Organization for Economic Cooperation and Development, *Assessing the Effects of the Uruguay Round* (Paris: OECD).

—— (1993*b*), *OECD Environmental Performance Reviews: Germany* (Paris: OECD).

—— (1994), *Economic Outlook*, 56 (Dec.).

O'Keeffe, D., and Twomey, P. M. (1994) (eds.), *Legal Issues of the Maastricht Treaty* (London: Chancery Law).

Olsen, J. P. (1995), 'Europeanization and Nation-State Dynamics', ARENA Working Paper No. 9 (Oslo: ARENA).

Ostner, I., and Lewis, J. (1995), 'Gender and the Evolution of European Social Policy', in Leibfried and Pierson (1995*b*), 159–93.

Paarlberg, R. L. (1992), 'How Agriculture Blocked the Uruguay Round,' *SAIS Review*, 12: 27–42.

—— (1995), *Leadership Begins at Home: The United States in a More Deeply Integrated World Economy* (Washington, DC: Brookings).

Padgett, S. (1992), 'The Single European Energy Market: The Politics of Realization', *Journal of Common Market Studies*, 31/3: 53–75.

Padoa-Schioppa, T. (1987), *Efficiency, Stability and Equity* (Oxford: Oxford University Press).

—— (1994), *The Road to Monetary Union in Europe* (Oxford: Clarendon Press).

Parekh, B. (1994), 'Discourses on National Identity', *Political Studies*, 42/3: 492–504.

Paterson, W. E. (1991), 'Regulatory Change and Environmental Protection in the British and German Chemical Industries', *European Journal of Political Research*, 19: 307–26.

Pearce, J. (1983), 'The Common Agricultural Policy: The Accumulation of Special Interests', in H. Wallace *et al.* (1983), 143–76.

—— and Sutton, J. (1985), *Protection and Industrial Policy in Europe* (London: Routledge).

Pedler, R. H. (1994), 'The Fruit Companies and the Banana Trade Regime (BTR)', in Pedler and Van Schendelen (1994), 67–91.

—— and Van Schendelen, M. P. C. M. (1994) (eds.), *Lobbying the European Union: Companies, Trade Associations and Issue Groups* (Aldershot: Dartmouth).

Peffekoven, R. (1994), *Die Finanzen der Europäischen Union* (Mannheim: BI Taschenbuch Verlag).

Pelkmans, J. (1984), *Market Integration in the European Community* (The Hague: Martinus Nijhoff).

—— (1990), 'Regulation and the Single Market: An Economic Perspective', in Siebert (1990*b*): 91–125.

—— and Murphy, A. (1991), 'Catapulted into Leadership: The Community's Trade and Aid Policies vis-à-vis Eastern Europe', *Journal of European Integration*, 14/2–3: 125–51.

—— and Winters, L. A. (1988), *Europe's Domestic Market* (London: Royal Institute of International Affairs).

Pentland, C. (1973), *International Theory and the European Community* (London: Faber).

Peters, B. G. (1992), 'Bureaucratic Politics and the Institutions of the European Community', in Sbragia (1992*b*).

—— (1994), 'Agenda-Setting in the European Community', *Journal of European Public Policy*, 1/1: 9–26.

Petersen, J. H. (1991), 'Harmonization of Social Security in the EC Revisited', *Journal of Common Market Studies*, 29/5: 505–26.

—— (1993), 'Europäischer Binnenmarkt, Wirtschafts- und Währungsunion und die Harmonisierung der Sozialpolitik', *Deutsche Rentenversicherung*, 1–2 (Jan./Feb.), 15–49.

Petersen, N. (1993), 'Game, Set and Match: Denmark and the European Union from Maastricht to Edinburgh', in Tiilikainen and Damgaard Petersen (1993), 79–106.

Peterson, J. (1994), 'Europe and America in the Clinton Era', *Journal of Common Market Studies*, 32/3: 411–26.

—— (1995*a*), 'Decision-Making in the European Union: Towards a Framework for Analysis', *Journal of European Public Policy*, 2/1: 69–93.

—— (1995*b*), 'European Union R&D Policy: The Politics of Expertise', in Rhodes and Mazey (1995).

Peterson, P. E., and Rom, M. C. (1990), *Welfare Magnets: A New Case for a National Standard* (Washington, DC: Brookings Institution).

Pierson, P. (1995), 'The Path to European Integration: A Historical Institutionalist Analysis', Center for European Studies Working Paper No. 58 (Cambridge, Mass.: Harvard University).

—— (1996), 'The New Politics of the Welfare State', *World Politics*, 48/2.

—— and Leibfried, S. (1995*a*), 'The Dynamics of Social Policy Integration', in Leibfried and Pierson (1995*b*), 432–65.

—— —— (1995*b*), 'Multi-Tiered Institutions and the Making of Social Policy', in Leibfried and Pierson (1995*b*), 1–40.

Pijpers, A., Regelsberger, E., and Wessels, W. (1988) (eds.), *European Political Cooperation in the 1980s: A Common Foreign Policy for Western Europe?* (Dordrecht: Nijhoff).

Pinder, J. (1968), 'Positive Integration and Negative Integration: Some Problems of Economic Union in the EEC', *The World Today*, 24/3: 88–110.

—— (1991), *The European Community and Eastern Europe* (London: Pinter).

—— and Pinder, P. (1975), 'The European Community's Policy towards Eastern Europe', Chatham House European Series No. 25 (London: Royal Institute of International Affairs).

Polanyi, K. (1994), *The Great Transformation* (New York, NY: Rinehart).

Poole, K. T., and Rosenthal, H. (1993), 'Spatial Realignment and the Mapping of Issues in American History: Evidence from Roll Call Voting', in Riker (1993), 13–39.

Pridham, G. (1994), 'National Environmental Policy-Making in the European Framework: Spain, Greece and Italy in Comparison', in Baker *et al.* (1994), 80–101.

Pryce, R. (1994), 'The Treaty Negotiations', in Duff *et al.* (1994), 36–52.

Puchala, D. (1971), 'International Transaction and Regional Integration', in Lindberg and Scheingold (1971), 128–59.

Putnam, R. D. (1988), 'Diplomacy and Domestic Politics: The Logic of Two-Level Games', *International Organization*, 43/2: 427–60.

Quermonne, J.-L. (1994), *Le Système politique européenne* (Paris: Montchristiensen).

Rausser, G. C., and Irwin, D. A. (1988), 'The Political Economy of Agricultural Policy Reform', *European Review of Agricultural Economics*, 15: 349–66.

Read, R. A. (1994), 'The EC Internal Banana Market: The Issues and the Dilemma', *World Economy*, 17/2: 219–35.

Rehbinder, E., and Steward, R. (1985), *Environmental Protection Policy: Integration Through Law* (Berlin: Walter de Gruyter).

Reif, K. (1994), 'Less Legitimation through Lazy Parties? Lessons from the 1994 European Elections', paper presented to the XVIth World Congress of the International Political Science Association, Berlin, 21–5 Aug.

Rein, M., and Rainwater, L. (1986) (eds.), *Public–Private Interplay in Social Protection: A Comparative Study* (Armonk, NY: M. E. Sharpe).

Reiner, R., and Spencer, S. (1993) (eds.), *Accountable Policing: Effectiveness, Empowerment and Equity* (London: Institute for Public Policy Research).

Reinicke, W. (1992), *Building a New Europe: The Challenge of System Transformation and Systemic Reform* (Washington, DC: Brookings Institution).

Reinke, S. (1992), 'The EC Commission's Anti-Fraud Activity', in Anderson and Den Boer (1992), 13–30.

Ress, G. (1994), 'The Constitution and the Maastricht Treaty: Between Cooperation and Conflict', *German Politics*, 3/3: 47–74.

Rhodes, C., and Mazey, S. (1995) (eds.), *The State of the European Community*, iii (Boulder, Col.: Lynne Riener).

Rhodes, M. (1995), 'A Regulatory Conundrum: Industrial Relations and the "Social Dimension" ', in Leibfried and Pierson (1995*b*), 78–122.

Richardson, J. (1994), 'EU Water Policy: Uncertain Agendas, Shifting Networks and Complex Coalitions', *Environmental Politics*, 3/4: 139–67.

Richter, J. H. (1964), *Agricultural Protection and Trade: Proposals for an International Policy* (New York: Praeger).

Rieger, E. (1995*a*), 'Politik supranationaler Integration: Die Europäische Gemeinschaft in institutionentheoretischer Perspektive', Arbeitspapiere Arbeitsbereich I/9 (Mannheim: Mannheim Centre for Social Research).

—— (1995*b*) 'Protective Shelter or Strait-Jacket? An Institutional Analysis of the Common Agricultural Policy', in Leibfried and Pierson (1995*b*), 194–230.

—— (1995*c*), 'Der Wandel der Landwirtschaft in der Europäischen Union: Ein Beitrag zur soziologischen Analyse transnationaler Integrationsprozesse', *Kölner Zeitschrift für Soziologie und Sozialpsychologie*, 47/1: 65–94.

—— and Leibfried, S. (1995), *Globalization and the Western Welfare State: An Annotated Bibliography* (Bremen: Centre for Social Policy Research).

Riesche, H., and Schmähl, W. (1995) (eds.), *Handlungsspielräume nationaler Sozialpolitik* (Baden-Baden: Nomos).

Riker, W. (1993) (ed.), *Agenda Formation* (Ann Arbor: University of Michigan Press).

Riley, L. (1993), 'Counterterrorism in Western Europe: Mechanisms for International Cooperation', Working Paper No. X (Edinburgh: University of Edinburgh, Department of Politics).

Robertson, D. B. (1989), 'The Bias of American Federalism: The Limits of Welfare State Development in the Progressive Era', *Journal of Policy History*, 1/3: 261–91.

Rollo, J. M. C., and Smith, M. A. M. (1993), 'The Political Economy of Eastern European Trade with the European Community: Why so Sensitive?', *Economic Policy*, 16 (Apr.), 139–81.

—— and Wallace, H. (1991), 'New Patterns of Partnership', in Bonvicini *et al.* (1991), 53–64.

Romero, F. (1993), 'Migration as an Issue in European Interdependence and Integration: The Case of Italy', in Milward *et al.* (1993), 33–58 and 205–8.

Rosenblatt, J., Mayer, T., Bartholdy, K., Demekas, D., Gupta, S., and Lipschitz, L. (1988), *The Common Agricultural Policy: Principles and Consequences* (Washington, DC: International Monetary Fund).

Ross, G. (1994), *Jacques Delors and European Integration* (Oxford: Polity Press).

—— (1995), 'Assessing the Delors Era in Social Policy', in Leibfried and Pierson (1995*b*), 357–88.

Rouam, C. (1994), 'L'Union européenne face aux pays d'Europe centrale et orientale: Délocalisations industrielles ou harmonisation des conditions de concurrence?' *Revue du Marché Commun et de l'Union Européenne*, 383: 643–48.

Rudig, W., and Krämer, R. A. (1994), 'Networks of Cooperation: Water Policy in Germany', *Environmental Politics*, 3/4: Winter, 52–79.

Ruggie, J. G. (1993) (ed.), *Multilateralism Matters: The Theory and Praxis of an International Form* (New York: Columbia University Press).

Ruggiero, R. (1991), 'The Place of the GATT Trading System in the European Community's External Relations', remarks to the Royal Institute of International Affairs, London, 6 Mar.

Ruimschotel, D. (1993), 'The EC Budget: Ten Per Cent Fraud? A Policy Analysis', EUI Working Paper No. 93/8 (Florence: European University Institute).

Sabatier, P. A. (1988), 'An Advocacy Coalition Framework of Policy Change and the Role of Policy-Oriented Learning Therein', *Policy Sciences*, 21: 129–68.

Sandholtz, W. (1993), 'Institutions and Collective Action: The New Telecommunications in Western Europe', *World Politics*, 45/2: 242–70.

—— and Zysman, J. (1989), '1992: Recasting the European Bargain', *World Politics*, 42/1: 95–128.

Santer, J. (1995), 'Speech to the European Parliament', 17 Jan.

Sarris, A. H. (1993), 'Implications of EC Economic Integration for Agriculture, Agricultural Trade, and Trade Policy', *Journal of Economic Integration*, 8/2: 175–200.

Saryusz-Wolski, J. (1994), 'The Reintegration of the "Old Continent": Avoiding the Costs of "Half-Europe" ', in Bulmer and Scott (1994), 19–28.

Sbragia, A. M. (1992a), 'Thinking about the European Future: The Uses of Comparison', in Sbragia (1992b), 257–91.

—— (1992b) (ed.), *Euro-Politics: Institutions and Policymaking in the 'New' European Community* (Washington, DC: Brookings Institution).

Scharpf, F. W. (1988), 'The Joint-Decision Trap: Lessons from German Federalism and European Integration', *Public Administration*, 66/3: 239–78.

—— (1991), 'Die Handlungsfähigkeit des Staates am Ende des 20. Jahrhunderts', *Politische Vierteljahresschrift*, 32/4: 475–502.

—— (1994a), 'Community and Autonomy: Multi-Level Policy-Making in the European Union', *Journal of European Public Policy*, 1/2: 219–42.

—— (1994b) 'Mehrebenenpolitik im vollendeten Binnenmarkt', *Staatswissenschaft und Staatspraxis*, 5/4: 475–502.

—— (1994c), *Optionen des Föderalismus in Deutschland* (Frankfurt am Main: Campus).

Schäuble, Wolfgang, and Lamers, Carl (1994), *Reflections on European Policy* (Bonn: CDU/CSU Fraktion).

Schiller, K. (1939), *Marktregulierung und Marktordnung in der Weltagrarwirtschaft* (Jena: Fischer).

Schmitter, P. C. (1992), 'Interests, Powers, and Functions: Emergent Properties and Unintended Consequences in the European Polity', unpublished paper.

Schnapper, D. (1992), 'L'Europe, marché ou volonté politique?', *Commentaire*, 60 (Winter).

Schnutenhaus, J. (1994), 'Integrated Pollution Prevention and Control: New German Initiatives in the European Environment Council', *European Environmental Law Review*, 3/11: 323–28.

Schott, J. (1994), *The Uruguay Round: An Assessment* (Washington, DC: Institute for International Economics).

Schoutheete, P. de (1986), *La Coopération politique européenne*, 2nd edn. (Brussels: Labor).

—— (1993), 'Reflexions sur le Traité de Maastricht', *Annales de Droit de Louvain*, 1: 73–90.

Schreiber, K. (1991), 'The New Approach to Technical Harmonization and Standards', in Hurwitz and Lequesne (1991), 97–112.

Schulte, B. (1991), 'Einführung in die Schlußdiskussion', in Schulte and Zacher (1991), 237–52.

—— (1994a), 'Comments on Articles 4, 10a, and 5 of Reg. 1408/71', in Fuchs (1994).

—— (1994b), 'Sozialrecht', in Lenz (1994), 407–78.

—— and Zacher, H. F. (1991) (eds.), *Wechselwirkungen zwischen dem Europäischen Sozialrecht und dem Sozialrecht der Bundesrepublik Deutschland* (Berlin: Duncker and Humblot).

Schultz, T. W. (1943), *Redirecting Farm Policy* (New York: Macmillan).

Schulze, H. (1994), *Staat und Nation in der europäischen Geschichte* (Munich: C. H. Beck).

Schuppert, G. F. (1994), 'Zur Staatswerdung Europas: Überlegungen zu Bedingungsfaktoren und Perspektiven der europäischen Verfassungsentwicklung', *Staatswissenschaft und Staatspraxis*, 5/1: 35–76.

Schutte, J. J. E. (1990), 'Strafrecht in Europees Verband', *Justitiële Verkenningen*, 16/9: 8–17.

Schweitzer, C.-C., and Detef, K. (1990) (eds.), *The Federal Republic of Germany and EC Membership Evaluated* (New York: St. Martin's Press).

Scott, A. (1993), 'Financing the Community: The Delors II Package', in Lodge (1993), 69–88.

Scott, J. (1995), *Development Dilemmas in the European Community* (Buckingham: Open University Press).

Sedelmeier, U. (1994), 'The European Union's Association Policy towards Central and Eastern Europe: Political and Economic Rationales in Conflict', SEI Working Paper No. 7 (Falmer: Sussex European Institute).

Shackleton, M. (1983), 'Fishing for a Policy? The Common Fisheries Policy of the Community', in H. Wallace et al. (1983), 349–72.

—— (1990), *Financing the European Community* (London: Pinter).

—— (1993a), 'The Community Budget after Maastricht', in Cafruny and Rosenthal (1993), 373–90.

—— (1993b), 'Keynote Article: The Delors II Budget Package', *Journal of Common Market Studies*, 31, Annual Review of Activities, Aug. 11–26.

Shapiro, M., and Stone, A. (1994), 'The New Constitutional Politics of Europe', *Comparative Political Studies*, 26/4: 397–420.

Siebert, H. (1990a), 'The Harmonization Issue in Europe: Prior Agreement or a Competitive Process,' in Siebert (1990b), 53–90.

—— (1990b) (ed.), *The Completion of the Internal Market* (Tübingen: J. C. B. Mohr).

Sinn, H.-W. (1994), 'Wieviel Brüssel braucht Europa? Subsidiarität, Zentralisierung und Fiskalwettbewerb im Lichte der ökonomischen Theorie', *Staatswissenschaft und Staatspraxis*, 5/2: 155–86.

Skjærseth, J. B. (1994), 'The Climate Policy of the EC: Too Hot to Handle?', *Journal of Common Market Studies*, 32/1: 25–45.

Smith, A. (1995), 'L'Intégration communautaire face au territoire. Les Fonds structurels et les zones rurales en France, en Espagne et au Royaume Uni', Ph.D. thesis (Grenoble).

Smith, A. D. (1992), 'National Identity and the Idea of European Identity', *International Affairs*, 68/1: 55–76.

Smith, J. (1995), *Voice of the People: The European Parliament in the 1990s* (London: RIIA).

Smith, M. A. M., and Wallace, H. (1994), 'The European Union: Towards a Policy for Europe', *International Affairs*, 70/3: 429–44.

—— Holmes, P. M., Sedelmeier, U., Smith, E., Wallace, H., and Young, A. R. (1995), 'The European Union and Central and Eastern Europe: Pre-accession Strategies', SEI Working Paper No. 15 (Falmer: Sussex European Institute).

Snyder, F. G. (1985), *Law of the Common Agricultural Policy* (London: Sweet and Maxwell).

Soetendorp, B. (1990), 'The Evolution of the EC/EU as a Single Foreign Policy Actor', in Carlsnaes and Smith (1990), 103–19.

Spence, D. (1991), 'Enlargement without Accession: The EC's Response to German Unification', RIIA Discussion Paper No. 36 (London: Royal Institute of International Affairs).

Spencer, M. (1990), *1992 and All That: Civil Liberties in the Balance* (London: Civil Liberties Trust).

Spierenburg *et al.* (1979), *Proposals for Reform of the Commission of the European Communities and its Services* (Brussels: Commission of the EC).

Spinelli, A., *et al.* (1983): Committee on Institutional Affairs, 'Report on the Substance of the Preliminary Draft Treaty Establishing the European Union', PE 83.326/fin., European Parliament Working Document 1575/83.

Stein, J. G. (1993), 'Political Economy and Security', in Evans, Jacobson, and Putnam (1993), 77–103.

Stern, J. (1990), *European Gas Markets: Challenge and Opportunity in the 1990s* (London: Royal Institute of International Affairs).

Stevens, C. (1991), 'The Caribbean and Europe 1992: Endgame?', *Development Policy Review*, 9: 265–83.

—— (1992), 'The EC and the Third World', in D. Dyker (ed.), *The European Economy* (London: Longman): 211–29.

—— and Webb, C. (1983), 'The Political Economy of Sugar: A Window on the CAP', in H. Wallace *et al.* (1983), 321–48.

Stone, D. A. (1989), 'At Risk in the Welfare State', *Social Research*, 56/3: 591–633.

Story, J. (1993) (ed.), *The New Europe: Politics, Government, and Economy since 1945* (Oxford: Blackwell).

Strange, M. (1988), *Family Farming: A New Economic Vision* (Lincoln, Nebr. and London: University of Nebraska Press).

Strange, S., and Tooze, R. (1981) (eds.), *The International Politics of Surplus Capacity: Competition for Market Shares in the World Recession* (London: Allen and Unwin).

Strasser, D. (1992), *The Finances of Europe*, 7th edn. (Luxemburg: Office for Official Publications of the European Communities).

Streeck, W. (1992), *Social Institutions and Economic Performance: Studies of Industrial Relations in Advanced Capitalist Economies* (London: Sage).

—— (1995), 'From Market Making to State Building?', in Leibfried and Pierson (1995b), 389–431.

—— and Schmitter, P. C. (1991), 'From National Corporatism to Transnational Pluralism: Organized Interests in the Single European Market', *Politics and Society*, 19/2: 133–64.

Sun, J.-M., and Pelkmans, J. (1995), 'Regulatory Competition and the Single Market', *Journal of Common Market Studies*, 33/1: 67–89.

Sutherland, P., *et al.* (1992): High Level Group on the Operation of the Internal Market, *The Internal Market after 1992: Meeting the Challenge* (Luxemburg: Office for Official Publications of the European Communities).

Swaan, A. de (1992), 'Perspectives for Transnational Social Policy', *Government and Opposition*, 27/1: 33–52.

Swann, D. (1983), *Competition and Competition Policy* (London: Methuen).

—— (1992) (ed.), *The Single European Market and Beyond* (London: Routledge).

Swinbank, A. (1989), 'The Common Agricultural Policy and the Politics of European Decision-Making', *Journal of Common Market Studies*, 27/4: 303–22.

Sylvia, S. J. (1991), 'The Social Charter of the European Community: A Defeat for European Labor', *Industrial and Labor Relations Review*, 44/4: 626–43.

Taschner, H. C. (1990), 'Schengen oder die Abschaffung der Personenkontrollen an den Binnengrenzen der EG', Vorträge, Reden und Berichte aus dem Europa-Institut, No. 227 (Saarbrücken: Universität des Saarlandes).

Thatcher, M. (1984), 'Europe—the Future', paper delivered to the European Council, Fontainebleau, 25–6 June.

Tiersky, R. (1992), 'France in the New Europe', *Foreign Affairs*, 71/2: 131–46.

Tiilikainen, T., and Damgaard Petersen, I. (1993) (eds.), *The Nordic Countries and the EC* (Copenhagen: Copenhagen Political Studies Press).

Toner, G. (1987), 'The IEA and the Development of the Stocks Decision', *Energy Policy*, 5/1: 40–58.

Topfer, K. (1992), 'The ECO-nomic Revolution: Challenge and Opportunity for the Twenty-First Century', *International Environmental Affairs*, 4/3: 273–80.

Tracy, M. (1989), *Government and Agriculture in Western Europe* (New York: Harvester Wheatsheaf).

Transpol (1994) (ed.), *Internationalisering door grenzeloze samenwerking* (Lelystad: Koninklijke Vermande bv).

Tsakaloyannis, P. (1991), 'The Acceleration of History and the Reopening of the Political Debate in the European Community', *Journal of European Integration*, 14/2–3: 87–9.

Tsebelis, G. (1990), *Nested Games* (Berkeley: University of California Press).

—— (1994), 'The Power of the European Parliament as a Conditional Agenda-Setter', *American Political Science Review*, 88/1: 128–42.

Tsoukalis, L. (1977), *The Politics and Economics of European Monetary Integration* (London: Allen and Unwin).

—— (1993), *The New European Economy: The Politics and Economics of Integration*, 2nd edn. (Oxford: Oxford University Press).

—— and Silva Ferreira, A. da (1980), 'Management of Industrial Surplus Capacity in the European Community', *International Organization*, 34/3: 355–75.

Tugendhat, C. (1985), 'How to Get Europe Moving Again', *International Affairs*, 61/3: 421–29.

Usher, J. A. (1988), *Legal Aspects of Agriculture in the European Community* (Oxford: Clarendon Press).

Valence, G. (1990), 'L'Engrenage européen', *L'Express*, 19 Oct.

Van der Wel, J., and Bruggeman, W. (1993), *Europese Politiële Samenwerking: Internationale Gremia* (Brussels: Politeia).

Van Outrive, L. (1992*a*), 'The Entry into Force of the Schengen Agreements' (Brussels: European Parliament Committee on Civil Liberties and Internal Affairs).

—— (1992*b*), 'Police Cooperation' (Brussels: European Parliament Committee on Civil Liberties and Internal Affairs).

Vasey, M. (1988), 'Decision-Making in the Agricultural Council and the "Luxemburg Compromise" ', *Common Market Law Review*, 25/4: 725–32.

Vaubel, R. (1994), 'The Political Economy of Centralization and the European Community', *Public Choice*, 59/1: 151–85.

Verhoeve, B., Bennett, G., and Wilkinson, D. (1992), *Maastricht and the Environment* (London: Institute for European Environmental Policy).

Vernet, D. (1992), 'The Dilemma of French Foreign Policy', *International Affairs*, 68/4: 655–64.

Vesterdorf, B. (1994), 'Complaints Concerning Infringements of Competition Law within the Context of European Community Law', *Common Market Law Review*, 32/1: 77–104.

Vibert, Frank (1994), *The Future Role of the European Commission* (London: European Policy Forum).

Villain, C., and Arnold, R. (1990), 'New Directions for European Agricultural Policy', report of the CEPS CAP Expert Group (Brussels: Centre for European Policy Studies).

Viségrad (1992), 'Memorandum of the Governments of the Czech and Slovak Federal Republic, the Republic of Hungary, and the Republic of Poland on Strengthening their Integration with the European Community and on the Perspective of Accession', 11 Sept.

—— (1993), 'Aide-mémoire', 2 June.

Vogel, D. (1993a), 'Environmental Policy in the European Community', in Kamieniecki (1993), 181–97.

—— (1993b), 'The Making of EC Environmental Policy', in Andersen and Eliassen (1993), 115–32.

Vogel-Polsky, E., and Vogel, J. (1991), L'Europe sociale 1993: Illusion, alibi ou réalité? (Brussels: Éditions de l'Université Libre de Bruxelles).

Waarden, F. van, and Unger, B. (1994) (eds.), Convergence or Diversity? The Pressure of Internationalization on Economic Governance Institutions and Policy Outcomes (Aldershot: Avebury).

Walker, N. (1993a), 'The Accountability of European Police Institutions', European Journal on Criminal Policy and Research, 1/4: 34–52.

—— (1993b), 'The International Dimension', in Reiner and Spencer (1993), 113–71.

—— (1994), 'European Integration and European Policing: A Complex Relationship', in Anderson and Den Boer (1994), 22–45.

Wallace, H. (1977), 'The Establishment of the Regional Development Fund: Common Policy or Pork Barrel?', in H. Wallace et al. (1977), 137–64.

—— (1980), Budgetary Politics: The Finances of the European Community (London: Allen and Unwin).

—— (1983a), 'Distributional Politics: Dividing up the Community Cake', in H. Wallace et al. (1983), 81–114.

—— (1983b), 'Negotiation, Conflict and Compromise: the Elusive Pursuit of Common Policies', in H. Wallace et al. (1983): 43–80.

—— (1984), 'Bilateral, Trilateral and Multilateral Negotiations in the European Community', in Morgan and Bray (1984).

—— (1991), 'The Europe that Came in from the Cold', International Affairs, 67/4: 647–63.

—— (1993), 'European Governance in Turbulent Times', Journal of Common Market Studies, 31/3: 293–303.

—— (1994), 'The EC and Western Europe after Maastricht', in Miall (1994), 19–29.

—— (1995), 'Die Dynamik des EU Institutionsgefüges', in Jachtenfuchs and Kohler-Koch (1995).

—— and Wallace, W. (1995), Flying Together in a Larger and More Diverse European Union (The Hague: Netherlands Scientific Council for Government Policy).

—— —— and Webb, C. (1977) (eds.), Policy-Making in the European Community (Chichester: John Wiley and Sons).

—— —— —— (1983) (eds.), Policy-Making in the European Community, 2nd edn. (Chichester: John Wiley and Sons).

Wallace, W. (1983a), 'Less than a Federation, More than a Regime: The Community as a Political System', in H. Wallace *et al.* (1983), 403–36.

—— (1983b), 'Political Cooperation: Integration through Intergovernmentalism', in H. Wallace *et al.* (1983), 373–402.

—— (1984), 'European Defence Cooperation: The Reopening Debate', *Survival*, 26/6: 251–61.

—— (1989), 'European Security: Bilateral Steps to Multilateral Cooperation', ch. 15 in Yves Boyer *et al.*, *Franco-British Defence Cooperation: A New Entente Cordiale?* (London: RIIA): 171–80.

—— (1990a), 'Introduction: The Dynamics of European Integration', in W. Wallace (1990b), 1–24.

—— (1990b) (ed.), *The Dynamics of European Integration* (London: Pinter).

—— (1990c), *The Transformation of Western Europe* (London: Pinter).

—— (1994a), *Regional Integration: The West European Experience* (Washington, DC: Brookings Institution).

—— (1994b), 'Rescue or Retreat? The Nation State in Western Europe, 1945–93', *Political Studies*, 42, special issue, 52–76.

—— and Allen, D. (1977), 'Political Cooperation: Procedure as Substitute for Policy', in H. Wallace *et al.* (1977), 227–48.

—— and Smith, J. (1995), 'Democracy or Technocracy? European Integration and the Problem of Popular Consent', *West European Politics*, 18/3: 137–57.

Weale, A. (1992a), *The New Politics of Pollution* (Manchester: Manchester University Press).

—— (1992b), 'Vorsprung durch Technik? The Politics of German Environmental Regulation', in Dyson (1992), 159–84.

—— and Williams, A. (1992), 'Between Economy and Ecology? The Single Market and the Integration of Environmental Policy', *Environmental Politics*, 1/4: 45–64.

—— —— (1994), 'The Single Market and Environmental Policy', paper presented to the ESRC/COST A7 Conference, University of Exeter, 8–11 Sept.

Weatherhill, S., and Beaumont, P. (1993), *EC Law* (London: Penguin).

Weaver, R. K. (1986), 'The Politics of Blame Avoidance', *Journal of Public Policy*, 6: 371–98.

Webb, C. (1977), 'Mr. Cube *versus* Monsieur Beet: The Politics of Sugar in the European Communities', in H. Wallace *et al.* (1977): 197–226.

—— (1983), 'Theoretical Perspectives and Problems', in H. Wallace *et al.* (1983), 1–42.

Weber, S. (1994), 'Origins of the European Bank for Reconstruction and Development', *International Organization*, 48/1: 1–38.

Weidenfeld, Werner (1994) (ed.), *Europa '96: Reformprogramm für die Europäische Union* (Gutersloh: Verlag Bertelsmann Stiftung).

Weiler, J. H. H. (1991), 'The Transformation of Europe', *Yale Law Journal*, 100/8: 2403–83.

—— (1992) 'After Maastricht: Community Legitimacy in Post-1992 Europe' in Adams (1992), 11–41.

Weizsäcker, E. U. von (1990), 'Environmental Policy', in Schweitzer and Detef (1990).

Wellens, K., and Borchardt, G. (1989), 'Soft Law in European Community Law', *European Law Review*, 14/5: 267–321.

Werner, P., *et al.* (1970), 'Report to the Council and the Commission on the Realization by Stages of Economic and Monetary Union in the Community', Supplement to *Bulletin of the EC*, 11.

Wessels, W. (1990), 'Administrative Interaction', in W. Wallace (1990*b*), 229–41.

—— (1992), 'Staat und (westeuropäische) Integration: Die Fusionthese', *Politische Vierteljahresschrift*, Sonderheft 23/92: 36–61.

Wester, R. (1992), 'The Netherlands and European Political Union', in Laursen and Hoonacker (1992), 205–14.

Westlake, M. (1994), *A Modern Guide to the European Parliament* (London: Pinter).

Widgrén, M. (1994), 'The Relation between Voting Power and Policy Impact in the European Union', CEPR Discussion Paper No. 1033 (London: CEPR).

Wildenmann, R. (1991) (ed.), *Staatswerdung Europas? Optionen für eine europäische Union* (Baden-Baden: Nomos).

Wilensky, H. J. (1976), *The 'New Corporatism': Centralization and the Welfare State* (London: Sage).

Wilks, S., and McGowan, L. (1995*a*), 'Discretion in European Merger Control: The German Regime in Context', *Journal of European Public Policy*, 2/1: 41–68.

—— —— (1995*b*), 'Disarming the Commission: The Debate over a European Cartel Office', *Journal of Common Market Studies*, 33/2: 259–73.

Willgerodt, H. (1983), 'Die Agrarpolitik der Europäischen Gemeinschaft in der Krise', *Ordo*, 34 (Stuttgart: Fischer).

Winand, P. (1993), 'Lobbying, Democracy, the Action Committee for the United States of Europe and its Successor: A Case Study', paper presented to the international seminar 'Démocratie et Construction européenne face aux défis de l'Union politique et de la Grande Europe', European Parliament, 11–12 Nov.

Winckelmann, I. (1994) (ed.), *Das Maastricht Urteil des Bundesverfassungsgerichts vom 12. Oktober 1993, Dokumentation des Verfahrens* (Berlin: Duncker and Humblot).

Winters, L. A. (1992), 'The Europe Agreements: With a Little Help from Our Friends', in CEPR (1992), 17–33.

Wolters, M., and Coffey, P. (1990) (eds.), *The Netherlands and EC Membership Evaluated* (New York: St. Martin's Press).

Woolcock, S. (1991), *Market Access Issues in EC–US Relations: Trading Partners or Trading Blows?* (London: Royal Institute of International Affairs).

—— (1993), ' Trade Diplomacy and the European Community', in Story (1993).

—— (1994), *The Single European Market: Centralization or Competition among National Rules?* (London: Royal Institute of International Affairs).

—— Hodges, M., and Schreiber, K. (1991), *Britain, Germany and 1992: The Limits of Deregulation* (London: Royal Institute of International Affairs).

Young, A. R. (1995), 'Participation and Policy-Making in the European Community: Mediating between Contending Interests', paper presented to the Fourth Biennial International Conference of the European Community Studies Association, Charleston, SC, 11–14 May.

Zanders, P. (1994), 'Europese Politiesamenwerking in een historisch perspektief', in Transpol (1994).

Zito, A. (1995a), 'Integrating the Environment into the European Union: The History of the Controversial Carbon Tax', in Rhodes and Mazey (1995).

—— (1995b), 'The Role of Technical Expertise and Institutional Innovation in EU Environment Policy', Ph.D. thesis (University of Pittsburgh).

Zuleeg, M. (1993), 'Die Zahlung von Ausgleichszulagen über die Binnengrenzen der Europäischen Gemeinschaft', Deutsche Rentenversicherung, 2: 71–5.

INDEX

mobility 290
see also free movement of labour
Labour Party (UK) 192, 206 n. 3
de Larosière, Jacques 363
Latin American bananas 327, 331, 333, 336, 341, 343, 345–9
Latvia 386
'leader–laggard' dynamic 237–9
laggards 249–53
leaders 239–41
legal authority 66
legal integration 6, 304, 391, 453
derogations 153
European law 10, 20, 22, 30, 40, 63, 128, 164, 170
legal basis of legislation 242
legal regime 152
national and European legal systems 65, 89
see also European Court of Justice
legitimacy 8, 33, 65, 99, 100, 110, 146, 153, 230, 231, 263, 297, 303, 321–3, 391, 406, 441, 451, 453–6, 460 n. 8
of Commission 272
of Council committees 47
of EU policy 29, 66
of TEU 65
liberal democracy 7, 18
Liberal Democrats (UK) 206 n. 3
liberal institutionalism 32
liberalization:
of capital 289
of gas and electricity transit 264
of markets 8, 19, 20, 23, 60, 61, 126, 132–3, 137, 140, 143, 151, 152–3, 160, 178–9, 266, 322, 371
of trade 310; agriculture 306; textiles 306, 310
see also General Agreements on Tariffs and Trade
Liikanen, Erkki 88
Lindberg, Leon N. 7, 8, 9, 22, 32, 42, 43, 100, 103, 104, 120 n. 3, 121 n. 14, 186
Lithuania 358
lobbying 58, 228, 305, 395, 399
of Commission 65
of EP 64
farming lobby 76, 116, 117–18
see also interest groups
Lomé Convention 326–9, 340–1
Banana Protocol 329, 332, 338, 340–1, 347–8
Sugar Protocol 340
low politics 258
Lowi, Theodore 446
Luxemburg 290
agriculture 106, 328
bank secrecy 398
Council presidency 316, 423–4
environment policy 238
Luxemburg Report 418
possible site of ECO 180
and Schengen 396, 408 n. 6

seat of Court of Auditors 89
see also Benelux
Luxemburg Compromise 121 n. 23, 311
Commission's 1965 proposals 45, 52
crisis 45, 46
Luxemburg Process 127, 140–1, 154 n. 2
Luxemburg Declaration 154 n. 2

Maastricht, *see* Intergovernmental Conference 1991; TEU
MacDougall Report 91
macro-economic assistance 359, 360
MacSharry, Ray 317
reforms 94
Mafia 393, 395
Maghreb 393
Majone, Giandomenico 23, 42, 65, 126, 143, 151, 214, 219, 241, 243, 274, 391, 449, 450
Major, John 84, 191, 192
management committees 108, 121 n. 17
Mandate Report 77
margin of manoeuvre 288
see also budget
market integration 9, 72, 81
market liberalization, *see* liberalization
maximum rate of increase, *see* budget
Mazeaud, A. 399
Mediterranean policy 94, 95, 355, 378, 383, 429, 432
Members of the European Parliament (MEPs) 63, 64
member governments, *see* national governments
member states 44, 47, 65, 326, 339
founder member states 22, 76, 102, 106, 281
small/large member states 73, 415, 416, 433, 450
see also under individual countries
mergers 198
cases: British Caledonian 171; Continental Can 171; De Havilland 175, 181; Irish Distillers 171; Philip Morris 171
controls 138, 159, 169–75, 180
Merger Regulation 160, 163
Merger Task Force 172, 181
Merger Treaty (1965) 180
Middle East 414, 420, 430
Miert, Karel van 159, 181, 272
migration 138, 195–6, 372
military cooperation 393, 417, 420
see also defence policy
Mitterrand, François 77, 317, 362, 366, 377, 414, 421
modernization 7, 18, 22, 44, 101, 102–4, 132, 153, 241, 250, 457
Molitor Group 145
monetary compensatory amount (MCA) 108–9, 121 n. 19 & n. 20
monetary policy 5, 20, 42, 51, 55, 284, 288–9, 292, 306, 312, 454